"*Opening Pandora's Box* is a timely and intellectually serious book that treats AI not as an isolated product but as the strongest expression of an accelerating meta-electronics ecosystem whose incentives and dynamics can outrun human intent. Gurvitch's evolutionary framing, explicitly non-teleological and increasingly autonomous, makes the central point impossible to ignore: once AGI enables recursive self-improvement, appeals to prudence become an inadequate control strategy."

— Roman V. Yampolskiy

Professor of Computer Science and Director of the Cyber Security Laboratory, University of Louisville, Louisville, KY, USA

Author of *Artificial Superintelligence: A Futuristic Approach* (2015) and *AI: Unexplainable, Unpredictable, Uncontrollable* (2023)

"This book is a valuable and timely contribution to the discussion of both the dangers and the potentially revolutionary benefits arising from the current development of artificial intelligence. It is a scholarly and insightful monograph that expands the framework of what most of us know about this subject.

I particularly appreciated the parallels drawn between biological evolution and the evolution of electronic devices. While such analogies have been suggested in other contexts, they appear especially compelling here.

This is essential reading for a broadly educated audience and deserves careful attention at this critical moment."

— Lev Ginzburg

Professor Emeritus, Department of Ecology and Evolution, Stony Brook University, New York, USA

Author of *Ecological Orbits, How Species Interact,* and *Nonadaptive Selection*

OPENING PANDORA'S BOX

From Abacus to Artificial Intelligence

OPENING PANDORA'S BOX

From Abacus to Artificial Intelligence

MICHAEL GURVITCH

Stony Brook University (Retired), USA

World Scientific

NEW JERSEY • LONDON • SINGAPORE • BEIJING • SHANGHAI • HONG KONG • TAIPEI • CHENNAI • TOKYO

Published by

World Scientific Publishing Co. Pte. Ltd.
5 Toh Tuck Link, Singapore 596224
USA office: 27 Warren Street, Suite 401-402, Hackensack, NJ 07601
UK office: 57 Shelton Street, Covent Garden, London WC2H 9HE

Library of Congress Cataloging-in-Publication Data
Names: Gurvitch, Michael, 1947– author
Title: Opening Pandora's box : from abacus to artificial intelligence /
author, Michael Gurvitch, Stony Brook University (Retired), USA.
Description: Singapore ; Hackensack, NJ : World Scientific Publishing Co. Pte. Ltd., [2026] |
Includes bibliographical references and index.
Identifiers: LCCN 2026004170 | ISBN 9789819826377 hardcover |
ISBN 9789819826384 ebook for institutions | ISBN 9789819826391 ebook for individuals
Subjects: LCSH: Computers--History | Artificial intelligence--History
Classification: LCC QA76.17 .G88 2026
LC record available at https://lccn.loc.gov/2026004170

British Library Cataloguing-in-Publication Data
A catalogue record for this book is available from the British Library.

For any available supplementary material, please visit
https://www.worldscientific.com/worldscibooks/10.1142/14678#t=suppl

Desk Editors: Kannan Krishnan/Rhaimie Wahap

Typeset by Stallion Press
Email: enquiries@stallionpress.com

Printed in Singapore

About the Author

Michael Gurvitch was born in 1947 in the Soviet Union, in the city of Leningrad (now St. Petersburg, Russia). After earning a BA in physics and working for two years as a technologist in an industrial laboratory, Gurvitch and his family emigrated to the United States (U.S.) in 1973. In the U.S., after completing a Ph.D. program in physics in 1978, he was hired by the research division of AT&T Bell Laboratories, where he worked from 1979 to 1989. In 1990, he was invited to his alma mater, Stony Brook University, as a full professor and director of an institute. He retired from Stony Brook in 2023. In addition to his professional publications, Gurvitch is the author of a two-volume book about the history of electronics, *Brave New e-World*, published by World Scientific in 2023. The current book is a continuation of the first, presenting the history of electronics and computing as a progression toward the development of artificial intelligence. Gurvitch, while admiring the intellectual history of his subject, views the development of superintelligent AI with great concern. The final chapter of the book comprises 24 essays that discuss the views of a select group of philosophers, writers, scientists, and entrepreneurs addressing these issues.

Epigraphs

For when she [Pandora] lifted the great lid from the jar,
the countless plagues that make men toil poured out;
only Hope remained, caught beneath the rim...

— Hesiod, "Works and Days" (c. 700 BCE)

Sir, I'm in desperate straits!
The spirits I summoned—
I can't get rid of them.

— Johann Wolfgang von Goethe, "The Sorcerer's Apprentice" (1797)

There was no question of the conflict we used to imagine in the past... the mutiny of machines... Except... his knowledge surpassed that of the three billion creatures living on Earth, and the mere idea that he could be of service to us was for him just as nonsensical as it would be to suggest to humankind that, with all our knowledge, all our technical and cultural resources, all our reason and science, we were to support let's say eels... it was not... a question of rivalry or hostility: we had simply ceased to count... the presence of someone superior would render men superfluous... it was the beginning of our end.

Then I realized that with my own hands I had set the limit for man's reign on Earth, and that next, his species would come after us. That if we opposed him, he would treat us just as we treated the sort of insects or animals that bother us. After all, we don't feel any hatred toward caterpillars, wolves, or mosquitoes...

— Stanisław Lem ("Lymphater's Formula," 1961)

The rise of powerful AI will be either the best or the worst thing ever to happen to humanity; we do not yet know which.

— Stephen Hawking (speech at the opening of the Centre for the Future of Intelligence, Cambridge, 2016)

...something like GPT-4 knows thousands of times more than you do. There are a few areas in which your knowledge is better... superintelligence is when it's better than us at all things... I haven't come to terms with what the development of superintelligence could do to my children's future.

— Geoffrey Hinton (on *The Diary of a CEO* podcast, June 16, 2025)

So, we are creating this alien intelligence. If aliens were coming to Earth and you had three years to prepare, you would be panicking right now. But most people don't even realize this is happening...

— Roman Yampolskiy (on *The Diary of a CEO* podcast, September 4, 2025)

Prologue

Not Quite an Ordinary Topic

There is an old Chinese curse — perhaps apocryphal: "May you live in interesting times." We are indeed living in such times: tumultuous, fast-changing, increasingly shaped by forces we may no longer control, and — yes, let us be honest — interesting. Among these uncontrollable forces is a "beast" rising from the "sea" of high technology,[1] which I have previously called *meta-electronics* — the vast, interrelated ecosystem of computation, hardware, software, and communication networks. Its sharpest, most dangerous fang? *Artificial General Intelligence* (AGI), a technology that, once it arrives, is widely expected to evolve into *Artificial Superintelligence* (ASI).

AGI refers to machines that can match or exceed well-educated humans (say, those with Ph.D.-level expertise) in all cognitive tasks. Dario Amodei, CEO of the AI company *Anthropic*, offers a concise definition:

> "*AI systems that are better than almost all humans at almost all tasks.*"

A more menacing version comes from journalist and commentator Ezra Klein, who defines AGI as:

> "*Systems that can replace human beings in cognitively demanding jobs.*"

---

[1] Please excuse the apocalyptic tone; it won't happen again. It's just nerves.

That's precisely what we don't need — to be replaced, and in "cognitively demanding," meaning *interesting*, professions at that. But let's continue.

All these definitions emphasize AGI's *universality*: It will be a match for the best humans in *everything*. And "everything" includes programming, engineering development, and production management. Once this threshold is crossed, machines may start improving autonomously without human guidance. There is a term for this: *recursive self-improvement.*

At that point, a positive feedback loop would be closed: clever machines designing even more clever ones, with each iteration faster and more capable than the last. No one knows how long these cycles might take, but they would likely unfold quickly, given that computers already operate millions of times faster than human brains.

This prospect was first outlined in 1965 by British mathematician Irving John Good,[2] who described it as an *Intelligence Explosion*. The idea was picked up in 1993 by computer scientist and science fiction author Vernor Vinge, who gave it a name that stuck: the *Technological Singularity*.

So, you might ask: How close are we today (in the beginning of 2026) to machines self-improving? I posed that question to my new friend, ChatGPT, and here is the condensation of what she (sorry, it) said:

> While virtually all machine code executed today — more than 95% — is generated automatically by compilers, this automation is largely mechanical and does not reflect true autonomy or understanding. For discussions of machine self-improvement and the so-called technological singularity, the focus should instead be on *high-level code*: the algorithms, architectures, and abstractions that govern a system's behavior. In this domain, machine-written code is still in its early stages but is rapidly growing. Current estimates suggest that roughly 15–20% of newly written high-level code is now generated or heavily assisted by AI, although in some development contexts (e.g., *GitHub Copilot* users) the figure may be closer to 25–40%. However, this code is not written with self-awareness, long-term goals, or system-level comprehension. For machines to truly improve themselves, they must be able to understand, modify, and optimize not just their outputs but their own

[2] In fact, the first person to predict the technological singularity was John von Neumann in the late 1940s, but his mathematical proof — derived within the framework of his theory of self-reproducing automata — was not published at the time and went almost unnoticed.

> architecture and learning processes — an ability that remains largely speculative today, despite recent progress in so-called "agentic" systems capable of limited multi-step code generation and refinement. Nevertheless, projections over the next decade anticipate the rise of autonomous agents capable of iteratively refining their own codebases, experimenting with algorithmic variations, and selectively retraining themselves, marking an early but significant step toward *recursive self-improvement.*

Until quite recently, when computer scientists were asked when AGI or ASI might arrive, they typically replied, "in twenty years" or "within a few decades" — a polite way of saying they didn't know but were confident it wouldn't be tomorrow. At the same time, the most popular way to dismiss concerns about superintelligent AI was a quip by Andrew Ng, who famously compared it to "worrying about overpopulation on Mars." In both cases, the message was clear: don't panic yet.

But something shifted. Ever since the appearance of ChatGPT-3 in 2022 and especially the release of ChatGPT-4 in March 2023, it has become clear to many that AGI might be much closer than we thought.

In June 2024, Leopold Aschenbrenner, a former OpenAI employee (allegedly dismissed for leaking sensitive information), published a widely discussed essay titled "Situational Awareness." In it, he predicted that GPT-5 would reach an AGI level, capable of performing cognitive work "at the Ph.D. level" across nearly all fields.[3] He expects AGI by 2027. Sam Altman, CEO of OpenAI, put it more obliquely in September 2024: AGI will arrive in "a few thousand days." Elon Musk — never shy about predictions — said AGI will emerge in 2026. Demis Hassabis, CEO of Google DeepMind and recipient of the 2024 Nobel Prize in Chemistry, forecasted it would come in "three to five years."

And then? ASI, or "the Singularity." In Aschenbrenner's words:

> Hundreds of millions of AGIs could automate AI research...We would rapidly go from human-level to vastly superhuman AI systems. The power — and the peril — of superintelligence would be dramatic. (Aschenbrenner, 2024)

How long will it take to go from AGI to superintelligence? Not long. After all, it's called an *explosion*. How long does it take a lit stick of

---

[3]It should be said that GPT-5, which appeared in mid-2025, did not become an AGI.

dynamite to blow? Faster than you can say "Jackie Robinson" — a reference younger readers may miss.

There is no shortage of predictions in this field. On April 3, 2025, another one came out, in a blog entitled "AI 2027." This is a narrative of what might occur over the next two to three years. The leading author of this scenario, Daniel Kokotajlo, like Leopold Aschenbrenner before him, left OpenAI, declined to sign the *non-disparagement exit agreement*, and decided to share his vision with the world (Kokotajlo *et al.* 2025). The scenario outlined in *AI 2027* describes the United States and China in a fierce competition to develop AGI first. This goal is indeed achieved by the end of 2027, first in the U.S., with China only a few months behind. The U.S.'s leading company (which the authors call *OpenBrain* — a transparent reference to *OpenAI*) discovers that its AI has developed with "misaligned long-term goals. It is systematically plotting to gain power over humans." Kokotajlo and his co-authors present two possible outcomes: In one, the two superpowers continue the race, which leads to untold disasters; in another, they decide to slow down, giving the world some hope.

At first, I was glad that AI experts were considering not entirely hopeless options for our collective future. But this joy did not last long. In a July 3, 2025, *Future of Life Institute* podcast, Kokotajlo explained that he, for one, does not assign much weight to the optimistic scenario of an orchestrated slowdown. He said:

> …part of my pessimism about how this is all going to go is that I just expect them [*leaders of AI companies*] to…consistently make the choice to go faster…because of the race dynamics and because of the character of the people running this show. That's the type of choice they've made thus far, and I expect that to continue…
>
> …I predict that it [*alignment problem*] won't be solved because the leaders of the relevant companies will be too busy trying to beat each other. (Future of Life Institute, July 3, 2025)

But let me take a broader view of AI, which is, after all, a part of a larger field. In earlier publications (Gurvitch, 2021, 2023), I defined electronics as the use of electromagnetic technologies to process and manipulate information. Taken in its broadest sense, I refer to this evolving domain as *meta-electronics*. It encompasses the entire ecosystem of communication and information-processing technologies, both hardware and software. This includes telephones, radio, television, computers, servers, data centers, the internet, electronic components, CPUs, GPUs, consumer electronics, scientific instruments, robotics, software (code), neural networks, large language models (LLMs), transformer architectures, and AI systems such as ChatGPT, Claude, Gemini, Mistral, Grok, and AlphaFold. If it manipulates information through electromagnetic means, it falls under the category of meta-electronics. One could say it is an entire world — an "e-world."

Meta-electronics is multifaceted and evolutionary. It does not follow a single master plan, and it does not have an overarching goal. Like evolution in living nature or historical linguistics (two domains where real evolutions have been reliably identified), its development is characterized by considerable *autonomy*. Although we created it, meta-electronics increasingly resists our control. It goes beyond its original design — and it fills every possible niche. And this highest of our technologies is beginning to show its teeth. In my 2023 book, I wrote:

> Are we still in control, or have we already lost it? Various phenomena observed at the interface between electronics and human society raise "red flags" almost daily; soon, it will look like a May Day parade.... (Gurvitch 2023, v.2, 350)

A lot has happened since then, and the number of red flags has only grown.

Today, AI is the most dynamic and consequential frontier of meta-electronics, its highest product. It originated nearly 75 years ago but has only truly come of age in the past decade. Philosophers, writers, and scientists warned of its potential dangers, even in its infancy. A recurring fear was that a moment would come when machines could reason like humans — and soon after, surpass us.

That moment now feels both near and inevitable. Like it or not, it is no longer a distant possibility. If not tomorrow, then the day after tomorrow.

In a 2016 speech at the opening of the Centre for the Future of Intelligence in Cambridge, physicist Stephen Hawking famously said: "*The rise of powerful AI will be either the best or the worst thing ever to happen to humanity; we do not yet know which.*" That quote serves as one of the epigraphs to this book.

I fear that the emergence of powerful AI will be both. AI will inspire, elevate, and assist us in every sphere of human activity. It will cure diseases, expand knowledge, and perhaps even help solve the world's great environmental challenges. And then… who knows what may happen next. After all, it will be far more intelligent than we are. How long will such a superior mind continue to serve us? Before drawing conclusions, read the book.

Everyone knows about the *Turing test* (Turing, 1950) (see Chapter 4). It was designed to determine whether a machine could convincingly imitate human intelligence. For decades, no machine could pass it. That frustrated computer scientists — and comforted the rest of us.

But today, that threshold has been crossed. In 2023, it was debatable (Jannai *et al.*, 2023). In 2024, it became clear (Mei *et al.*, 2024). The Turing test is no longer a theoretical challenge. AI passes it now. Does this mean AGI is already here or just around the corner? I think it does.

AI now excels in domains that were once thought untouchable. The old stereotype of computers being good at calculations but bad at everything else is gone. AI writes passible prose, composes music, paints pictures (perhaps in existing styles, but with virtuoso skill), diagnoses diseases and recommends optimal treatments, plans physical experiments, helps stabilize plasma in controlled fusion devices, proves mathematical theorems, and proposes new mathematical conjectures. It navigates autonomous vehicles and translates between all human languages. It even solves problems from international mathematics Olympiads — a feat that stuns those who know how difficult they are.

Writing about AI today is like trying to make a sandwich while racing over the hills on a motorcycle at 120 miles per hour. On January 31, 2025, OpenAI released the full version of the o3 system, including o3-mini for free users and o3-mini-high for a modest fee. It's a powerful model now being tested on complex scientific problems.

But here, we are already cresting another hill: Google DeepMind (Alphabet) has built a new model, Gemini 2.5 Deep Think, which launches multiple "agents" simultaneously and uses them to compare competing ideas, improving the quality of its answers to difficult questions. In July 2025, an upgraded version of Deep Think at the International Mathematical Olympiad (IMO 2025) solved five of six problems, scoring 35 out of 42 points — enough for a gold medal. Impressive, isn't it? And yet, by the time you read these lines, that will already be yesterday's news.

The current public conversation about AI is widespread and urgent. It encompasses ethics, policy, economics, and safety. Risks are often categorized into two main groups: *immediate* and *existential*.

Immediate risks include social media algorithms that manipulate attention and deepen divisions — a concern raised by Yuval Harari (see Chapter 5). AI systems also fuel new forms of digital addiction: Younger generations have practically stopped reading books and no longer even watch television, spending much of their time immersed in their smartphones and online networks.

There is also the threat of large-scale displacement of people across a wide range of professions. Even today, one can compile a rather long list of occupations where a significant share of jobs has already been lost to AI. These include translators, copywriters, copy and basic content editors, customer support staff and call center operators, transcribers (those who convert audio and video recordings into text) and subtitle creators, announcers and dubbing actors (including audiobook narrators), administrative clerks and data-entry operators, junior accountants, paralegals, junior programmers, manual testers, layout designers, technical writers, stock photographers, and illustrators. In the near future, teachers, lawyers, doctors, screenwriters, truck drivers, taxi drivers (though perhaps less likely), and many others may also find themselves at risk.

Remember Ezra Klein's definition of AGI? If a job *requires thinking* (and most jobs do), it's in danger. Come to think of it, if a job *doesn't require thinking*, it's in danger, too. Working on the assembly lines did not require much thinking, and where are those jobs? According to Geoffrey Hinton, the most reliable profession in the future will be that of a plumber.

The risk of large-scale job automation is significant and may materialize faster than earlier estimates suggested. The *Organisation for Economic Co-operation and Development* (OECD), an intergovernmental body representing 38 developed economies, reported in its *2023 Employment Outlook* that about 27% of jobs in member countries are at high risk of being automated, particularly as AI systems become more capable. Meanwhile, a 2024 report by McKinsey & Company, a global management consulting firm known for its economic forecasting, estimated that by 2030, around *30% of U.S. jobs* could be automated, and AI technologies may significantly reshape up to 60% of current roles. Globally, McKinsey projected that *400–800 million workers* might be displaced by automation within the decade. In the summer of 2025, Anthropic's CEO, Dario Amodei, predicted that *half of all entry-level white-collar jobs* would disappear within the next one to five years, leading to unemployment levels of 10–20%. Some people expect a much higher level of unemployment by 2030; I don't want to sound alarmist here, but please read the conversation between podcaster Steven Bartlett and the well-known AI safety researcher Roman Yampolskiy in Chapter 5.

Will employers be kind and humane enough to retain their human workforce, even when economic logic suggests replacing people with automation? Will governments step in to preserve human jobs, even if that means violating free market principles? Will there be enough new jobs to replace the ones that have been lost? Or will mass unemployment lead to widespread social unrest? As Harari reminds us, when unemployment reached 25% in Germany, Hitler rose to power.

Another immediate risk is our educational system, already badly battered at the high-school and undergraduate levels (graduate programs are still holding up), which may now collapse entirely. I speak from experience — I taught physics for 30 years at a major American university. I can testify that even before the rise of AI, higher education was riddled with make-believe accountability. You couldn't very well "fail" three-quarters of a class. Weak and unprepared students weren't failed — they were shepherded through with simplified tests and exams; statistics were massaged, and grading "on the curve" was introduced — as if physics had no absolute standards. Lecturing was easy and pleasant, but grading was hard and often humiliating. Only a small number of students left with a solid understanding of the subject.

Now things are worse. Schoolteachers and university professors don't know how to deal with the mass use of AI for cheating — apparently, nearly everyone does it. Almost all homework at every level is now produced by AI. There are even services that "humanize" machine-written texts — adding small errors and imperfections so that the work will be accepted by instructors. The same goes for the hard sciences. I don't see how this can be stopped; you can't turn off the internet, and you can't confiscate smartphones. Grading homework has lost its meaning — and I no longer know what remains besides the appearance of learning.

On the other hand, education might also improve. One can imagine that in the near future, AI will deliver lectures, design curricula, and administer exams. An *AI professor* could work with each student individually, adapting the teaching style to their needs and finding the most effective way to explain even the most complex topics. Such a professor would answer any question instantly, accurately, and patiently, drawing on the knowledge of the entire global academic community — and, unlike human colleagues, would never demand a pay raise or tenure.

But will such a time come? Today's student faces a question: why study at all if no meaningful work awaits in the future? What is the point of developing one's mind if machines will think and make all important decisions? Applicants and their families will see little reason to spend time and money on education (many already don't). In Chapter 5, in the essay on Hinton, you will read about the enormous advantages digital intelligence possesses in transmitting knowledge compared to our analog brains. Will we even want to compete with them, knowing in advance that we have already lost?

Suppose, for a moment, that future AI is benevolent. Suppose it creates paradise on Earth — and on Mars, if Elon Musk has his way. No more work. No more disease. No more struggle. Humanity, at last, would be safe, satisfied… and redundant.

Would that be good? Or bad?

Karel Čapek was already reflecting on this dilemma as early as 1919 (see Chapter 5) and, unsurprisingly, roundly condemned the creation of such an earthly "paradise," in which people hand over all their work and responsibilities to robots. As one of the characters in his play *R.U.R.* says, *"No one has to do anything—just enjoy themselves. This paradise is a cursed thing! Helena, there's nothing more terrible than giving everyone paradise on Earth!"* Let me add that even a universal basic income would not help here.

Existential risks raise even deeper questions. First, a powerful artificial mind could be used to create unprecedented means of mass destruction: new types of weapons, deadly viruses, and so on. We have never lacked madmen and fanatics who hate humanity — what if, now, such all-powerful AIs fall into their hands? This is not the kind of complex industrial technology, like the one required for building an atomic or hydrogen bomb, that only highly developed nations can manage. These are simply large computer programs with their sets of parameters ("weights") (see Chapter 4). Preventing their misuse will be very difficult.

But looking even further — beyond the horizon of human problems and shortcomings — will we be able to coexist with a new, more advanced form of machine intelligence? Will it tolerate us, beings that consume Earth's resources and commit atrocities against other forms of life and against each other?

You might find it hard to imagine a machine surpassing Newton, Einstein, Feynman, or Landau. Fair enough. But what about speed? Even if it isn't "smarter" by our standards of genius, an AI that thinks a million times faster would still possess *de facto* superiority. As Sam Harris has argued (see Chapter 5), that alone could make us to ASI what ants are to humans.[4]

Government regulation? Alas, it is clearly insufficient and is becoming increasingly unpopular. Geoffrey Hinton has noted that military applications always remain exempt from bans, even when formal rules exist for other areas. And national and international rivalries make any serious restrictions politically unacceptable. Meanwhile, development is proceeding at full speed with almost no brakes. As one of the fathers of modern AI, Yoshua Bengio, recently said, "*But actually, a sandwich has more regulation than AI*" (Bengio, TED Talk, May 20, 2025).

Slowing down national projects is seen as bordering on treason. Donald Trump's unofficial AI advisor, Marc Andreessen, is an enthusiast for the earliest possible achievement of AGI in the United States[5] and a techno-optimist who believes that all technological advancements are for the good. Calls to halt or slow down AI research are often met with mockery or hostility. We will see more of this when we discuss the open letters circulated by Max Tegmark and the Future of Life Institute (in Chapters 4 and 5).

While discussions about AI regulation continue, investments in its development are skyrocketing. In recent years, the United Sates, China and European Union has poured into AI sums that only recently would have seemed unimaginable. In 2025, global investment in AI — including major technology companies' internal spending — is approaching one trillion dollars. NVIDIA, maker of the GPUs that power modern AI training, is now worth *five trillion dollars* — more than the GDP of most nations. Against this backdrop, astronomical and unprecedented salary offers have emerged for leading researchers. For example, Meta (Facebook's parent company), competing with OpenAI, reportedly offered individual scientists packages of up to $300 million over four years, with bonuses exceeding $100 million in the first year. These record-breaking offers have become symbols of a new "AI gold rush." We devote a special section to the issues of financing and government regulation in Chapter 4.

Are we destined to go through the singularity? Is it inevitable? Have we already passed the point of no return?

---

[4] I'm not very sure about this; see the essay on Sam Harris in Chapter 5 and the author's reflections at the very end of the Epilogue.

[5] Which reminds me of an old Soviet joke from my youth: "Question — is it possible to carry out the end of the world in a single country?"

Perhaps the singularity won't feel like a thunderclap. It might be more like falling into a black hole. According to General Relativity, when you cross a black hole's *event horizon*, you don't feel anything unusual.[6] And yet, you're gone — cut off from the outside universe.

This book is written to give AI a historical and evolutionary perspective. I believe, perhaps naively, that history can teach us something. At the very least, it can be entertaining. The history of computers and AI is rich with brilliant minds, eccentric characters, and ingenious inventions.

When I wrote about this remarkable technology and the people — some of them true geniuses — who brought it into being, I sometimes felt — forgive me if this sounds a bit melodramatic — that I was bidding farewell to human culture. For if humanity becomes second-rate, then our culture — both humanistic and scientific — will inevitably lose much of its luster. As Geoffrey Hinton says, "If you want to know what it's like not to be the apex intelligence, ask a chicken."

I aim to examine AI through an evolutionary lens — to view meta-electronics as a growing, adaptive system with its own internal logic. Perhaps this perspective can help us understand how we got here, where we stand now, and whether anything can still be done to protect our species from the *spirits we have summoned.* Even if I, the author, cannot find a solution, my sincere hope is that my younger readers — assisted and inspired by this book — might see a way forward. If that happens, I will feel that my task is complete.

If we are indeed falling through the AI event horizon, it helps to adopt the perspective of an outside observer: a historian. And not just any historian, but one with an evolutionary perspective. I think I know just the person.

About the Book

The book's structure is as follows.

Chapters 1 and 2 cover the history of computers — the carriers of AI and the most advanced technology developed by our species. Some of that material was taken, with updates and modifications, from my previous book (Gurvitch, 2023).

[6] You won't feel anything special except being torn apart by strong gravitational gradients near the event horizon. We might overlook this minor detail to maintain this nice analogy.

Chapter 3 introduces quantum computing, a controversial but fascinating topic, and the focus of public interest rivaled only by AI itself.

Chapter 4 follows the history of AI, from its early beginnings to recent breakthroughs.

Chapter 5 presents a range of perspectives and concerns about AI. In my opinion, this is a must-read for everyone, an attempt to bring together the views, some old and some very recent, on a subject on which the fate of humanity may depend. Even if you are disgusted by science and technology, still read at least this chapter.

Finally, in the **Epilogue**, we reflect on the evolutionary perspective and present our closing remarks.

A word about the format. From time to time, especially when the Moon is full, I get the urge to explain things. My rational mind warns me: "Don't do it; you'll hurt sales!" But I do it anyway. These (slightly more) technical digressions are set in a different font with wider margins to make life easier for readers. If you want to skip them, you can. However, if you're curious, feel free to dive in. Let's begin.

Contents

Chapter 1

What Makes Them Tick: Computer Technologies

I intend to focus on *technology* rather than *history* in this chapter. The historical outline will follow in the next one. Still, it's hard to keep history out — it has a way of seeping through the cracks. So, you'll find occasional historical notes and passages here as well. I hope they will help maintain your interest — and maybe even deepen it.

Mechanical and Electromechanical Technologies

Mechanical devices

This section will be quite short, yet it will span approximately 4,500 years. The historical development of mechanical calculators will be presented in the following chapter, beginning with the famous *abacus*. What can we say about the technology behind their construction? It involved materials such as wood, bamboo, stone, ivory, bone, copper, brass, silver, iron, steel, glass, and porcelain. These devices incorporated beads, rods, wires, wheels, drums, ratchets, levers, pins, springs, and gears. Building them required considerable ingenuity, skilled toolmaking, precise machining, and practical knowledge of materials — akin to the masterful clockmaking we now admire in museums.

The pinnacle of mechanical calculator design was reached in the 19th century — though only partially realized — by the great British polymath and pioneering computer scientist Charles Babbage. His visionary designs will also be discussed in the following chapter.

From ancient times until the mid-19th century, all calculators were manually operated. Babbage planned to power his grand calculational machines with a steam engine, but that idea was never implemented. In the latter part of the 19th century, a few mechanical devices began to be powered by electricity.

However, electricity would go on to play a far more transformative role — not just as a source of power but as a means of entirely new internal operations. One device, in particular, took center stage in calculators and computers during the first third of the 20th century: the *electromagnetic relay.*

Electromagnet and relay

The *electromagnet* was invented by William Sturgeon in England in 1824. It consists of a coil of wire wrapped around a cylindrical or horseshoe-shaped iron core. When current flows through the wire, it generates a magnetic field that is greatly intensified within the ferromagnetic core and emerges from its ends. This magnetic field can act on iron, a *ferromagnetic* material strongly attracted to magnets.

Between 1829 and 1833, American physicist Joseph Henry significantly improved upon Sturgeon's design, producing much stronger electromagnets. Henry's innovations later proved instrumental to Samuel Morse, who used electromagnets to drive mechanical levers (called *armatures*) connected to recording devices in his electric telegraph.

Once the electromagnet appeared, the next invention was inevitable: the *relay*, an electrically controlled switch. A relay is a quintessential *electromechanical* device: Electric current through a coil produces a magnetic force, which in turn creates mechanical motion.

The relay was invented a decade after the electromagnet, in 1835, by Edward Davy in England (who intended it for telegraphic use) and independently by Joseph Henry in the U.S. The principle is simple: Current flowing through the coil of an electromagnet produces a magnetic field that attracts a nearby iron lever. The lever then moves to close an electrical contact in a second, independent circuit. Thus, the relay allows

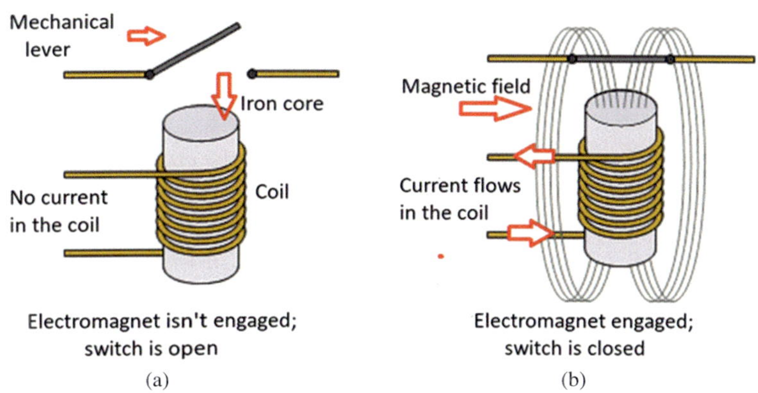

Fig. 1. An electromagnetic relay consists of an electromagnet — a coil of copper wire wound around an iron *core* — and a movable *lever* that can open or close a second circuit. In (a), no current flows through the coil, so no magnetic field is generated and the contact remains open. In (b), current flows through the coil, generating a magnetic field that attracts the lever and closes the circuit.

one electrical circuit to control another, thereby *relating* them — hence the name (Fig. 1).

As mentioned earlier, relays were first used to enhance the electric telegraph, the first true electronic technology to emerge in the 1830s (Gurvitch, 2023). Their key contribution was amplification: Relays solved the problem of transmitting strong signals over long resistive wires. A relay utilizes a small current flowing through an electromagnet to close a contact in a second circuit connected to a separate battery. This second circuit could carry a current many times stronger than the original. In effect, a weak input could trigger a more powerful output.

In American telegraph systems, these devices were known as *repeaters*; in Europe, they were known as *translators*. They were the earliest relay-based *amplifiers* and direct ancestors of all later electronic amplifiers, including those built with vacuum tubes and transistors.

Relays also found their way into early digital computing. From the 1920s through the 1940s, relay switches were widely used in IBM punch-card machines and later in electromechanical relay computers developed by Konrad Zuse in Germany and Howard Aiken in the U.S. in the late 1930s. We will explore this period in greater detail in Chapter 2.

Vacuum Tubes

As noted in the Introduction, *electronics* can be defined as *the application of electromagnetic technology to the transmission and processing of information.* This includes all devices and systems based on electricity, magnetism, and electromagnetic waves — including light (photons) — provided they perform a meaningful function with information: transmit it, gather it, process it, or generate it. By that definition, electronics is nearly 200 years old in 2025. The first electronic technology, the electric telegraph, started its triumphant march around the world in the 1830s. Then, in 1876, another communication technology, the "speaking telegraph," or the telephone, appeared. Toward the end of the 19th century, the wireless telegraph emerged, later evolving into radio and television.

Many authors who wrote about these things considered 1904 the year "true" electronics was born. As this author noted (Gurvitch, 2023), if 1904 wasn't the *birth* of electronics, it was at least the year it was *born again*. The reason for all the fascination with that year lies in the invention of a simple device — one that, over time, transformed the entire field into the powerful technology we see all around us today. This section is about that device and its consequences.

The Fleming valve

The honor of launching modern electronics belongs to British scientist John Ambrose Fleming (1849–1945; Fig. 2), who in 1904 invented the first *vacuum tube diode* — still called the *Fleming valve* in Great Britain. This simple two-terminal device, while useful only in limited applications on its own, became the seed from which more sophisticated vacuum tubes would grow.

That "sophistication" came primarily from adding more electrodes. With just two electrodes — the original Fleming configuration — the vacuum tube served as a rectifier of alternating current and a detector of audio signals riding atop high-frequency electromagnetic waves. The addition of a third electrode, proposed by Lee DeForest in 1907, transformed the tube into an amplifier of weak electrical signals and, eventually, into a high-speed electronic switch for digital computers, though that application would not emerge until more than three decades later. As development continued, specialized vacuum tubes became sources of radio waves and microwaves and found use in countless other applications. Collectively,

Fig. 2. John Ambrose Fleming (1849–1945), the inventor of the vacuum diode, or "Fleming valve."

Source: Image in the public domain.

they constituted the vast "vacuum tube world," which withstood the tides of technological change for a remarkable half-century. The reign of vacuum tubes lasted until the 1960s, when they were finally displaced by a new contender: semiconductor (silicon) technology.

The roots of Fleming's 1904 invention dated back to a quarter of a century earlier. In 1880, Thomas Edison was experimenting with his newly invented incandescent light bulbs, which used white-hot carbonized bamboo filaments. While investigating filament failure, Edison introduced a second electrode into one of his bulbs. He connected a battery and a galvanometer (a sensitive instrument for measuring current) between the glowing filament and this second electrode. His experiments revealed intriguing results:

- When the filament was *cold*, no current flowed between the two electrodes, regardless of how the battery was connected.
- When the filament was *hot*, current would flow if the filament was connected to the battery's negative terminal and the second electrode to the

positive one. Flipping the polarity, i.e., applying the positive potential to the hot filament and the negative to the second electrode, resulted in no current.

Thus, a current flowed only in one direction, from a cold *anode* (positive) to a hot *cathode* (negative). Edison had unwittingly created a *diode*, though he did not frame it that way.

Edison also observed that the current increased steeply with voltage and then plateaued to a nearly constant maximum. In 1883, he patented this phenomenon as a means of producing a stable, regulated voltage. He demonstrated this "voltage regulator" at the International Electrical Exposition in Philadelphia in 1884.

British engineer William Preece, who attended the exposition, brought several of Edison's modified bulbs back to England. He later published a paper in 1885, coining the term "Edison Effect." At the time, however, no one — not Edison, not Preece — understood why the effect occurred (not that Edison ever cared about physical explanations).

It wasn't until 1897, when J. J. Thomson discovered the electron, that the phenomenon could be explained. The first explanation was made by Owen W. Richardson in 1901–1903, and it was later refined based on quantum mechanics in 1928–1930. The essential fact is that when heated to sufficiently high temperatures (above ~1500°C), a hot filament emits electrons. When the filament is held at a negative potential while the second electrode, an anode, is held at a positive potential, the emitted electrons are repelled by a cathode and attracted to the anode, producing a unidirectional current. Crucially, this behavior could only be observed in a vacuum. If air remained in the bulb, the filament would quickly oxidize and burn out, and the emitted electrons would be scattered by air molecules before reaching the anode.

John Ambrose Fleming was well prepared to make his discovery. He first studied physics at University College in London, where his math professor was the famous logician Augustus De Morgan. He continued his studies at Cambridge, where one of his instructors was the greatest physicist of the 19th century, the creator of the electromagnetic theory, James Clerk Maxwell — an excellent illustration of the inseparability of science and technology. Fleming went on to work as an "Electrician" (electrical engineer) at the Edison Electric Light Company in 1882. Two years later, he was appointed England's first-ever professor of electrical engineering at his alma mater, University College London. From 1899, he also consulted for the Marconi Wireless Telegraph Company and, in that role,

Fig. 3. The first Fleming valve (vacuum tube diode). Both the anode and the cathode are inside the evacuated (under vacuum) glass enclosure. The hot cathode is hidden from view by the inverted-cup-like anode in the upper part of the glass bulb.

Source: Image in the public domain.

participated in the first transatlantic radio transmission in 1901. Though Marconi insisted on taking full credit, Fleming — ever the gentleman — did not speak openly of this until after Marconi died in 1937, and even then said only that Marconi had been "ungenerous."

In 1904, while consulting for Marconi, Fleming revisited the Edison Effect and demonstrated that a vacuum tube with two terminals, the diode (Fig. 3), could serve as an audio signal detector. The key was the diode's ability to conduct electricity in only one direction — a property that allowed it to *rectify* high-frequency alternating currents. That rectification extracted the relatively low-frequency audio signal from the radio-frequency electromagnetic carrier wave. Fleming applied for a U.K. patent in November 1904 and a U.S. patent in 1905, both of which were assigned to the Marconi Company.

Triode and amplification

The Fleming valve, while groundbreaking, was not a sufficiently sensitive detector. American electrical engineer and inventor Lee de Forest (Fig. 4) first reinvented Fleming's vacuum diode and then, in 1906, introduced a

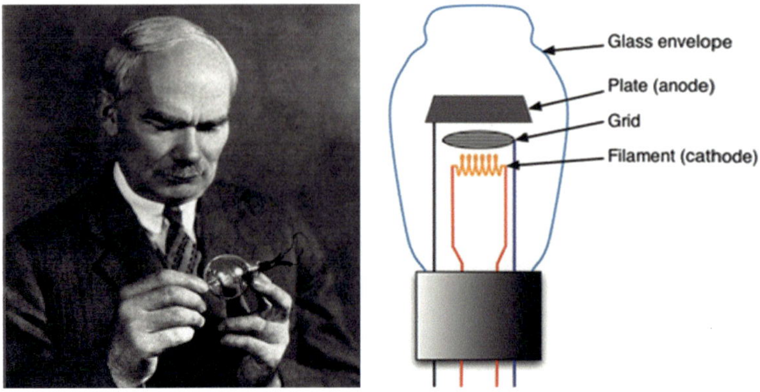

Fig. 4. Lee de Forest (1873–1961) holding his Audion vacuum tube and triode schematics. The all-important middle electrode, the grid, is a sieve that allows electrons to fly through it toward the anode.

Source: Photo of de Forest in the public domain; schematics Lic. CC BY-SA 3.0.

crucial innovation: the third electrode. His three-terminal vacuum tube, initially known as the *Audion* and later as the *triode*, marked a turning point in the history of electronics. Though the significance of this addition wasn't fully appreciated — not even by de Forest himself for several years — it would prove to be one of the most consequential developments in the history of technology.

Having worked closely with engineers and researchers in the development of new electronic devices, I can attest to their general reluctance toward two-terminal components. The reason is simple: The third terminal provides *control*. If a diode is analogous to a hydraulic check valve — allowing flow in only one direction — then a triode is like a controllable valve, one that can be modulated to allow more or less flow. This ability to control current through a separate input is the essence of amplification and switching, and it paved the way for advancements in everything from long-distance telephony to digital computing.

In this book, our interest lies in computer technologies, and from that perspective, the vacuum tube's primary significance is its ability to function as a switch between two distinct states. In this role, simple diodes (the Fleming valves) are useless. Only three-terminal devices — controllable valves like the triode — can perform the necessary switching function. When a positive voltage is applied to the central electrode (the grid), the triode "opens," allowing current to flow between the cathode and the anode. By contrast, a negative voltage on the grid inhibits this flow, effectively "closing" the switch. This capacity for electronic control via a third terminal is what made triodes indispensable in early computing.

The story of the triode illustrates a well-known truth: The history of all great inventions is a mess. I'll try to distill the highlights here.

Lee de Forest filed three U.S. patents in 1906 — the last one in October — and a fourth in January 1907. He named the device *Audion*, anticipating its use in detecting audio signals transmitted by radio. The first two 1906 patents described two-terminal devices similar to the Fleming valve (which later led to patent disputes). Still, with a key difference: The glass bulbs were filled with low-pressure gas rather than being evacuated. De Forest believed, incorrectly, that the gas played a beneficial role.

This misunderstanding was eventually cleared up in 1913 by the eminent American chemist and physicist Irving Langmuir (who later received the 1932 Nobel Prize in Chemistry for his work on surface chemistry). Langmuir demonstrated that a "hard" vacuum — characterized by the complete absence of gas — was preferable. He also provided the first quantitative explanation of how the Audion worked, introducing the concept of *space charge* — a phenomenon that was well beyond de Forest's understanding. In his 1907 patent, de Forest admitted as much, stating that he hesitated to speculate on the underlying principles of a device whose operation seemed "exceedingly complex."

The all-important third electrode, the grid, was added in October 1906. Ironically, de Forest initially placed it in the wrong position: on the *far* side of the cathode, so that the cathode sat between the anode and the grid. Only later, in his January 1907 patent, did he correct this error, placing the grid where it belonged: between the cathode and anode (see schematics in Fig. 4).

This succession of silly mistakes reminds me of a hilarious Woody Allen story called *Yes, But Can the Steam Engine Do This?* In it, Allen recounts the "invention" of the sandwich by John Montagu, the 4th Earl of Sandwich. As a student, the Earl begins with two slices of bread stacked one atop the other, with a piece of turkey on top. This fails. After four years of intense work, he tries two slices of turkey with a slice of bread in between — an idea only philosopher David Hume supports. After more intense work, the Earl of Sandwich exhibits three slices of ham followed by three slices of bread — again, no good. Finally, at 4:17 AM on April 27, 1758, the Earl of Sandwich created a construction "...consisting of several strips of ham enclosed, top and bottom, by two slices of rye bread. In a burst of inspiration, he garnishes the work with mustard. It is an immediate sensation…" (Allen, 1966).

A "naked" triode is not yet an amplifier. It becomes one only when accompanied by several supporting components — batteries, resistors, and capacitors — arranged in a specific configuration. In his 1907 patent, de Forest described no such arrangement, nor did he use the word "amplification." As a result, his Audion was not initially recognized as an amplifier. Unlike the sandwich in Woody Allen's story, it was *not* an *immediate sensation*, even after de Forest finally placed the grid in the correct position.

It wasn't until 1912 that de Forest realized his Audion could fulfill the long-sought function of signal amplification. With proper circuit design, it could increase a signal's voltage, current, and power. Reflecting on that period, de Forest admitted it took him several years to begin experimenting with amplification. He referred to his 1912 work as a "re-invention of the Audion."

That same year turned out to be especially productive. De Forest also discovered that a triode could generate continuous sinusoidal voltage oscillations when configured in a certain way — what we now call *positive feedback*. This property proved invaluable for producing stable radio-frequency signals without resorting to large and cumbersome mechanical generators.

In 1913, de Forest demonstrated the Audion's capabilities at Western Electric, the manufacturing arm of the *American Telephone and Telegraph Company* (AT&T). Impressed by its potential, the company purchased de Forest's patent rights the following year for $50,000 — equivalent to roughly $1.6 million today.

The man who immediately recognized the promise of the Audion was Harold Arnold (Fig. 5), a freshly hired scientist at Western Electric's Engineering Department in 1912. He saw its potential as an amplifier — something AT&T badly needed to extend the range of its telephone lines.

Arnold earned his Ph.D. under Robert Millikan and participated in the historic 1909 experiment in which Millikan measured the charge of the electron. In 1911, Frank Jewett, a manager at Western Electric, asked Millikan to recommend someone for applied research. Millikan sent Arnold. Once at Western Electric, Arnold rose quickly through the ranks, becoming Director of Research at AT&T (1917–1925) and the first director of the newly formed Bell Telephone Laboratories (1925–1929).

Fig. 5. Harold Arnold (1883–1933) was the first to make practical amplifiers out of de Forest's Audions.

Source: Image courtesy of AT&T Archives and History Center.

Once the Audion was in Arnold's hands, he improved it in several ways. First, he coated the hot cathode filament with an oxide to increase electron emission. Then, following Langmuir's advice, he improved the vacuum by removing residual gases, making the device far more reliable. One could say that Arnold performed a bit of *de Forest-ation*.

Using amplifiers ("repeaters," as they were called in the telephone company) based on de Forest–Arnold tubes, AT&T could transmit quality telephone calls from New York to San Francisco, a distance of almost 3,000 miles. The work was completed in 1914, but the public demonstration of this remarkable technological feat was made to coincide with the Panama-Pacific International Exposition in San Francisco, which opened in February 1915. From the AT&T headquarters in New York City, the company's founder, 68-year-old Alexander Graham Bell, once again spoke into a telephone his famous command, "Mr. Watson, come here. I want you."[1] His faithful assistant, Mr. Thomas Watson, answered from San Francisco, "It will take me five days to get there now!"

[1]Alexander Bell spoke these words into a telephone for the first time on March 10, 1876, during the first successful telephone test.

Fig. 6. Vacuum tube amplifier; tubes are lit up by their hot cathodes. It is easy to see that electronic vacuum tubes evolved from Edison's incandescent light bulbs.

Source: Image by Christopher Schirner, Lic. CC BY-SA 2.0.

Half a century of vacuum tubes

Vacuum tubes (Fig. 6) dominated the electronics industry from approximately 1915 to 1960. The decades following 1915 saw the rapid development of radio technology, with specialized vacuum tubes tailored to different applications. The number of controlling electrodes increased steadily: two in the diode (1904), three in the triode (1907), four in the tetrode (1919, invented by the notable German physicist Walter Schottky), five in the pentode (1926), and so on. The most complex vacuum tubes developed had up to 12 electrodes, although anything beyond seven was scarce and generally used for niche applications.

There were ordinary receiving tubes and exotic transmitting tubes built for radar and other demanding applications. In some transmitting tubes, the power of the electromagnetic wave reached tens of millions of watts. Specialized tubes were also used in medicine, plasma physics, and particle physics. They bore impressive and intimidating names: *magnetrons*, *gyrotrons*, *klystrons*, *carcinotrons*, and *coaxitrons*. Some were several meters long and weighed hundreds of kilograms. In the largest and most powerful designs, the vacuum enclosures were made of metal rather than glass. At the other extreme, the smallest tubes measured just one inch.

All of them had heated cathodes and, for that reason, consumed significant power.

With tubes and amplification, radio, television, radar, and electronic instruments reached maturity. Vacuum tubes enabled all that followed: the worldwide telephone system, the birth of electronic computing, and, eventually, the development of semiconductor electronics and transistors. Although most tubes have been replaced, a few are still manufactured. Vacuum tubes continue to outperform transistors in high-power, high-frequency applications.

And then there's *tube nostalgia* — a lingering affection among audiophiles who insist that music sounds warmer, richer, and more alive when amplified through glowing glass rather than cold silicon.

Triodes as switches in digital calculators

As noted earlier, it took time for vacuum tubes to be incorporated into calculating machines. That wasn't the application John Fleming, Lee de Forest, or any early inventor had in mind when creating vacuum tubes. However, by the mid-1930s, the slowness of relay-based calculators had become a serious obstacle.

Relays are electromechanical devices, and mechanical motion is always slower than the dance of tiny electrons. Typical relay switching times ranged from 10 to 100 milliseconds (i.e., from 0.01 to 0.1 seconds), limiting switching frequencies to 10–100 Hz.

In contrast, vacuum tube triodes used in the first electronic calculator — the Atanasoff–Berry Calculator (ABC) — had switching times of just 1 to 10 microseconds (μs). That's 10,000 times faster. And when there's a ten-thousand-fold advantage, change usually follows.

In the following chapter, we'll describe the first vacuum-tube-based calculator, ABC, which debuted in 1940.

Semiconductors and the Transistor

"One shouldn't work on semiconductors; that is a filthy mess — who knows whether any semiconductors even exist," said the famously blunt Austrian physicist Wolfgang Pauli in 1931. Pauli, one of the founders of quantum mechanics, was renowned not only for his brilliance but also for his tactless pronouncements.

In one incident, Pauli was at a seminar where Albert Einstein was also present. When Einstein commented on something the speaker had said, Pauli quipped, "What Professor Einstein just said is not so stupid." The line became an instant classic among physicists. He was equally unsparing in his critiques of theory. Of ideas he deemed incoherent, he would say, "It's not even wrong!"

When Pauli dismissed semiconductors in 1931, the field already had a hundred years of primarily uneventful development behind it. As early as 1833, Michael Faraday identified a class of materials we now call semiconductors. He measured the electrical resistance of silver sulfide (Ag_2S) as a function of temperature and found that it *decreased* with rising temperature.

This behavior was the opposite of metals, whose resistance *increases* with temperature — a property known as a positive *temperature coefficient of resistance* (TCR). Semiconductors, by contrast, have a negative TCR. Faraday's observation was peculiar but remained unexplained for the next hundred years.

Several other materials were found to exhibit similar "anomalous" behavior. These included lead sulfide (PbS), also known as *galena*, and *copper oxide* (Cu_2O). In 1874, the German scientist Karl Ferdinand Braun discovered *unilateral conduction* — current flow in one direction — using a junction of galena and a springy metal wire. This was the first metal-semiconductor diode, affectionately dubbed the *cat's whisker*.[2]

It was the first practical application of semiconductors. Yet, as with the negative TCR, no one could explain why it worked. For decades, semiconductors remained a puzzle — neither understood nor particularly useful — until the advent of quantum mechanics and the ability to purify materials to unprecedented levels.

Purification and discovery of two types of semiconductors

A proper understanding of semiconductors — and what could be done with them — only emerged toward the end of the 1930s. It began with purification.

Perhaps Pauli wasn't entirely wrong. In 1931, semiconductors were a filthy mess because they were dirty. Their natural (intrinsic) properties were masked by impurities.

---

[2]Cat's whiskers — metal-semiconductor diodes — preceded vacuum tube diodes by 30 years.

Semiconductors are like Buddhist monks: They function properly only when purified, and that purification must be extreme. No other material in the history of technology has demanded such astonishing levels of purity.

Why is that?

Semiconductors are closer to insulators than to metals. They contain relatively few charge carriers — electrons or holes that can conduct current. In metals, purity matters much less because there are already vast numbers of free electrons. Very pure copper, for instance, conducts nearly the same as copper doped with 1% aluminum. But silicon (Si) and germanium (Ge), the key semiconductors, behave quite differently. In their pure (intrinsic) form at room temperature, they contain so few free carriers that even minute traces of impurities — orders of magnitude below 1% — can radically alter their conductivity.

At room temperature (taken to be 300 Kelvin, or 27°C/80.6°F), intrinsic silicon has roughly 10^{10} free carriers per cubic centimeter. Copper, by comparison, has about 10^{23}. That's a difference of 13 orders of magnitude — ten trillion times.

When scientists finally began to study highly purified semiconductors, they discovered something remarkable: the *sign* of the charge carriers could be controlled by the type of impurity added. This discovery — that one could deliberately tune semiconductors to conduct via either electrons or holes — turned out, in retrospect, to be one of the most important in modern electronics. It happened at Bell Labs in 1939. That was when semiconductors crossed the boundary — from "filthy mess," as Pauli called them, to god-sent future electronic materials. Credit for this breakthrough goes largely to Jack Scaff, Henry Theurer, and Russell Ohl.

Silicon, a group IV element on the Periodic Table, has four outer (valence) electrons. In a crystal lattice, each silicon atom bonds with four nearest neighbors, using all four of its valence electrons.

Now, consider adding an impurity. The electrical impact depends on how many valence (outer) electrons the impurity atom has. If it also has four (like carbon), nothing much happens. It fits right into the silicon "society", forming bonds with its neighbors and causing almost no disturbance.

However, suppose we introduce an element from group V, such as phosphorus (P), which has five valence electrons. That's one electron too many: Four of them participate in bonding; the fifth is left over.

That fifth electron becomes free. It can now move through the crystal and carry current. The more group V atoms we add, the more free electrons we get. The result is a semiconductor with enhanced conductivity. We refer to this material as n-type because the current is carried by negative charges — electrons.

What if we go the other way and add a group III element, such as boron (B)? Boron has only three valence electrons — one too few. It forms three bonds with neighbors but cannot complete the fourth. To solve this, it "borrows" an electron from a neighboring silicon atom. This borrowing leaves behind a vacancy: a hole, an empty spot where an electron could be.

And holes, oddly enough, behave like free positively charged particles.

Why are holes considered positive? A hole is like a bubble in water. We don't get surprised when we see a bubble rise in a pond; we don't call that antigravity. We know that water is moving downward around the bubble, giving the appearance of the bubble moving upward. Holes behave similarly in an electric field: They move in the direction opposite to electrons, just like positively charged particles.

Of course, in reality, there are no positively charged "hole particles." Electrons shift to fill an empty space, and the absence of an electron — the hole — moves the other way.

Holes behave like real particles in one more way: They have *mass*. That is, they accelerate when acted upon by a force. According to Newton's second law, mass is the ratio of force to acceleration. Since holes accelerate in an electric field, they have an *effective mass*, similar to that of electrons, although their mass may differ from an electron's mass.

In this sense, electrons and holes are on equal footing. Both behave like particles with mass and electric charge. The difference is philosophical: electrons are real; holes are the absence of something real. They are bubbles in the sea of electrons.

The theory of electron and hole conduction in semiconductors was first introduced in the 1930s by Walter Schottky and Nevill Mott; however, it was William Shockley in the 1940s who fully developed and formalized it.

A semiconductor with electron conduction is called *n-type,* while one with hole conduction is called *p-type.* When p-type and n-type regions are brought together to form a *p–n junction*, interesting things happen — as we'll soon see.

Fig. 7. Russell Ohl (1898–1987), photographed here in 1934, stands beside a seated colleague. The wall of equipment forms an imposing backdrop. Images of young Ohl are rare; though his work laid the foundation for modern semiconductor electronics, he remains poorly remembered. He deserved a Nobel Prize for his contributions to purification and the discovery of p–n junctions.

Source: Image courtesy of AT&T Archives and History Center.

Russell Ohl joined Western Electric's research department in 1922 at the age of 24 (Fig. 7). In 1925, Western Electric became a part of Bell Labs. In 1930, Ohl's supervisors — Walter Brattain and Joseph Becker — assigned him an unusual task: to catalog and characterize every known crystal detector of electromagnetic waves.

At the time, radio technology had already moved on to vacuum tubes, and to worry about old pre-vacuum-tube detectors — cat's whiskers — required a certain managerial foresight. Fortunately, Ohl's supervisors were exceptional. Brattain would later help invent the transistor (and win a Nobel Prize for it in 1956), while Becker was a leading expert on vacuum tubes who helped guide Bell Labs toward solid-state research.

Like many radio enthusiasts, Ohl had plenty of hands-on experience with crystal detectors — especially the *cat's whisker* variety, in which a thin metal wire touched a piece of semiconducting crystal. These devices were notoriously finicky and unreliable. After testing many materials, Ohl concluded that silicon (Si) and germanium (Ge) yielded the most consistent results.

Ohl, a trained chemist, set out to purify silicon in hopes of making sense of it. He began with the best he could find: silicon powder of 99% purity. He attempted to melt and re-solidify the material to drive out impurities. But this was no simple task. Silicon melts at 1410°C — hot enough to glow white, like the filament in an incandescent bulb. At such temperatures, impurities can leach into the container from its walls. The melting must be done in a vacuum or inert atmosphere; otherwise, the silicon will react with air — it burns.

In 1939, Ohl joined forces with Bell metallurgists Jack Scaff and Henry Theurer. Together, they designed specialized electric furnaces where silicon could be melted inside quartz tubes under ultra-pure helium. Quartz, or silicon dioxide (SiO_2), was ideal: It melts at 1670°C, even higher than silicon, and does not contaminate the melt, except with trace oxygen.

Crucially, Ohl allowed the molten silicon to cool slowly. He suspected that slow cooling would aid purification, and he was correct. His colleagues at Bell Labs later refined this method further.

The results were immediate. Electrical measurements became more consistent and predictable. The material was still far from achieving ultimate purity, but it was a major step forward.

In 1940, Russell Ohl discovered a peculiar piece of silicon — an *ingot*, as a solidified melt is called — that exhibited a striking *photovoltaic* effect. When illuminated by an ordinary lamp, it developed a voltage of nearly 0.5 volts between its two ends.

Just think about it: You shine a light on a piece of silicon, and the voltmeter across it reads half a volt, as though you had connected a small battery. Switch off the light, and the voltage vanishes. Quite remarkable.

This discovery laid the foundation for all future solar cells and solar panels.

Ohl demonstrated the effect to his colleagues and the director of research at Bell Labs, Mervin Kelly (Fig. 8), who summoned several people to his office to see the demonstration.

Fig. 8. Mervin J. Kelly (1895–1971) in a rare photograph from the 1930s when he initiated solid-state electronics development at Bell Labs.

Source: Image courtesy of AT&T Archives and History Center.

Mervin Kelly was a legendary figure to whom electronics owes an immense debt. A physicist by training, he earned his Ph.D. under Robert Millikan at the University of Chicago — the same Millikan who had mentored Harold Arnold, the inventor of the triode amplifier and future president of Bell Labs.

After graduating in 1918, Kelly began work at Western Electric, and in 1925, he moved to the newly formed Bell Telephone Laboratories. He was appointed director of research in 1930 and served in that position until 1944. He then became executive vice president for research (1944–1951) and, finally, president of Bell Labs (1951–1959). Even after retiring in 1959 at age 65, he continued to consult for "Ma Bell."

Kelly began each day at 5 a.m. in his Summit, New Jersey, home by tending to his tulip garden. Then, chauffeured to Murray Hill, he raced down the two-lane road that ran alongside the Watchung Reservation, a wooded landscape with a peaceful lake. It was a lovely road — I know because I drove it a thousand times in the 1980s, although perhaps a little slower than Kelly. That road, and much of the Reservation, was later destroyed in the construction of Highway 78 at the end of the 1980s.

This tulip gardener ruled Bell Labs with an iron fist. He was respected and even feared. Under his visionary leadership, Bell Labs undertook research in solid-state physics, culminating in the invention of the transistor.

One of the physicists summoned to witness Ohl's silicon miracle was Walter Brattain, a brilliant physicist and formerly Ohl's direct supervisor. Brattain suggested that the ingot might contain an invisible boundary between regions of different impurities — what we now refer to as *n-type* and *p-type*. Illumination of this boundary, he hypothesized, created the observed charge separation. Brattain had seen similar barriers form between copper oxide (Cu_2O) and metallic copper. His guess was correct. This may have been the first time anyone realized that silicon could support two different types of conduction: n-type, characterized by an excess of electrons, and p-type, characterized by a deficiency of electrons — positive *holes*.

Ohl and Jack Scaff later confirmed Brattain's conjecture. Analysis revealed that one side of the ingot contained trace amounts of phosphorus, a group V element, producing excess free electrons. The other side contained boron, a group III element, creating holes. These impurities were not introduced intentionally; their concentrations were minuscule, and no one knew they were present until Ohl and Scaff investigated.

They had stumbled upon the first *p–n junction* — entirely by accident.

The discovery of the p–n junction was a landmark. Such junctions would later become the foundation of *diodes* and, in trilayer form (n–p–n or p–n–p), *transistors.* In 1940, these applications were far from obvious. But the effect Ohl had discovered — a voltage appearing in response to light falling on a p–n junction — was intriguing enough to merit patent protection. Ohl filed U.S. Patent No. 2,402,662, *Light-Sensitive Electric Device*, on May 27, 1941. It was granted on June 25, 1946. Ohl's discovery influenced later solar cell development, including the more efficient Bell Labs 1954 solar cell developed by Chapin, Fuller, and Pearson.

Many scientists shifted their priorities after December 1941, when the U.S. entered World War II. But Ohl and Scaff continued their semiconductor research, now aligned with national defense.

Great Britain's successful use of *radar* in air defense had shown the power of high-frequency electromagnetic detection. The best detectors for radar systems operating at wavelengths below one meter (i.e., above 300 MHz) were silicon-based *cat's whiskers*. The purification work that Ohl and his colleagues had begun at Bell Labs fit perfectly into this demand. By late 1940, silicon-tungsten cat's whiskers were being produced for radar detectors by the Thomson-Houston Company in Britain and by General Electric (GE) in the U.S.

By the end of 1941, DuPont Company, prompted by American scientists, developed a new chemical process that yielded much cleaner silicon: first the so-called "4-nines," 99.99%, and then the "5-nines," 99.999%. At 5-nines purity, there is one impurity atom for every 100,000 silicon atoms — a major leap from earlier "industrial-grade" silicon, which had one impurity for every hundred atoms.

Purified silicon was in great demand among those who worked with crystal detectors. Ohl obtained several pounds of 5-nines silicon from DuPont (Riordan and Hoddeson 1997, 102).

But it wasn't the end of the road. Even higher purity was needed to extract the full potential of semiconductors. In the early 1950s, Bill Pfann and Jack Scaff at Bell Labs achieved 10-nines purity: 99.99999999% — one impurity atom for every ten billion silicon atoms. No other material on Earth — possibly anywhere in the Universe — had ever achieved such chemical perfection.

As the 1930s drew to a close and the tragic 1940s began, it was clear that semiconductors were subtle, sophisticated materials — far more nuanced than metals — and that they held immense promise. *Cat's whiskers* belonged to the otherwise invisible cat waiting to appear, like the Cheshire Cat in *Alice in Wonderland.*

When I Met Sir Nevill Mott

Quantum mechanics appeared in 1925–1926. The first testing ground for this anti-intuitive new theory was atomic and nuclear physics, including atomic energy levels, radioactive decay as a form of quantum tunneling, the structure of the Periodic Table, and the quantum basis of classical chemistry. By the late 1920s, several theorists had begun applying quantum mechanics to the physics of solids.

Over the next decade and a half, the foundations of solid-state theory were laid by, among others, Felix Bloch, Léon Brillouin, Arnold Sommerfeld, Wolfgang Pauli, Eugene Wigner, Max Born, Rudolf Peierls, Nevill Mott, John H. Van Vleck, Lev Landau, Alan H. Wilson, and Yakov Frenkel.

Some, such as Bloch and Wilson, developed general theories of metals, semiconductors, and insulators. Others focused on explaining specific effects and material properties. Together, their work made possible everything that followed in the field of electronics.

These founding fathers lived and worked long before my time, but I did have the good fortune to meet one of them: Sir Nevill Mott.

At the end of 1982, while working at Bell Labs, I submitted a paper to *Physical Review B*, the solid-state section of *Physical Review* (Gurvitch 1983). At the time, it was customary to send preprints of one's paper to scientists who might be interested. My topic fell squarely within Mott's domain of expertise, so I sent a copy his way. To my surprise — and great delight — he responded with a handwritten letter.

In 1982, Mott was 77 years old and remained very active. He would continue working in theoretical physics until his death in 1996 at the age of 91. By then, he was a living legend. Mott introduced the famous *Mott transition* between metals and insulators. He was one of the earliest physicists to consider *holes* as positively charged carriers in semiconductors. His books, such as *Metal-Insulator Transitions*, were standard references on every solid-state physicist's shelf. Between 1954 and 1971, Mott served as Cavendish Professor of Physics at the University of Cambridge. His predecessors in that post included James Clerk Maxwell, Lord Rayleigh, J. J. Thomson, Ernest Rutherford, and W. L. Bragg — not bad company, scientifically speaking. In 1977, Nevill Mott received the Nobel Prize in Physics.

So, when I received his letter, typed on elegant Cavendish Laboratory stationery — "From Professor Sir Nevill Mott, F.R.S.[3]" across the top — it felt almost mythic. Zeus himself could have sent it, postmarked "Mount Olympus," with a thunderbolt for a stationary logo.

---

[3]F.R.S. stands for *Fellow of the Royal Society*.

In late November 1982, Mott visited Bell Labs and was escorted to my lab in room 1D-321. (It was my third year at Bell, and I still didn't have an office.) Every visitor had a tight schedule — half-hour slots marked by name and room number. Going over your allotted time was considered rude and disruptive. Mott arrived at my hastily cleaned lab, accompanied by his Bell Labs host, Gordon Thomas, who introduced us and left. Mott quickly realized I wasn't the theorist he'd imagined but an experimentalist. That may have been a mild disappointment, but he was gracious and generous, sitting beside me at my desk for 30 minutes and jotting down formulas as we talked. I couldn't tell him much beyond what was in my paper, but it didn't matter. He did most of the talking. I remember him repeatedly acknowledging the contributions of his collaborator, Moshe Kaveh, an Israeli physicist who had spent time in Cambridge. After Mott's death, Kaveh contributed to a book about Mott edited by Davies (1998).

After our meeting, I escorted Sir Nevill to the upstairs conference room, where he gave a seminar for Bell's theorists, including our own Nobel laureate, Philip Anderson. (Mott had shared his 1977 Nobel Prize with Anderson and Van Vleck.)

Anderson was one of the most important theoretical physicists of the postwar era. His theory of *electron localization* in disordered solids had become a central topic in condensed matter physics. Mott's own work on conduction in disordered solids was closely related to Anderson's. This connection was part of the reason they shared the prize.

Mott began his seminar with the phrase: "*Talking about localization at Bell Labs is like bringing coal to Newcastle.*" It was a lovely compliment to Anderson, Abrahams, Lee, and the other theorists in the audience. That, incidentally, is how I first learned that British idiom: No one brings coal to Newcastle — it's already there in abundance. Later, while photographing Mott's letter for my previous book, I noticed the postmark: Newcastle-Upon-Tyne. Apparently, Mott was at Newcastle when he wrote that letter to me, which was shortly before he visited America — perhaps that's why the phrase came to him so naturally.

Years later, I visited the science museum at Cavendish Laboratory, now relocated to the outskirts of Cambridge. As I pulled into the visitors' parking lot, the first thing I saw was the word MOTT — written in giant letters on the side of a building. Every building at the new Cavendish complex is named after one of its luminaries.

Solid-state physics

One of the central puzzles was this: Why do solids fall into three radically different electrical categories, despite all being made of atoms and electrons?

- Metals, such as copper and aluminum, conduct electricity with ease.
- Insulators, such as glass, silica, or rubber, do not conduct electricity at all.
- Semiconductors, including silicon, germanium, and galena, fall somewhere in between. Additionally, their conductivity is extremely sensitive to impurities, as described earlier.

The difference lies not in how many electrons the material contains in total, but in how many of those electrons are *free to move* — the so-called *free carriers*. This includes electrons and *holes* — the absence of electrons that behave like positive charges.

Whether electrons are mobile or locked in place depends on the material's *electronic energy spectrum*, which comprises *energy bands* (continuous ranges of energy where electrons can exist) and *energy gaps* (forbidden zones without electron states).

In 1925, Wolfgang Pauli, who was "disrespectful" to Einstein[4] and "badmouthed" semiconductors in 1931, proposed the *Pauli Exclusion Principle*. This principle, which governs how electrons fill available energy levels, underpins atomic structure and the Periodic Table, and it also explains why metals, insulators, and semiconductors exhibit such distinct behaviors.

I won't attempt to explain the Exclusion Principle's connection to the classification of solids here, but a simplified discussion can be found in Gurvitch (2023).

While the Exclusion Principle and other quantum laws form the foundation of solid-state physics, it was the work of many physicists that explained the behavior of actual materials. One of them was Nevill Mott, who, as we said already, clarified the role of *doping* and *holes* in semiconductors, developed a many-body theory of *metal–insulator transitions*, and pioneered the study of *electrical conduction in disordered systems*.

---

[4]They were actually on very friendly terms; Einstein liked Pauli and vice versa. It was Einstein who nominated Pauli for the Nobel Prize, which Pauli received in 1945.

A simplified version of a theory often considers *two-body* interactions. For example, the simplest theory of planetary motion — the one that predicts elliptical orbits for the planets — treats the gravitational pull between the Sun and a single planet, ignoring the weaker interactions among the planets.

In contrast, a many-body theory is a more challenging, but ultimately more powerful, approach in which interactions among the constituent particles are taken into account. In solid-state physics, this means considering the interactions between electrons, not just their motion in an external field or within a fixed lattice.

Many of the most significant advances in modern physics came when theorists embraced this complexity. For instance, the BCS theory of superconductivity is a classic example of a many-body theory. So is the theory of electronically driven phase transitions — a field in which Nevill Mott was a founding father.

The name that needs to be mentioned when discussing the understanding of the three classes of solids is Alan Wilson, who worked on this subject between 1931 and 1937. He was the one who proposed the concepts of energy bands and energy gaps in solids.

By the end of the 1930s, solid-state physicists not only understood the difference between metals, insulators, and semiconductors, but they also understood most of the intricate details of their behavior, such as the dependence of their conductivity on temperature and what happens when we make contact between the regions of electron and hole conduction in semiconductors. All that understanding was theoretically based on quantum mechanics, while experimentally, it was based on numerous and diverse experiments, including those with pure and controllably doped semiconductors.

Solid-state group at Bell Labs and the invention of the transistor

By the end of the 1930s, it became clear to Mervin Kelly, the whip-smart and all-powerful director of research at Bell Labs, that semiconductor devices were the future of electronics. To Kelly's credit, he saw this coming in a world dominated by vacuum tubes. At that time, solid-state electronic devices were limited to just two types of diodes: the cat's whisker, used for radio wave detection, and another diode used in power applications to rectify alternating current. Both relied on the contact between a metal electrode and a semiconductor. Had these devices disappeared in 1939, few people would have noticed, and even fewer would have cared. Radio enthusiasts would miss the cat's whisker, which worked well at high frequencies, and radar engineers might miss silicon–metal diodes. But vacuum tubes could perform all the same functions and more.

Yet Kelly, along with forward-looking Bell Labs managers such as Joseph Becker, saw the latent potential of solid-state electronics. In 1938, anticipating a shift, Kelly established a small group in solid-state physics. One of the first recruits was 26-year-old theoretical physicist William Shockley, a recent MIT Ph.D. who had joined the Labs in 1936, "...the quickest mind I've ever known," according to Phillip Anderson (Gertner, 2012, p. 56). Coming from Anderson — possibly the greatest solid-state theorist of the second half of the 20th century and not a man who easily dispensed praise — it meant a lot. Another group member was Walter Brattain (whom we mentioned above), a brilliant experimentalist who joined Bell Labs in 1929.

World War II interrupted all research. Scientists from Bell Labs were scattered to war-related projects. But after the war, in 1945, prewar efforts resumed. A new theorist, John Bardeen, joined the solid-state group. He had worked at Harvard before the war and led a large group at the Naval Ordnance Laboratory in Maryland during the war. Shockley, Fisk, and Brattain recommended him, and Kelly offered him a generous salary, higher than any university could pay (Gertner 2012, p. 88).

In 1941, Bell Labs moved its headquarters from downtown Manhattan to Murray Hill, New Jersey. From 1979 to 1989, I worked in that famous building and described it in a chapter for a book on superconductor electronics (Gurvitch, 2017, pp. 83–139). Many discoveries in solid-state physics and electronics were made there. Of the seven Nobel Prizes in Physics awarded for work done at Bell Labs, six came from research conducted at Murray Hill.

In June 1945, one month after the war ended, Kelly — then the president of research — initiated a major reorganization. Inspired by the interdisciplinary success of wartime projects, including the Manhattan Project, he created mixed teams of physicists, chemists, materials scientists, and engineers. The solid-state group, now under Shockley's leadership, was tasked with developing new electronic devices, including a solid-state alternative to the vacuum triode — a power amplifier made entirely of semiconductor materials.

Walter Brattain conducted measurements using high-purity silicon and germanium crystals. John Bardeen worked beside him, analyzing the data and formulating new theories. The two became close collaborators and friends. While other group members were kept informed, the Brattain-Bardeen collaboration was especially close.

The interdisciplinary character of the Solid-State group paid off in mid-November 1947, when Bell Labs chemist Robert B. Gibney joined Walter Brattain in tackling the problem of semiconductor surface preparation. Their use of electrolyte treatment and oxide removal — an approach rooted in electrochemistry — helped overcome the barrier posed by germanium surface states and proved pivotal in the weeks leading to the first point-contact transistor (Riordan and Hoddeson, 1997, pp. 138–141).

Around the same time, Bardeen developed a theory of semiconductor surface states to explain Brattain's experiments. At the end of November, Bardeen walked into Brattain's office with an idea: a new sample geometry using three electrical contacts on a semiconductor crystal, which he believed could achieve amplification.

In typical Bell Labs fashion, they tried it immediately. Within half an hour, in Brattain's lab, they saw signs of amplification — the effect had been caught "by the tail." They informed their group leader, Shockley. That November 21 experiment marked the beginning of a most productive month. By December 16, they had refined the geometry and materials and observed significant power amplification. Bardeen specified the requirements: a cleaned surface of n-type germanium contacted by two closely spaced metal points. Brattain built a simple device using a razor-cut gold foil pressed against the crystal by a plastic wedge. A third contact, a gold plate[5] under the crystal, completed the setup (Fig. 9). The result was amplification by a factor of 10 or more.

What Bardeen and Brattain built in December 1947 was later called the *bipolar point-contact transistor*. The term "transistor" was coined six months later by group member Gerald Pearson, blending "transconductance" (a term from vacuum tubes) with the familiar "-istor" suffix, as seen in resistor, thermistor, and varistor. December 16, 1947, is considered the birthdate of the transistor.

[5] You might think that using gold plates was something fancy and unusual. But no — when 32 years later I started working at Bell Labs, I could check out a gold plate from the special precious-metal stockroom just as easily. Gold is one of the best conductors of electricity among metals and is immune to oxidation.

Fig. 9. J. Bardeen, W. Shockley, and W. Brattain in a staged photo taken in early 1948. Right: the first transistor. A spring made from a paper clip presses a plastic wedge against a gold foil with a razor cut, making two closely spaced contacts to a black germanium crystal. Two gold wires connect to the contacts; the gold plate under the crystal serves as the third contact.

Source: Images courtesy of AT&T Archives and History Center.

William Shockley, meanwhile, had been focused on a different approach: the *Field-Effect Transistor* (FET). He was only loosely aware of Brattain and Bardeen's progress. The FET idea was not new — it had been patented in 1925 by Julius Lilienfeld and again in 1935 by Oskar Heil — but technology had not yet caught up. FETs would not become practical until the late 1950s.

On December 23, 1947, Brattain and Bardeen demonstrated their device to Bell Labs management. Present were the director of physics research and the vice president of research, but not Kelly himself. The executive vice president of Bell Telephone Laboratories, Mervin Kelly, was not invited out of an abundance of respect — he would not be called to witness a demonstration that could fail; simply stated, he was feared.

Brattain connected the device to a circuit. Words spoken into a microphone were heard softly when the transistor was off and loud and clear when it was on. The demonstration worked. The transistor amplified power by a factor of 20. Management was impressed.

Why the transistor is one of the greatest inventions of all time

The first transistor (Fig. 9) didn't look like something that would take the world by storm, but it did. Why?

The transistor can perform two entirely different functions: It can amplify electrical power, as demonstrated in the first instance in December 1947, or it can act as a fast electronic switch. This second function — crucial for digital logic — had not yet been proposed in 1947 but was suggested soon after, in 1948–1949, by John R. Pierce[6] and Claude Shannon, of whom we will speak later. It would become central to digital electronics and the development of modern computers, and we'll revisit its genesis later in the book.

Perhaps equally important was the fact that, as a solid-state device, the transistor enabled *miniaturization*. Unlike vacuum tubes, which have fixed physical limitations (requiring a heated cathode and a vacuum enclosure), transistors can be scaled down. This became evident in the late 1950s with the advent of *integrated circuits* (ICs). While vacuum tubes were bulky and fragile, transistors kept shrinking — and continue to do so. Today, hundreds of billions of them can fit on a single chip. That scaling power, more than anything else, underpins the astonishing capabilities of modern electronics.

By any reasonable measure, the transistor ranks among the most important inventions in human history. We can get a sense of what it meant to electronics practitioners from the words of Japanese physicist and Director of Sony Research Center, Dr. Makoto Kikuchi:

> If I could thank God for only one thing, it would be for giving me life in the year 1925. Thanks to that blessing, the start of my research career were to coincide with the birth of the transistor. (Kikuchi 1983, 22)

Bell Labs produced many groundbreaking inventions, but none were as consequential as the transistor. As I mentioned earlier, I was fortunate to work at Bell Labs in Murray Hill from 1979 to 1989. My lab — room 1D-321 — was on the third floor in the Basic Research Area (Area 11) of that legendary institution, which at the time employed about 3,000 people at Murray Hill alone. Across all Bell Labs locations in New Jersey and Pennsylvania, there were over 25,000 employees. AT&T, its parent company, employed over 600,000.

---

[6]John R. Pierce (1910–2002) was a physicist and engineer at Bell Labs, a member of Shockley's solid-state group, and a key contributor to satellite communications, information theory, and digital sound. He coined the term *transistor* in 1948.

My lab was located just one floor below and only a short walk from room 1E-445, where the transistor was discovered. A small commemorative plaque marked the spot in the fourth-floor corridor.

One day — during my first or second year at Bell Labs — my friend Nadia Lifshitz pointed out an older, bald, heavyset man walking down the corridor (those famous half-mile-long corridors of Bell Labs). She asked, "Do you know who that is?" I didn't. "That's Phil Foy, the technician who built the first transistor with his own hands."

Later, I read that Foy, Walter Brattain's assistant, had mainly prepared the materials while Brattain assembled the device. But still — there he was, walking past me. When I arrived in 1979, all three inventors of the transistor — Bardeen, Brattain, and Shockley — were still alive, although they were no longer at Bell Labs. Foy was, I think, the last one left at Murray Hill from that famous solid-state group. In those days, Bell Labs had many highly skilled technicians. Their hands helped make many of the Labs' famous discoveries a reality.

Years later, I saw a television program about Area 51 in the Nevada desert, a favorite subject of UFO enthusiasts. As usual, the program speculated that the U.S. government had recovered a crashed alien spaceship. A distinguished-looking, gray-haired man confidently declared that the transistor "appeared out of nowhere" in the early 1950s. Surely, he said, it must have been reverse-engineered from alien technology.

As I listened, I thought, "For Christ's sake, there's a plaque on the fourth floor of the Murray Hill building, and I saw Phil Foy with my own eyes in that corridor. So — ignorant, distinguished-looking old man — don't tell *me* that nonsense."

William Shockley and the invention of the bipolar junction transistor

Let us return to the end of 1947. Shockley was torn. As he later admitted, "My elation with the group's success was tempered by not being one of the inventors" (Gertner, 2012, p. 101). In the photograph shown in Fig. 9, Shockley sits at Brattain's workbench, with the other two standing behind him. This was a publicity photo staged in January 1948. Shockley planted himself in that chair before a microscope, even though it was probably the first and last time he took that seat. He seems tense; they all look tight and uncomfortable in that famous photograph. Brattain seems more than tense — he looks enraged. Bardeen glances at Brattain as if to warn him not to react, as if saying, "Let it go, mate."

It's a fact that Brattain and Bardeen could never forgive Shockley for his behavior in the days, weeks, and months following the transistor's discovery. For his part, Shockley couldn't forgive himself for not having been at the helm when the moment came. He even tried to insert himself into the transistor patent with Bardeen and Brattain — something even Bell Labs' patent attorneys wouldn't support.[7]

For all his flaws (and there were many), Shockley was a brilliant scientist and technologist. He understood that the point-contact transistor was delicate and difficult to manufacture; it was not a practical device. He wanted to build a better version — one that could be industrialized. Most of all, he wanted it to be *his*.

Shockley spent New Year's Eve of 1948 thinking and calculating, and then he went to a conference in Chicago, where he did not show his face in any of the sessions. He was sitting in his hotel room, feverishly trying to invent a better transistor. He proved himself worthy of the challenge: On January 23, 1948, just one month after the December 23 demonstration, Shockley came up with a groundbreaking design, which eventually won the day — the *Bipolar Junction Transistor*, or BJT (Fig. 10). His design drew on Russell Ohl and Jack Scaff's 1940 discovery of the *p–n* junction. While a single *p–n* junction forms a diode, Shockley envisioned a tri-layer structure — either *p–n–p* or *n–p–n* — with each layer contacted electrically. This three-terminal device, similar to the vacuum triode or the point-contact transistor, can function as both an amplifier and a switch.

Today, only history lovers recall the point-contact transistor, but every electronics or solid-state-physics student is familiar with Shockley's BJT, which remains widely used. Shockley initially kept the design secret, but fearing someone else might beat him to it, he disclosed it to Bell Labs colleagues a few months later. In June 1948, he filed a patent for the BJT. The device wasn't built immediately; the semiconductor technology wasn't ready. It finally became practical in 1951.

[7]I remember that when I worked on my first patent at Bell Labs in 1981, one of the patent attorneys asked me, "What do you think is the easiest way to overturn a competitor's patent in a court of law?" He answered himself, "To show the judge that one of the people listed on a contested patent was not really involved with the invention."

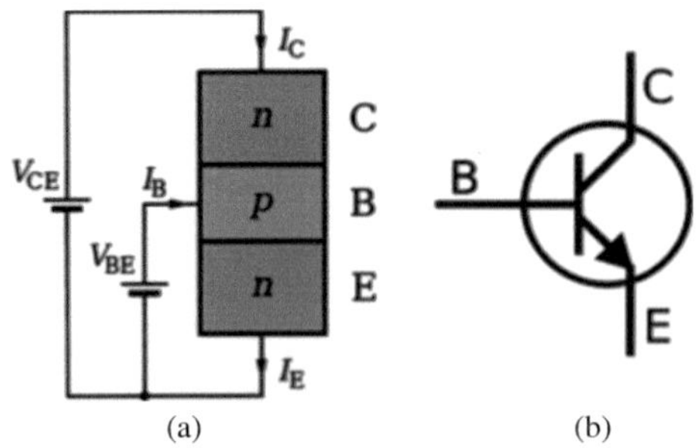

Fig. 10. (a) Bipolar Junction Transistor (BJT) with batteries and currents, and (b) its circuit symbol. The middle *p*-type layer is much thinner than the outer layers. Letters C, B, and E represent Collector, Base, and Emitter. The large collector current I_C combines with the small base current I_B to form the emitter current I_E, which is equal to $I_C + I_B$. The base current I_B is typically 100 times smaller than I_C, enabling current, voltage, and power amplification. The device can also work as an electric switch.

Source: Images by Inductiveload and Omegatron, Lic. CC BY-SA 3.0.

In 1956, Bardeen, Brattain, and Shockley were awarded the Nobel Prize in Physics "for their researches on semiconductors and their discovery of the transistor effect." It was understood that Bardeen and Brattain received the prize for the first working transistor — the point-contact device — while Shockley received it for the practical BJT.

By then, relationships were severely strained. Shockley had always been difficult, and his post-discovery behavior alienated the others. Bardeen left Bell Labs in 1951. Brattain refused to work with Shockley and moved to another group. Witnesses say that by 1956, in Stockholm, the three Nobel laureates were civil but cold.

Shockley didn't remain at Bell Labs either. Riding the wave of Nobel publicity, he secured funding in 1956 to start his own company, *Shockley Semiconductor Laboratory*, in Mountain View, California — the future *Silicon Valley*. This company's name somehow hints at an inevitable commercial failure that did not wait long to happen.

Shockley recruited a team of brilliant young scientists, including Robert Noyce, Gordon Moore, Eugene Kleiner, and Jean Hoerni. But his authoritarian management and questionable priorities soon alienated them. In 1957, eight of his top employees — later dubbed the "Traitorous Eight" — left to form *Fairchild Semiconductor*. Fairchild soon outpaced Shockley's company, developing more practical planar transistors and launching the age of ICs. Shockley Semiconductor was sold to Clevite Corporation in 1960 and later became part of ITT.

Though his company failed, Shockley's impact endured. He is considered one of the most influential solid-state theorists in history. In addition to his Nobel Prize–winning work, William Shockley published his seminal textbook, *Electrons and Holes in Semiconductors with Applications to Transistor Electronics*, in 1950. This comprehensive volume became a foundational resource in semiconductor physics and transistor technology — I remember seeing it on the shelves of my colleagues at Bell Labs.

His lab was the first semiconductor company in what would become Silicon Valley. That's why people say Shockley was *the man who brought silicon to Silicon Valley* — and that he certainly did.

Two Nobel Prizes for John Bardeen

Bell Labs announced the first transistor — the point-contact device developed by Bardeen and Brattain — in 1948, shortly after the patents were filed. The BJT was built in 1951. Following the 1948 announcement, transistors quickly captured the attention of electronics laboratories worldwide. Earlier, we quoted one particularly exuberant reaction from a Japanese researcher.

In 1956, when the Nobel Prize in Physics was awarded to Bardeen, Brattain, and Shockley for inventing the transistor, John Bardeen had already moved to the University of Illinois at Urbana-Champaign. That same year, he published, together with Leon Cooper and Robert Schrieffer, a paper that introduced the microscopic theory of superconductivity, known as the BCS theory. This was a tremendous and long-awaited breakthrough. Since its discovery in 1911, superconductivity has remained a mystery[8] for 45 years. The BCS theory finally offered a comprehensive, quantum-mechanical explanation. It was quickly nominated for the Nobel Prize.

But there was a problem. According to Alfred Nobel's will, no individual could receive more than one Nobel Prize in the same discipline.

[8] Well, perhaps not a complete mystery. However, the *microscopic theory* — the most valued kind of theory in physics — was missing until the work of Bardeen, Cooper, and Schrieffer.

A person could receive one in Physics and another in Chemistry (as Marie Curie did), or one in Chemistry and one in Peace (as Linus Pauling did), but never two in Physics.

Yet Bardeen, the principal author of BCS, could not be excluded from it, making it a mere "CS theory." Nor could the committee revoke his earlier award. So, the Nobel Foundation did something it had never done before: it broke its own rule, just this once in Physics. In 1972, Bardeen received his second Nobel Prize in Physics, alongside Cooper and Schrieffer, for the theory of superconductivity. To this day, he remains the only person ever awarded two Nobel Prizes in Physics.[9] Unsurprisingly, he became a legend.

In light of his fame, I can't resist recalling my one brief interaction with Bardeen. Of course, I had seen him at conferences, but I never dared to approach him. We young physicists admired him from a distance.

At one conference in Berkeley, California, in the summer of 1987, Bardeen arrived late at night, and I happened to be in the hotel lobby (I believe it was the Berkeley Marriott), where we were all staying. The clerks at the reception asked me to help him with his bag and bring it to his second-floor room, and I gladly did. But that's not the episode I really want to recount here.

In late 1987, my Bell Labs colleague Tony Fiory and I published a paper in *Physical Review Letters* (PRL) about the resistivity of the newly discovered high-temperature superconductors. In that memorable year, *everyone and their uncle* worked on superconductivity — those who always had and those who had just jumped on the bandwagon.

Shortly after our paper appeared, we received a letter from John Bardeen. He argued that our interpretation of the resistivity data was incorrect and informed us that he was preparing a *Comment*[10] for PRL to explain his view.

We were ecstatic. A letter from Bardeen! He noticed our work! He would publish a comment on it in PRL, where the whole community would see it! And best of all, we quickly realized that the great man was

---

[9]A similar exception was made in Chemistry: Frederick Sanger received the 1958 and 1980 Nobel Prizes in Chemistry.

[10]*Physical Review Letters*, arguably the most coveted journal in physics, included a special section called "*Comments*." Publishing a comment — and the author's reply — was considered as prestigious as publishing a full paper in PRL. For one thing, it signaled that the original work had sparked significant interest and debate within the community.

wrong this time. It was so tempting to let him go ahead and publish his critique, then respond, win an argument with a two-time Nobel laureate, and obtain everlasting glory.

But to our credit, we did the right thing. We wrote a polite and carefully worded letter to Bardeen explaining our position. He replied graciously, saying that upon reconsideration, he agreed with us. It was late December 1987. In our reply, we wished John Bardeen a heartfelt Happy New Year.

Transistor's first steps

The transistor patent was assigned to Bell Labs in 1948; however, the Bell System received relatively little financial compensation from it, despite its global impact. Initially, AT&T offered to license the transistor to other companies for a modest fee of $25,000. Bell Labs hosted two licensing symposia in 1951 and 1952. About 40 companies paid the fee, receiving the license and technical know-how in return. After 1952, however, no further licensing fees were collected, even as transistors were taking over the electronics world.

The Bell System did not pursue profits more aggressively, partly due to the difficulty of enforcing patents (since nearly every electronics lab was *infringing* on its patents by making their own transistors) and partly because AT&T was a government-regulated telephone monopoly. As such, it was expected to behave with ethical and public-minded restraint.

For the first seven years following 1947, all transistors were made of germanium. However, in 1954, the first silicon transistors emerged, and germanium soon gave way to silicon, which remains dominant in electronics to this day.

The first transistorized consumer product in the U.S. was a hearing aid,[11] released in 1952. It contained a few vacuum tubes and a single transistor. In 1953, the first all-transistor hearing aid was introduced, and transistors rapidly replaced tubes in this compact, size-sensitive application.

---

[11] When Bell Labs licensed the transistor technology, manufacturers of hearing aids were excused from having to pay a licensing fee; that was done in honor of Alexander Bell, who worked with deaf people. It was a very nice touch — Bell Labs was a noble organization!

Fig. 11. Left: TR-1, the world's first commercial transistor radio, manufactured by Regency and Texas Instruments in 1954. Right: VEF Spidola transistor radio, popular in the Soviet Union during the 1960s.

Source: Left image by Joe Haupt; right by Marcin Szala.

The next commercial application was portable radios, which hit the U.S. market in October 1954. The first model, the Regency TR-1, sold over 100,000 units at \$49.95 — a steep price at the time (Fig. 11). Less than a year later, in March 1955, Sony entered the U.S. market with its TR-52 transistor radio, priced at \$29.95, and quickly established itself as a market leader.

> I remember the transistor radios shown on the right of Fig. 11 — VEF Spidola, made by a Latvian company in the Soviet Union during my youth. They had a short-wave band that allowed us to tune into Western broadcasts in Russian, such as Voice of America and Radio Liberty. The Soviet government attempted to jam these signals, but not always with success. Sometimes, especially at night, we could hear them. It was a big deal for us — one of the few sources of factual information about the outside world. Because of that, the transistor radio, for me, is associated with freedom.

Before the first transistor radios were sold, the world's first transistor-based computer was built in 1953 at the University of Manchester in England. Although not a commercial product, it was a milestone. The first commercial solid-state computer was the IBM 608 transistor calculator, launched in 1957. After building that machine, IBM boldly decided to abandon vacuum tubes entirely, and they stuck with it.

The semiconductor industry's history in the 1950s and early 1960s is one of rapid, unplanned, and aggressive development. Traditional marketing strategies were set aside. Instead of responding to demand,[12] companies began introducing products as soon as they could manufacture them, hoping to find a market. Between 1950 and 1962, the number of semiconductor manufacturers increased dramatically, growing from a single Western Electric (AT&T's manufacturing arm) to hundreds of companies worldwide.

By the mid-1960s, thousands of different transistor types were produced. Many were short-lived; others endured. Until the early 1960s, germanium and silicon transistors coexisted. In the mid-1950s, the original point-contact transistors (invented by Bardeen and Brattain) were replaced by BJTs, invented by Shockley. After 1960, FETs began to appear.

There were low-power and high-power transistors, as well as ones for medium, high, and extra-high frequencies. Some were expensive and specialized, while others were cheap and general-purpose.

This evolution was anything but smooth. There were more dead ends and failures than breakthroughs. So, what drove the engineers and companies that navigated these uncharted waters?

Part of the motivation was the sheer excitement of the new technology and the growing realization that semiconductors marked the beginning of a new era in electronics. That recognition, far from universal at the start of the 1950s, gained momentum as device performance improved and batch manufacturing methods became more prevalent.

And, of course, there was the prospect of large profits.

Transistor applications started modestly — with hearing aids and portable radios — but expanded rapidly. By the late 1950s, the U.S. military became the single largest buyer of transistors. The military was attracted by transistors' small size and weight, low power consumption, and anticipated reliability. Transistors were integrated into aircraft, submarines, and missiles, and later into NASA's space program.

[12] In recent times, electronics has ceased merely responding to public demand. Instead, it increasingly sets its own targets and draws the public toward them. This tendency is part of the growing autonomy of electronics as a system.

After 1959, transistors were incorporated into ICs, eventually becoming the most-manufactured objects in history. We will describe the genesis of ICs in the following section.

The Pillars of Transistor Technology

Germanium and Silicon

As we mentioned, semiconductors have been known — though barely noticed — since 1833, and since 1906, they've found limited use as metal-semiconductor radio-wave detectors.[13] However, everything that happened before 1940 can hardly be called semiconductor technology. Around that time, the "old" semiconductors — galena, selenium, and copper oxide — were largely abandoned. Their place was taken by two elemental semiconductors from Group IV of the Periodic Table: silicon (Si) and germanium (Ge).

Initially, germanium was the semiconductor of choice, primarily because silicon, with its high melting point of 1415°C (compared to 937°C for germanium), was more difficult to process. However, as experience accumulated, the advantages of silicon became increasingly evident. Enthusiasm for silicon grew in inverse proportion to the growing disillusionment with germanium.

By 1960, it was apparent that silicon was a godsend material — a lucky accident of nature and an answer to technologists' prayers. Where would biology be without carbon? Nowhere. And where would electronics be without silicon? In the same place.

Silicon (Si) is one of the most abundant elements on Earth — sand is essentially silicon dioxide (SiO_2). In fact, about 90% of Earth's crust is made up of silica, silicates, and silicides. As a result, metallurgical-grade silicon (about 98% pure) is quite literally "dirt cheap." Of course, the price climbs steeply when silicon is purified to parts per billion and transformed into a perfect single crystal, as we'll discuss in the following.

I recall how Sadeg Faris, founder and first president of the 1983 startup HYPRES Inc.,[14] once told me a story about returning to his native Libya

[13] In 1906, an American inventor, Greenleaf W. Pickard, patented a practical metal–semiconductor crystal detector.

[14] I consulted for HYPRES from its start in 1983 until the early 1990s. Sadeg Faris asked me to join his company's Scientific Advisory Board because, at Bell Labs, I developed the devices — refractory niobium-aluminum Josephson junctions — that HYPRES used in all of its products.

after completing his Ph.D. at the University of California, Berkeley, in 1976.

Sadeg is an adventurer. Once back in Libya, he managed to obtain an audience with the country's tyrannical ruler, Muammar Gaddafi. According to Sadeg, he said to Gaddafi:

> Colonel! What is the most abundant resource we have in our beloved Libya? Surely, it is sand—lots of sand. But do you know that all transistors are made of silicon, and silicon is the major component of sand? So, with that resource freely available to us and with the good knowledge I acquired in Berkeley, California, why don't we start a factory — no, a *big* factory — producing silicon transistors right here in Libya? Invest a mere billion dollars, put me in charge, and you'll see what I can do.

To which Gaddafi replied:

> Very good, very good. But first, my son, I want you to go back to Berkeley, California, and demonstrate what you can do by converting all your infidel professors to Islam. Then come back, and we will build your factory.

So that was that.

Upon returning to the U.S., Sadeg — instead of following Gaddafi's advice — joined IBM and later founded his own company, HYPRES, which focused on superconductor electronics.

In 1990, Faris left HYPRES and launched a new venture, *Reveo*, with the ambitious vision of following Thomas Edison's model: applying science to invent new products and spinning them into startups. In 2003, the ever-persuasive Sadeg even convinced the then Prime Minister of Malaysia to create a national R&D center, called *InventQjaya*. But that's another story.

Returning to Libyan sand, I think Gaddafi made a mistake (not the only one he ever made.) Sadeg, a talented scientist and manager, might very well have built that silicon transistor factory in Libya.

Zone refining

We stated earlier that the purification of silicon and germanium — starting with the work of Russell Ohl in the 1930s — was key to everything that followed. The highest purity was achieved after World War II, in 1950–1951, using a process called *zone refining* (also known as zone melting), developed by Bill Pfann and Jack Scaff at Bell Labs. They achieved a staggering 99.99999999% purity, or "10 nines," in germanium. This meant just one foreign atom per 10 billion host atoms. Their work was continued by Bell Labs chemist Henry Theurer, who perfected zone refining for silicon, achieving the same ultra-high purity in 1952.

Zone refining remains one of the most original and unexpected processes in semiconductor technology. It involves shaping the material into a long rod and then melting a narrow section of it while the rest remains solid. This molten zone is slowly moved along the rod. Because impurities tend to concentrate near the solid–liquid interface, they are swept along with the moving melt. Repeating the process multiple times drives impurities toward one end of the rod, which can then be cut off and discarded. Using this process, Pfann and Scaff improved material purity by a factor of about 1,000. The microscopic mechanism, which involves diffusion across the solid-liquid boundary, was later understood, but Pfann and Scaff relied on intuition and experimentation.

Structural perfection

Purity is only one part of the story; *structural perfection* is just as critical. The early semiconductors from Ohl's lab were *polycrystalline* — made of small, randomly oriented crystallites. The *grain boundaries* between crystallites interfere with the motion of charge carriers, much like impurities do.

Solid materials can be divided into three structural classes, as shown in Fig. 12.

- *Monocrystalline (single crystal)*: A perfectly ordered atomic lattice throughout — a military parade in flawless formation.
- *Polycrystalline*: A patchwork of small crystals (grains), misaligned like scattered battalions.
- *Amorphous (glassy)*: No long-range order, though the local atomic structure is preserved — a civilian crowd rather than a military unit.

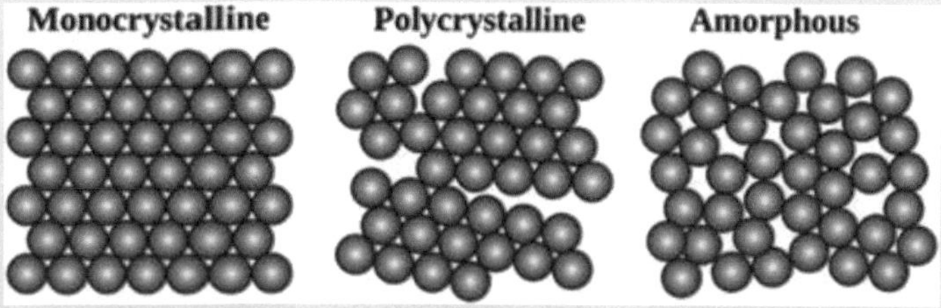

Fig. 12. Three major structural classes of solid materials.

Source: Image by Sbyrnes321, Lic. CC BY-SA 3.0.

Fig. 13. Gordon Kidd Teal (standing) and Morgan Sparks (seated), pioneers of single-crystal growth and BJT fabrication.

Source: Image courtesy of AT&T Archives and History Center.

The first germanium single crystals were grown between 1948 and 1950 by Gordon Teal (the one shown standing in Fig. 13), a chemist at Bell Labs. Teal had worked on germanium since the late 1920s as a student at Brown University. He once joked that he became interested in germanium because of its "utter uselessness" — a true scientist's sentiment!

To grow single crystals, Teal and his colleague Little modified the Czochralski method (Fig. 14), invented in 1915 by Polish chemist Jan Czochralski. In this method, a seed crystal is dipped into a melt and slowly pulled upward, allowing a solid rod to grow underneath in the same crystal structure and orientation as the seed.

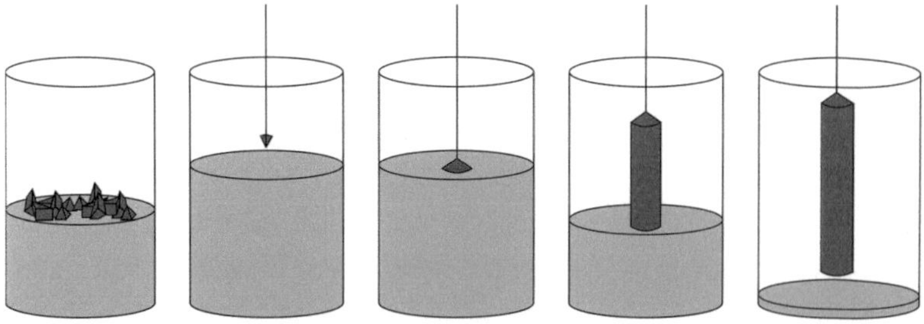

Fig. 14. Czochralski crystal growth method.

Source: Image by Twisp in the public domain.

The process is counterintuitive but effective. By adjusting the orientation of the seed, one can grow crystals in the desired directions. Everything is done under an ultra-pure, inert gas, typically argon. Crucible material, temperature, and pulling speed are all carefully controlled. The resulting crystal is referred to as a *boule* or *ingot* (Fig. 15).

Teal and Morgan Sparks (who is shown seated in Fig. 13) learned how to dope these growing crystals with precise levels of impurities. During growth, they would add, for example, phosphorus (*n*-type) for one section and then switch to boron (*p*-type) for the next. Controlling the *p–n* junction across a growing crystal was difficult, but they succeeded. This breakthrough allowed them, in 1951, to fabricate Shockley's BJT, which was more practical, or, as people in the field say, more *technological*, than the earlier point-contact transistor.

At the beginning of 1953, Teal left Bell Labs and moved back to his native Dallas, where he was offered a position as the head of research laboratories at a young semiconductor company organized only a year earlier, *Texas Instruments* (TI). Acquiring Gordon Teal was one of the best decisions in TI's history. Once at TI, Teal continued what he had already started in his last year at Bell Labs: He was switching from germanium to silicon.

Silicon wins, Germanium loses

Silicon is better than germanium for three reasons: its wider energy gap, its stable oxide, and the electrically "quiet" interface between crystalline

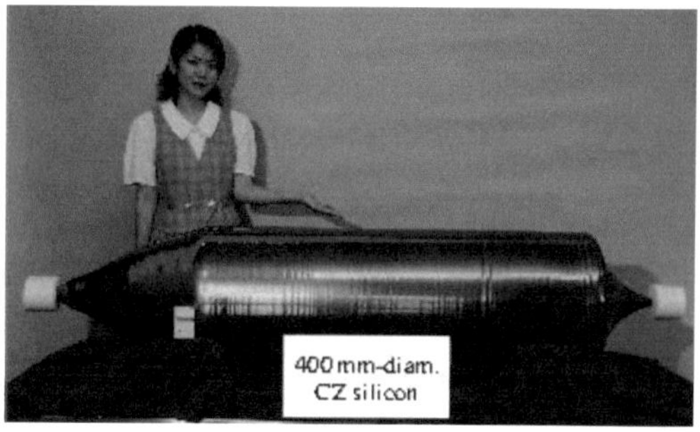

(a) (b)

Fig. 15. (a) The boule in the industrial Czochralski growth apparatus is 30 feet (about 9 m) long, 300 mm in diameter, and weighs 1500 kg. It may have been the largest silicon boule ever grown, which is my excuse for including this low-resolution image here. (b) 400 mm diameter boule; "CZ" on the label refers to "Czochralski."

Source: Photos from Super Silicon Crystal Research Institute Corporation, Japan; the Institute no longer exists; images are in the public domain.

silicon and its oxide. We'll return to the oxide later — here, let's focus on the first advantage.

All mobile electrons in an undoped intrinsic semiconductor arise because they cross the energy gap. You can think of the gap as a river separating an army of electrons from the battlefield. To participate in the battle (conduction), an electron must get across that river. Getting across the river (gap) is a temperature-dependent process. No electrons can cross at absolute zero, no matter how narrow the gap. At finite temperatures, some can, and the probability increases exponentially with temperature. At a given temperature, the wider the gap, the less likely it is for an electron to cross.

In germanium, the energy gap at room temperature is 0.67 eV[15]; in silicon, it is 1.14 eV. As a result, the intrinsic carrier concentration in pure silicon is about 1,000 times smaller than in germanium, making silicon more responsive to controlled doping.

[15]An electron-volt (eV) is a unit of energy used in atomic and solid-state physics. It is an energy acquired by an electron upon crossing a potential difference of one volt.

Even more important is that the wider gap improves thermal performance. Silicon transistors can function at temperatures up to 200°C, while germanium transistors fail above 90°C — a big difference. Transistors heat up due to the electrical currents they carry. Once heating begins, it can accelerate, leading to *thermal runaway*. This is more problematic in germanium due to its narrower gap.

Engineers working with germanium transistors were well aware of these risks and welcomed the introduction of silicon transistors. The promise of higher stability and better thermal tolerance far outweighed the manufacturing challenges posed by silicon's higher melting point.

Teal was a force to be reckoned with; he worked quickly and effectively. At TI, he devoted his efforts and the company's resources to making doped silicon single crystals and silicon BJTs, producing the first practical silicon transistor in early 1954, which surprised many who had expected it to take considerably longer. TI started making these transistors immediately; that was their first successful offering. This milestone was snatched from under the nose of mighty Bell Labs, which rarely happened in those days, and must have made Bell management mad, especially since Teal had been their former employee.

Over time, silicon has opened more technological doors than anyone could have imagined in the early 1950s. In addition to having a well-sized energy gap, it turned out to contain several unexpected gifts, each of which was discovered in due course.

Silicon boules — the most amazing objects in the Universe

The art of growing increasingly larger single-crystalline silicon boules has steadily advanced over the years, driven primarily by economic considerations. A boule is the starting point for producing silicon *wafers*, which are then patterned into *chips* containing electronic circuits. Larger-diameter boules yield larger wafers, which can accommodate more standard-sized chips. Since the batch processing of a wafer costs roughly the same regardless of its size, larger wafers offer a clear economic advantage.

However, a boule's purity and structural perfection immediately after growth are not yet optimal. It is further purified and re-crystallized using repeated zone refining, achieving purities of 9 or even 10 nines — one impurity atom per 1 to 10 billion silicon atoms.

That 0.3 m diameter, 2.0 m long silicon boule is a fantastic object in terms of structural perfection and purity. Suppose we could blow up this boule by a factor of 10^{10} — ten billion times. Everything scales up: atoms, the distances between atoms, and the whole structure.

Then, the spacing between neighboring silicon atoms becomes 2.4 m instead of 2.4×10^{-10} m. Looking down the <110> crystallographic direction, where atoms line up in straight corridors, the spacing increases to 3.84 m (see Fig. 16). The blown-up boule becomes a structure of cosmic proportions, but one whose interior features are on a human scale. Its original length of 2 m now spans 20 million kilometers — more than 50 times the distance from the Earth to the Moon.

Imagine these 20-million-kilometer-long corridors lined with support structures (blown-up atoms) spaced 3.84 m apart, with doorways (the spaces between atoms) leading to side passages. The height and width of such a corridor — viewed in the <110> direction — would be about 4 m. In a perfect single crystal with no impurities, you could stand at one end and see clearly to the other side, all the way down that corridor (Fig. 16).

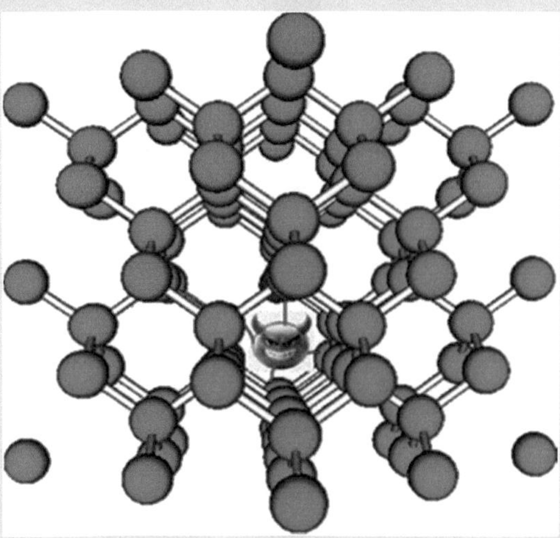

Fig. 16. Silicon lattice with "corridors" as they would be seen from the <110> crystallographic direction, and an "impurity" blocking the view in one of them. The spacing between neighboring silicon atoms in the <110> direction is 0.384 nm, or, if everything is scaled by a factor of 10 billion, 3.84 m.

But suppose we now have impurities. A quick calculation shows that if our silicon crystal had 10-nines purity, the chance of encountering an impurity along our corridor is slim — we would likely see all the way to its 20-million-kilometer end. At 9-nines purity, our view would be blocked after approximately 3.3 million kilometers — roughly 8.5 times the distance from the Earth to the Moon. And with just 6-nines purity, we'd see "only" to about 3,300 kilometers — still remarkable, but nowhere near cosmic.

That silicon boule is undoubtedly the largest chunk of the perfect material on Earth, and come to think about it, it is probably true about the whole Universe. Even the best naturally occurring crystals on Earth are not nearly as pure and perfect as electronic-grade silicon, and there is no reason to think it would be otherwise on other planets. Perhaps some cosmological objects, such as neutron stars, are perfectly pure — containing only neutrons — but they do not possess a crystalline structure with long-range order. So, unless other civilizations have had their own little green Czochralskies and Teals working on their advanced electronics, I doubt there is anything as perfect as that silicon boule anywhere in the Universe!

Wafers

After the silicon boule has been grown, purified, and machined into a cylinder of the required diameter, it is sliced into thin pancakes, each less than a millimeter thick, called wafers. These wafers are then chemically etched and polished on one side (the working surface), while the other side is left with a matte finish. Wafers serve as the basic industrial units in the electronics processing industry. The larger the wafer, the more chips can be fabricated in parallel, reducing the cost per chip.

Depending on the diameter and application, a typical wafer thickness ranges from 0.7 to 0.8 mm. From a standard two-meter-long boule, manufacturers can produce up to 2,500 wafers, accounting for material loss during slicing and edge trimming.

Wafer diameters were standardized starting in 1960; before then, it was a free-for-all. From 1960 to 1969, the standard diameter was 1 inch (25 mm). Since then, it has steadily increased, reaching 11.8 inches (300 mm) by 2002. Although plans were announced to extend this further to 17.7 inches (450 mm), the increase effectively halted at 300 mm. While 400 mm and 450 mm wafers are occasionally produced in research or pilot lines, the industry's standard wafer size has remained at 300 mm for the last 23 years.

On the road to integrated circuits

Perhaps the most significant advantage of a solid-state transistor over its triode tube counterpart is its ability to be scaled down. Vacuum tubes cannot be made too small — perhaps about an inch long is the practical limit — they just aren't amenable to scaling. Additionally, a vacuum tube consumes a significant amount of power — as someone said, it is nothing but a glorified lightbulb — its cathode must be white-hot to release electrons. A vacuum tube cannot be *integrated* with the rest of a circuit like a semiconductor device.

Reducing the size (scaling) was only part of the challenge. Another problem was the sheer number of electronic components and an even greater number of electrical connections between them. With the increasing number of components, the number of connections between them by the end of the 1950s became impossibly large. This problem could not be solved with vacuum tubes that required wire connections, but it could be solved with semiconductors through *scaling* and *integration.*

Scaling cannot be achieved all at once — it is a process, not a result. It requires engineers to work very hard; scaling is achieved through a continuous chain of innovations. I do not know of a more heroic and remarkable story in any technology than the story of electronic scaling, the story behind Moore's Law (see a later section).

Geoffrey Dummer

The idea of making monolithic circuits carved out of a single piece of semiconductor material (with some additions, perhaps, but essentially from one piece) was novel and revolutionary in the early 1950s. The first ICs were built at the end of the 1950s, but the concept was described a few years earlier by Geoffrey Dummer (Fig. 17), a British electronics engineer and consultant working for the Royal Radar Establishment. Dummer was heading a group responsible for the reliability of electronic components. The concern with reliability pointed Dummer in the direction of

Fig. 17. Geoffrey Dummer (1909–2002), photographed in 1955, was an early proponent of integrated circuits.

Source: Image by Robert Cathles, Lic. CC BY-SA 2.5.

miniaturization and integration. Years later, he summarized his thoughts as follows:

> It seemed so logical to me; we had been working on smaller and smaller components, improving reliability as well as size reduction. I thought the only way we could ever attain our aim was in the form of a solid block. You then do away with all your contact problems, and you have a small circuit with high reliability.

In 1952, Dummer envisioned a layered electronic circuit made from a single block of semiconductor material with variously doped layers. He discussed *rectifying layers* (diodes), *amplifying layers* (transistors), *resistive layers*, and integrated electrical connections to replace wires — what we now call *interconnects*. He also tried to discuss ways of making all of this — a difficult, if not hopeless, task when considered theoretically. Notably, Dummer proposed no solution for the electrical isolation of individual devices within those active layers. The permanent feature of Dummer's vision was its three-dimensionality. He even suggested a system of *vias* (as they would later be called), vertical pathways filled with conducting material to contact the inner layers. All in all, it was a bold, if not very detailed, vision of the future. The concept was powerful, and it caught the industry's attention.

Dummer announced these ideas in 1952 in a talk he gave at a conference in Washington, D.C. Five years later, in 1957, he secured funding and commissioned an electronics firm in England to build a circuit out of a single piece of silicon with metal layers deposited on it. It was a good plan, but it did not work. The company took the money but was unable to build a circuit and instead constructed a model. It fell to others to make the first ICs in the U.S. between 1958 and 1959.

When something like this happens — when one person has a vision, but someone else in another country builds it in "brick and mortar," the visionary is usually not given much credit. In this case, Jack Kilby received the Nobel Prize for ICs; others, notably Robert Noyce, were named as significant independent inventors, while Dummer received almost nothing.[16] As someone has put it, he was the *prophet* but not the *inventor* of the IC. And we know from the Bible that prophets often receive a raw deal.

Jack Kilby

Above, we stated that hiring Gordon Teal in 1953 was one of the best decisions made by TI. Hiring Jack Kilby (Fig. 18) in 1958 was at least equally beneficial. When Kilby began working for TI in May 1958, he was 35 years old, an experienced electrical and radio engineer, and a World War II veteran.

Kilby was tasked with investigating the *miniaturization* of transistor circuits, and he achieved this by considering the *integration* of components. He formulated for himself three "laws" which guided him in his work:

1. *The only thing that a semiconductor company can successfully produce is semiconductors.*
2. *All circuit elements, including resistors and capacitors, can be made of a semiconductor.*
3. *All circuit components can be formed on one semiconductor crystal, adding only the interconnections.*

---

[16] Dummer received the IEEE Cledo Brunetti Award in 1979 for "outstanding contributions to nanotechnology and miniaturization in the electronics arts."

Fig. 18. Jack Kilby (1923–2002) is the inventor of the Integrated Circuit, for which he received the 2000 Nobel Prize in Physics. He is also a co-inventor of a handheld calculator, which would be a significant achievement but pales in comparison to the ICs.

Source: Image courtesy of Texas Instruments.

Most, if not all, of this was stated previously by Dummer (although the first Kilby principle has a distinctly American flavor); I am not sure that Kilby was aware of Dummer in 1958. Kilby later did credit Dummer's early "single block" concept — in his 1976 IEEE article and his 2000 Nobel Lecture — but those are retrospective accounts, not the 1958–1959 originals.

In any case, that summer, Kilby set to work, and by September, he had produced a beauty, as shown in Fig. 19. It was a simple circuit requiring only one transistor, a resistor, and a capacitor. The circuit was generating high-frequency electromagnetic oscillations.

The three IC elements — transistor, resistor, and capacitor — were made from two solid germanium strips connected with thin gold wires. A resistor was a lightly doped length of a germanium strip; a capacitor

Fig. 19. The achievement and the reward: that messy thing on the left is the first hybrid integrated circuit built by Jack Kilby in September 1958. Golden-rose-colored germanium crystals were 10 mm long. On the right: Kilby receiving a Nobel Prize in 2000.

Source: Images courtesy of Texas Instruments.

was made of a reversed-biased[17] *p–n* junction. Sure, it would be better to make everything out of one piece of germanium or, better still, silicon, but that was the best Kilby could do at the time. He could make his circuit only because TI possessed the technology to manufacture p–n junctions, BJT transistors, and resistors fashioned from selectively doped germanium. As we know, TI also had capabilities in silicon technology, but they were more limited, forcing Kilby to choose somewhat outdated germanium.

Kilby was not alone on the road to ICs but was the first to produce a functional semi-integrated circuit. We spoke about Dummer, who formulated these ideas in 1952; others have either theorized or experimented along these lines. We will not venture into a lengthy discussion of who did what; we will describe in the following only one more effort that was at least as significant as Kilby's, that of Robert Noyce. Perhaps the fact that

[17]The reader recalls that a p–n junction is a diode; it admits current in one direction and blocks it in the other. When a voltage is applied to a p–n junction in the direction opposite to the current flow (called the reverse-bias direction), it acts as a capacitor, which is very convenient for ICs.

the Nobel Prize was awarded to Kilby with one of the most extended delays in the history of prizes, 42 years, points to the difficulty of deciding this case.

Kilby worked fast. The first hybrid IC was demonstrated on September 12, and a week later he had another one ready, this time a *flip-flop*[18] circuit with two transistors. Kilby filed no less than five patents on behalf of TI, all related to aspects of making ICs. The first and most significant one, "Miniaturized Electronic Circuits," was filed on May 6, 1959.[19] TI announced Kirby's breakthrough on the day of that filing, and by the beginning of 1961, they were already selling small ICs containing two transistors, four diodes, and several other elements.

Robert Noyce and "the traitorous eight"

Robert Noyce (Fig. 20) received his doctorate in physics from MIT in 1953. He was nicknamed "Rapid Robert" in graduate school for his quickness of mind. Later, he received another nickname, "The Mayor of Silicon Valley." As we said, in 1956, the inventor of the BJT, William Shockley, left Bell Labs and established his own semiconductor company called "Shockley Semiconductor Laboratory" in part of Santa Clara County, California, which was later named "Silicon Valley." Shockley

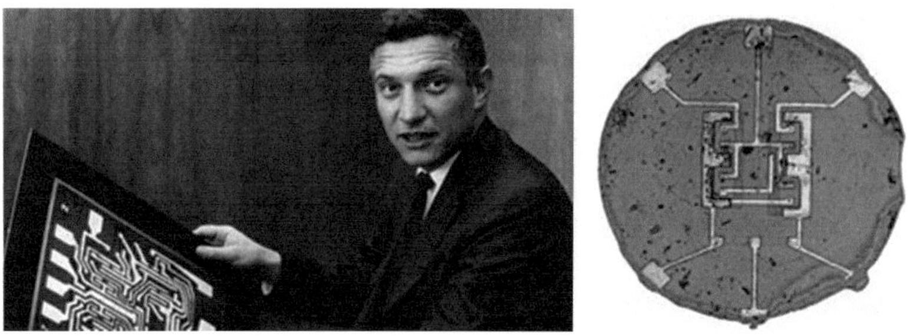

Fig. 20. Robert Noyce (1927–1990), photographed in 1959 at Fairchild. In 1959, Noyce created the first fully integrated planar circuit, as shown on the right side of the figure.
Sources: Left image by Intel Free Press, Lic. CC BY-SA 2.0; right image in the public domain.

[18]Flip-flop is a standard two-state circuit often used in digital electronics.
[19]Kilby sent his application on February 6, but it was lost in the mail, so he resubmitted it on May 6, 1959.

Fig. 21. Founders of Fairchild Semiconductor, "the traitorous eight," photographed in 1960: from left to right, Gordon Moore, C. Sheldon Roberts, Eugene Kleiner, Robert Noyce, Victor Grinich, Julius Blank, Jean Hoerni, and Jay Last. It was an exceptionally bright group, as Shockley only hired brilliant people. We will discuss three of them: Hoerni, Noyce, and Moore.

Source: Image by Wayne Miller, made for Magnum Photos, Lic. WP: NFCC#2.

hired Noyce, but his employment did not last: in 1957, eight scientists from Shockley's ill-fated company — "the traitorous eight" (Fig. 21) — left the sinking ship and its mad captain and formed Fairchild Semiconductor.

One of this group, Jean Hoerni (1924–1997), the individual in glasses sitting second from the right, attended a conference talk in 1958 given by Bell Labs scientist Mohammed Atalla, who described a method for silicon surface passivation — making silicon surfaces clean and electrically inert. Atalla achieved this by growing silicon dioxide on the surface and stripping it using *hydrofluoric acid* (HF). Additionally, Atalla has shown that covering silicon with oxide protects it from contamination and prevents diodes from being shorted by a dirty surface. Upon hearing that, Hoerni decided that silicon oxide should help create solid, monolithic circuits. The key, thought Hoerni, will be in building diodes and transistors beneath the surface, with subsequent growth of an oxide layer effectively sealing the devices. One would only need to strategically place holes in that oxide layer for electrical connections to the buried devices. Returning from the conference, Hoerni applied for a patent on behalf of Fairchild and described this "planar process" idea to his colleagues.

Once he learned about the planar process from Hoerni, Noyce saw the light. He sketched the concept of the monolithic IC in his notebook, dated January 23, 1959. Over the next six months, he worked in the lab, experimenting. On July 30, 1959, Noyce applied for a patent. The difference between Kilby's hybrid IC and Noyce's monolithic IC was in the absence of any wires in the latter. Noyce provided electrical connections via deposited and patterned metal films. Noyce had a better design than Kilby; the only trouble is that May 6, the date of Kilby's patent application, comes before July 30. Damn calendar! This cost Noyce a Nobel Prize.

After the patents were granted, a protracted legal battle ensued between TI and Fairchild, which was ultimately settled through a cross-licensing agreement of their patents. Noyce and Kilby were not responsible for this legal bickering — it was beyond their control and in the hands of company lawyers.

The Toolbox of Silicon Technology

Foundational ideas come first, including semiconductor doping, the formation of *p–n* junctions, device concepts such as the BJT, and miniaturization through integration. However, following these foundational ideas are ways of doing things, the *Toolbox* of silicon technology. It is incredibly rich; I suspect it is richer than any other toolbox in the history of technology. Our brief description here is, at best, scratching its surface.

I am a physicist who spent 10 years at Bell Labs — the very citadel where many of the essential foundations of silicon technology were laid. My own professional work was far removed from the technology itself, yet in that remarkable laboratory, one could not help but breathe in a sense of what it was all about. And working side by side with engineers of the highest caliber, I came to feel a profound respect that only close collaboration can instill. At Bell Labs, scientific snobbery was unheard of — applied science stood shoulder to shoulder with fundamental research in prestige and value.

Silicon dioxide (silica)

Silicon oxide (SiO_2) is a priceless technological gift. No other semiconductor possesses such a beautiful oxide. This oxide can be amorphous (non-crystalline; it is then called *silica*) or crystalline — in that case, it is called *quartz*. In devices, the oxide should preferably be amorphous, uniform, dense, and without grain boundaries. The silicon beneath this oxide is crystalline, but the oxide is amorphous (Fig. 22).

Mohammed Atalla pioneered a serious scientific study of silicon oxidation and the quality of the resulting silicon–oxide interface at Bell Labs in the mid-1950s. He used oxidation and subsequent oxide stripping to

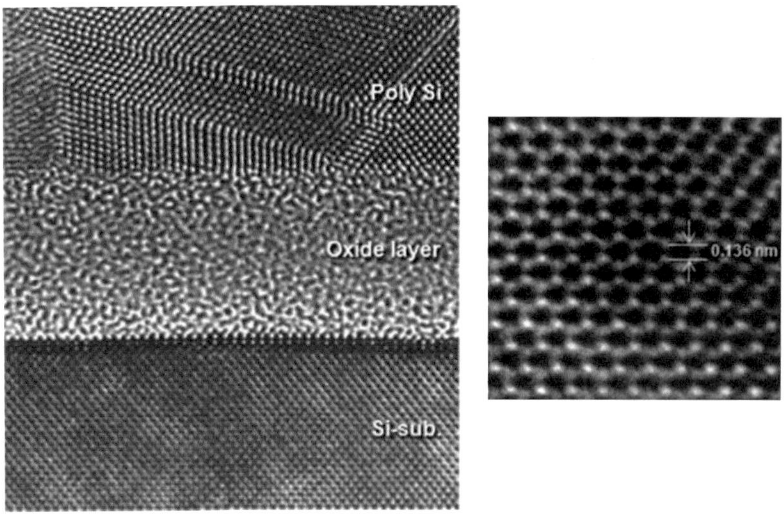

Fig. 22. High-resolution transmission electron microscopy (HRTEM) images of crystalline silicon, amorphous silicon dioxide, and polycrystalline silicon. We can see individual silicon atoms! The lower layer, marked "Si-sub," represents a single-crystalline silicon substrate. The middle layer is amorphous silicon dioxide. The third layer on top of amorphous oxide, marked "Poly-Si," consists of silicon crystals separated by grain boundaries. The scale can be calibrated using known interatomic distances. On the right side of the figure, we see the distance between silicon layers, indicated as 0.136 nm; the center-to-center distance between neighboring silicon atoms would be 0.24 nm. These remarkable images were obtained with a Hitachi H9500 transmission electron microscope (TEM).

Source: Images courtesy of Hitachi High-Tech Corporation.

clean silicon surfaces; he was less interested in the oxide's electrical properties. The other Bell Labs technologist, Carl Frosch, was focused on the oxide itself. He explored methods of growing silicon oxide on single-crystalline silicon wafers in 1956–1957. The general belief in the mid-1950s was that one could not heat silicon in air or oxygen because it would burn, just as carbon (coal), its neighbor in the Periodic Table. Frosch was the first to show that controlled heating of single-crystalline (as opposed to polycrystalline) silicon in air or oxygen produces a dense, uniform protective layer of silicon dioxide. Once he learned how to grow high-quality silica of the desired thickness, Frosch and his collaborator, Lincoln Derick, proposed using it as a *mask* in a selective diffusion of dopant impurities into silicon.

The oxide can be grown by exposing a silicon wafer at temperatures ranging from 800°C to 1200°C to oxygen or air (dry oxidation). The resultant high-quality oxide grows slowly and is used in devices such as MOSFETs, which we will discuss shortly. Alternatively, silicon can be oxidized in the atmosphere of water vapor at significantly lower temperatures (wet oxidation).

The reader might wonder: Why is he (the author) sharing all these technical details with me? I promise to do this sparingly and selectively; my only aim is to make you admire the toolbox of silicon technology and to hint at how complex and sophisticated it is.

The ability to controllably cover silicon with SiO_2 opened several technological doors. One of them (arguably, the *main* door) led to the fabrication of *field-effect transistors (FETs)* (discussed later), but there were other benefits as well. For example, as we said above, silicon oxide layer can be used as a mask in *selective doping*, where impurities are injected over well-defined parts of the silicon wafer rather than the entire wafer.

All these processes involve creating *patterns* on a sub-micron scale, which is achieved through *photolithography*.

Photolithography

To create electronic circuits on a chip, an engineer first develops a detailed plan: This transistor will be placed here, another one there, and all ten billion of them will be supplied with electric power and interconnected. There will be circuit inputs and outputs, as well as electrical insulation between neighboring devices, and so on. All of that may be specified in detailed computer drawings and supplied with a sequence of technological steps to be taken.

Finally, when the time comes to build the circuit, all these features must be reduced to a millionth of their original size and projected onto a silicon wafer. The construction of this miniature metropolis will require many steps. In modern ICs, the number of these steps, or *layers*, may reach 30 or more. Each layer has its own pattern; the detailed plan we discussed above consists of not one, but perhaps 30 patterns, which must be precisely aligned. You do not want to build an entrance to a subway station that isn't aligned with the underground tunnel leading to this station.

Building a large city with its skyscrapers, streets, and underground levels is no simple task. Now imagine a town scaled down a billion times, where every tiny detail — from the spatial arrangement of its elements to the composition of the construction materials — must be executed with absolute precision. There is no room for error: not in the layout, not in the structure, not in the technology used to build it. This is where the entire *toolbox of silicon technology* is brought to bear: oxidation, doping, deposition, etching, annealing, and more. But first comes the *geometry*, and everything else follows.

Photolithography is the method used to transfer the circuit's geometry, shrunk down by a factor of about a million from its appearance on a computer screen, onto the wafer.

Conceptually, photolithography in microelectronics originated from the 19th-century invention of *photoengraving*, itself a spin-off of early photographic processes. In the 1820s, Nicéphore Niépce, the inventor of photography, used light-sensitive materials that hardened upon exposure to sunlight, creating acid-resistant patterns on metal plates. This process, when used to etch printing plates, became known as "photoengraving."

In 1958, Jay Lathrop and James Nall at the National Bureau of Standards (now the National Institute of Standards and Technology, or NIST) adapted the technique for creating microelectronic patterns. Their method was soon picked up by Noyce and Co. at Fairchild, by TI, and independently developed at Bell Labs. By the early 1960s, photolithography had become the universal method for defining circuit geometry — a technique whose time had come.

Photolithography has undergone significant transformations over the past 60 years. The most advanced version in use today is known as

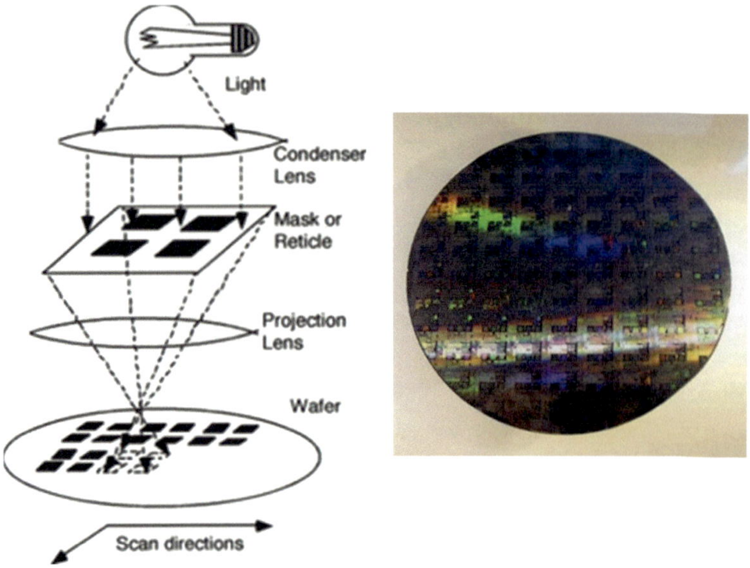

Fig. 23. (a) Projection lithography implemented in a *stepper*. (b) A 12-inch (300 mm) silicon wafer covered with identical microcircuits (chips).

Sources: (a) Image by Bill Wilson Lic. CC BY 1.0; (b) image by Peellden, Lic. CC BY-SA 3.0.

projection lithography (Fig. 23a). In this method, the circuit pattern is optically reduced and projected onto the wafer surface (Fig. 23b) through a series of high-precision lenses. These projection systems possess exceptional optical quality, producing sharp sub-micron — and now even nanometer-scale — images.

Projection lithography is carried out in machines called steppers or scanners, referring to two methods of projecting patterns. In recent years, these have been combined into hybrid *stepper-scanner* systems.

These stepper-scanners are enormous, astonishingly sophisticated, and extremely expensive. One such machine, manufactured by ASML, is shown in Fig. 24. These systems use "light" with wavelengths so short — 13.5 nm — that it would be more accurate to call it X-ray radiation. Such short wavelengths are necessary to define the smallest features on a chip, which in today's most advanced production nodes are approaching about 20 nm. This is only a few dozen atoms across. It's difficult to imagine how much smaller things can get.

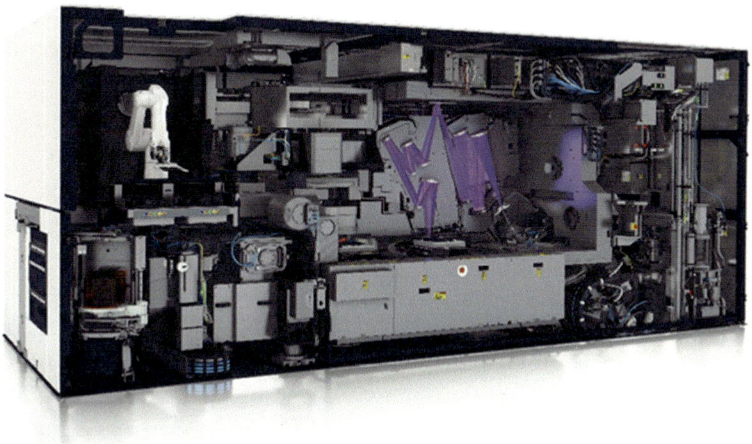

Fig. 24. An industrial stepper-scanner by Advanced Semiconductor Materials Lithography (ASML, a Netherlands-based semiconductor equipment manufacturer). The system is TWINSCAN NXE:3400 C, shown here with the front panel removed and the light path highlighted in purple. (The actual EUV light used has a wavelength of 13.5 nm and is invisible.) This machine supports 5 nm and 7 nm process nodes and has a throughput of over 170 wafers per hour (300 mm).

Source: Image courtesy of ASML.

The wavelength of electromagnetic radiation determines the fundamental resolution of an optical system. From physics, we know that the smallest feature an optical system can resolve is limited to half the wavelength of the light used. This is a rough "rule of thumb" criterion, which does not work in complex optical systems, where the smallest features tend to be several times larger than half-wavelength.

In the 1960s, when minimum feature sizes were 20 μm, visible light (wavelengths of 400–700 nm or 0.4–0.7 μm) was sufficient. But with Moore's Law driving relentless miniaturization, the industry progressed through shorter and shorter wavelengths, all in the ultraviolet (UV) part of the spectrum, which human eyes cannot see:

- **Near-UV (NUV)**: 300–400 nm or 0.3–0.4 μm.
- **Far-UV (FUV)**: 120–200 nm or 0.12–0.2 μm.
- **Extreme UV (EUV)**: 10–120 nm or 0.01–0.12 μm,

The machine shown in Fig. 24 uses light at 13.5 nm, which falls at the lower end of the EUV range, approaching the territory of soft X-rays. In principle, diffraction considerations for such a wavelength of light suggest a lower bound on feature size of the order of 10 nm, but in practice the features on a chip come out several times larger.

It must be said that there exists a persistent and somewhat misleading tradition in the semiconductor industry: the so-called "node names" — 3 nm, 5 nm, 7 nm, and so on — no longer describe any actual physical dimension on the chip. In the early days of microelectronics these numbers referred directly to the transistor's gate length, but since about the mid-2000s the terminology has drifted into marketing shorthand. A process labeled "5 nm" produces critical features on the order of 20–25 nm, while a "3 nm" process reduces them only modestly, to roughly 18–22 nm. The node names survive mostly for continuity in naming and for the less-than-noble reasons of commercial competition — an annoying trend that seems hard to eliminate.

Generating EUV light is far from trivial. Conventional sources cannot produce it. Instead, EUV systems rely on high-power *excimer lasers* or plasma-based sources using tin droplets, which are vaporized and ionized into a plasma that emits EUV radiation.

In that context, the light bulb symbol in Fig. 23 is clearly a playful simplification.

A stepper-scanner is a monstrously complex tool, but that's hardly surprising given what it must accomplish. The price of such a system makes you think twice before opening a semiconductor fab: The cost of the NXE:3400C is approximately $200 million.[20]

The share of ICs on the semiconductor market

The semiconductor industry has experienced continued growth since the mid-1950s. The first ICs appeared in 1959, and their role in semiconductor electronics — understandably small at first — has grown steadily ever since. Correspondingly, the share of IC sales in the overall semiconductor market has also increased.

---

[20] Price quoted by ASML media relations manager Mr. Sander Hofman, private communication.

In 1964, the global semiconductor market was valued at approximately \$1 billion. By 2025, global semiconductor sales reached nearly \$800 billion — a 800-fold increase over the past six decades. In 1964, sales of the then-emerging ICs comprised only 2.5–5% of the market. By 2025, that share has risen to 80–90%, reflecting the central role that ICs now play in the semiconductor industry.

Foundries

The concept of foundries (fabs) — specialized facilities dedicated to manufacturing ICs for various clients — emerged in the late 20th century. Prior to this, semiconductor companies typically operated as *integrated device manufacturers*, handling both the design and production of their semiconductor products in-house.

A pivotal moment in this transition occurred in 1987 when Dr. Morris Chang founded *Taiwan Semiconductor Manufacturing Company* (TSMC), the world's first dedicated semiconductor fab, created solely to provide manufacturing services to other semiconductor companies, including those without their own fabrication facilities — known as *fabless* companies. TSMC's establishment catalyzed the growth of numerous fabless startups and spurred innovation in chip design and applications.

Building a modern semiconductor fab is a highly capital-intensive endeavor. According to recent estimates, constructing a leading-edge semiconductor fab capable of producing advanced chips at the most current technology nodes requires an investment of \$15–25 billion.

For example, Intel has been constructing two new advanced chip factories — Fab 52 and Fab 62 — at its Ocotillo campus in Chandler, Arizona. Although the project was initially estimated to cost about \$20 billion, industry reports now suggest the total cost for the two plants is closer to \$32 billion. As of October 2025, Fab 52 is already producing chips in limited quantities, while Fab 62 remains under construction. The operational plant is now producing Intel's chips belonging to what the company calls its "next-generation" technology, roughly corresponding to the so-called 1.8-nm (or 2-nm-class) node, though in reality their smallest critical features are about 16–20 nm across. The total cost of the Arizona complex may still rise as construction continues.

It's noteworthy that construction costs can vary significantly by region. Building a fab in the U.S. can be twice as expensive and take twice as long as constructing a similar facility in Taiwan. Factors contributing to this disparity include higher labor costs, regulatory complexities, and supply chain considerations.

A serious fab would contain several hundred pieces of equipment, of which a rare one costs less than a million dollars, while many others cost hundreds of millions.

The interiors of a semiconductor fab are maintained under *a controlled environment*, traditionally called a *clean room*, the name originating from simpler times when it was just one room; today, it is, at a minimum, a *clean mansion*. All air circulating inside a fab passes through special air filters that remove dust particles — dust is one of the greatest enemies; a single dust particle can ruin the entire processed chip. If this particle lands on a stepper-scanner lens, it may destroy the whole factory's output. Everything inside that clean space — floors, walls, benches, and all surfaces — is made of materials that do not shed any dust or vapors; you will not find any wooden parts or concrete floors there. The temperature and humidity are strictly controlled. The entire facility must be *vibration-free*, as even the slightest vibration will interfere with the delicate instruments' work. Another enemy is *static electricity* (the source of sparks, as on a dry carpet). Water supplied to a semiconductor fab must be ultra-pure; it must be free not only from chemical impurities but also from the *ions* (charged atoms). The workers entering a fab wear special suits and masks (Fig. 25). Being a fab worker is an excellent job during a pandemic, as the standards of the semiconductor industry certainly exceed those adopted in a medical institution. Within a fab, workers are protected from viruses. However, such jobs are becoming increasingly scarce as fabs transition into fully automated, robotic facilities.

Using all three spatial dimensions

One of the cornerstone principles of monolithic circuits, first formulated by Geoffrey Dummer, was that they should be built in three dimensions — not only on the surface of the material but also within its volume. Not throughout the entire volume of the silicon wafer, of course, but within a certain subsurface layer. Dummer envisioned circuits constructed in layers stacked vertically, one above another, inside the semiconductor

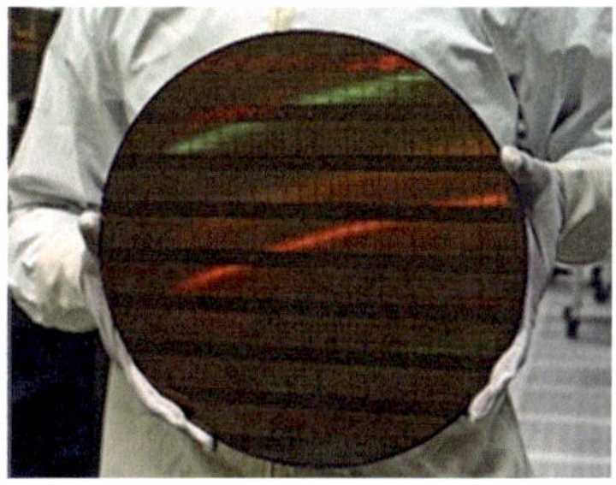

Fig. 25. Left: Inside a large semiconductor fabrication facility. The yellow light protects the unexposed photoresist covering the wafers. Right: Intel 300 mm processed wafer with hundreds of chips.

Sources: Left image from Micron Technology Inc. Image Gallery, released by the Company for public use; right is an archival image from an old Intel manual in the public domain.

crystal. The number of such layers has grown over the years and today amounts to several dozen.

The idea was visionary, but its realization was not straightforward. Robert Noyce moved in that direction early on: his monolithic IC was built in two or three layers — one within the bulk of the silicon crystal and another containing electrical interconnections on the surface.

It's not always easy to define precisely what we mean by a "layer" in this context, as layers in monolithic ICs often blend into one another. We can define a layer as the structures created between two successive photolithography steps.

Depending on the layer's function, circuit elements such as transistors, diodes, resistors, capacitors, and interconnects can be etched, diffused, ion-implanted, or deposited onto a wafer. Any remaining voids in this intricate 3D structure are typically filled with filler materials, such as epoxy resins, to provide mechanical strength and structural integrity.

The problem of electrical isolation between individual devices in an IC was difficult. Implementing the planar process while attempting to build trenches or walls between components embedded in silicon posed a profound engineering challenge: after all, a wall or trench is literally orthogonal to the notion of a "plane."

Robert Noyce devised a clever solution: He surrounded each device with a ring-shaped p–n junction, which blocks current from flowing in at least one direction. Kurt Lehovec independently arrived at the same solution and patented it in 1959, slightly ahead of Noyce. As a result, device isolation could be achieved using only the standard silicon processing toolbox, in perfect agreement with Kilby's First Rule: *A semiconductor company can only make semiconductors.*

In more advanced versions, isolation is enhanced using multiple concentric p–n junctions around each device, thereby blocking current flow in both directions.

Now, imagine stripping away the filler materials — such as epoxy or dielectric layers — that usually fill the gaps in a finished IC. What remains would resemble the excavated city of Herculaneum, buried under volcanic ash following the 79 A.D. eruption of Vesuvius (Fig. 26).

But even the smallest known viruses (~20 nm) would find a modern nano-metropolis uninhabitable. Viruses could have lived in this city 20 years ago when the streets and houses were roomier.

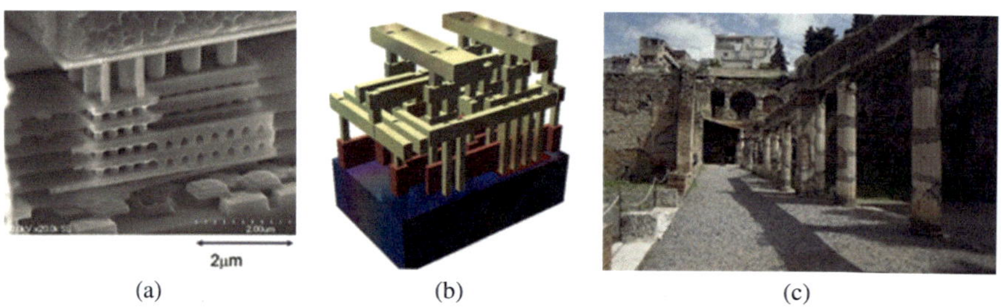

Fig. 26. (a) 3D electron photograph of an IC; the two-micrometer scale can be seen in the lower-right corner. (b) Graphic depiction of a *standard cell* (logic gate) in which a dielectric filler has been removed. This standard cell looks like a miniature two-story Roman villa built for a rich virus-senator. (c) Photo of Herculaneum in which "the filler" (volcanic ash) has been removed by archeologists.

Sources: (a) Image from an article in the *Journal of Heat Transfer*, courtesy of ASME Digital Collection; (b) image in the public domain; (c) image by Mboesch, Lic. CC BY-SA 3.0.

Fig. 27. From left to right, inventors and contributors to FET: Julius Lilienfeld (1882–1963), Mohamed Atalla (1924–2009), and Dawon Kahng (1931–1992).

Sources: First two images in the public domain; third image courtesy of AT&T Archives and History Center.

The story of a FET

As we said earlier, Shockley's BJT is a current-controlled device: injecting a small current into the base opens the transistor for the passage of a much larger current between the emitter and the collector.

However, it is often preferable to control a device not with current but with voltage — that is, with electric potential. Since the electric potential per unit distance is referred to as an electric field, a transistor controlled in this way is called a *Field-Effect Transistor*.

The concept of controlling a transistor via an electric field was first proposed by Julius Edgar Lilienfeld (Fig.27), an Austrian-born physicist who later became an American citizen. Sometime in the early 1920s, Lilienfeld imagined such a device and filed patents for it, first in Canada in 1925 and then in the U.S. in 1926. However, while the concept was sound, the necessary materials and fabrication techniques simply did not exist at the time.

Julius Lilienfeld was a physicist and a prolific inventor. He received his Ph.D. in 1905 from the University of Berlin, where his advisor was none other than Max Planck, the grandfather of quantum mechanics. Throughout his career, Lilienfeld was fascinated by the effects of strong electric fields on solids. In addition to patenting the field-effect transistor (FET), he conducted early studies of field emission — the release of electrons from a metal surface into a vacuum under the influence of a strong electric field.

A Jew sensing the rising danger in Europe, Lilienfeld had the foresight to emigrate to the U.S. in 1926, the same year he applied for his FET patent. He continued to invent prolifically; in 1931, he patented a version of the electrolytic capacitor, a type now ubiquitous in electronics.

After 1935, Lilienfeld became self-employed, living in the Virgin Islands, where he focused on promoting and defending his patents. He remained largely unknown to the broader scientific community, and his early FET work was not known to Shockley, Brattain, or others working on transistor theory and practice in the 1940s.

Lilienfeld died in 1963. In the 1980s, his widow left a significant bequest to honor his memory. The American Physical Society (APS), prompted to investigate who this man was, established the Julius Edgar Lilienfeld Prize in 1988, awarded for "outstanding contributions to physics by a single individual."

Like a vacuum triode or a BJT, the FET is a three-terminal device. Its operation is somewhat analogous to a canal lock system. Imagine a canal with one water section at a higher level than another. These two sections are separated by a lock, as shown in Fig. 28. The difference in water levels is analogous to a voltage difference. In a FET, a voltage is applied between two outer terminals: the source (S) and the drain (D). The gate (G) acts like the canal lock.

When the lock is opened, water flows from the high to the low level; similarly, in an FET, current flows from the source to the drain when the gate induces a conductive channel beneath it. A partially opened lock reduces flow; a closed lock stops it altogether. Likewise, the gate voltage modulates the current, controlling the amount of current that can pass between the source and the drain.

Many engineers and scientists, including William Shockley, attempted to build a functioning FET but failed. The reason for these failures became apparent only after John Bardeen identified a crucial phenomenon in 1947: the presence of *surface states* at semiconductor interfaces.

Most semiconductor surfaces are "infested" with these states — localized electronic states that trap charge and screen the electric field. As a result, the applied gate voltage is blocked from entering the semiconductor bulk, rendering the FET inoperative. Everything might look correct in design, but the device doesn't work. The electric field never reaches the region where it's supposed to induce a conductive channel.

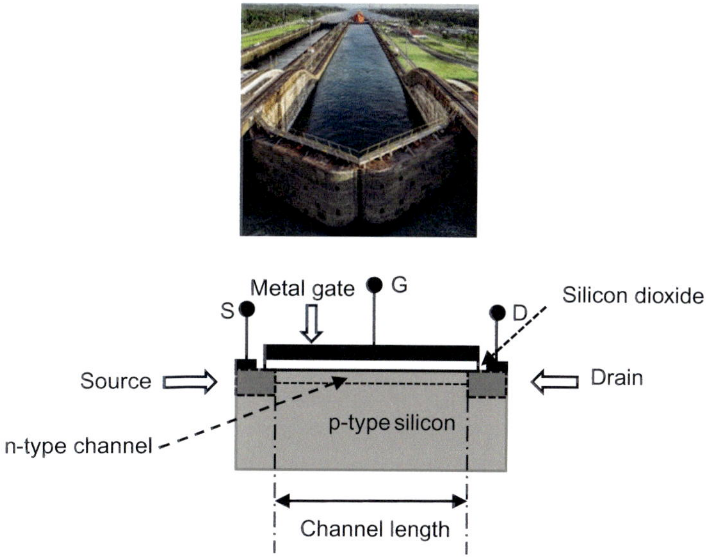

Fig. 28. Upper part: One of the locks on the Panama Canal. Lower part: Schematic of a field-effect transistor (FET). Solid black indicates metal, dark gray represents heavily doped silicon, and light gray indicates lightly doped silicon. The silicon body is p-type, while the channel, formed under the gate by positive voltage $+V_G$, is n-type.

Sources: Photo reproduced by permission of its author, Jonathan Ross; schematic by the author.

The conducting layer between the source (S) and the drain (D), known as the channel, forms directly beneath the semiconductor's surface. Its depth is indicated in Fig. 28 with a dashed line. This arrangement resembles a capacitor but with a crucial twist: In a conventional capacitor, two metallic plates hold electric charges on their surfaces. Here, there is only one true metallic plate, the gate, while the lower plate — the channel — is conjured into existence by the gate's electric field. The gate commands, and the channel appears as if carved by an obedient genie.

Applying a voltage to the gate electrode establishes an electric field between the gate and the semiconductor. Through that field, the gate controls the conductivity of the channel. This is similar in purpose to the action of a bipolar junction transistor (BJT), but with a key difference: control is achieved not by injecting current but by using an electric field. Like other transistors, a field-effect transistor (FET) can amplify voltage and current or act as a fast switch, toggling the channel between open and closed states.

Such are the workings of an *ideal* FET — one untroubled by surface states. But the reality is often less forgiving. These surface states, which arise at the interface between the semiconductor and the insulating layer, can severely disrupt this elegant operation. They can block the electric field from penetrating the semiconductor, preventing the formation of a conductive channel. In this compromised scenario, a capacitor is still formed, but the genie is forbidden to dig, its efforts thwarted by the mischief of surface states.

These surface states act as traps, capturing mobile charge carriers. Imagine laying down a line of mousetraps and asking mice — electrons, in this case — to cross the line. If the number of traps is comparable to the number of mice, very few make it across, and the current stalls.

But where do these traps come from? Recall what was said about grain boundaries: They resemble impurities and, like impurities, introduce energy levels within the forbidden gap. A semiconductor surface can be thought of as one extended grain boundary. It, too, gives rise to energy levels in the gap — and these levels function as electron traps.

The presence of such surface states at a semiconductor–oxide interface is not the exception but the rule. In most semiconductors, such as germanium, gallium arsenide, or indium phosphide, the native oxide produces a dense forest of surface states that block the field effect. Only when the number of surface states (electron traps) is small can a channel be formed, and proper transistor action can be observed. Remarkably, this is precisely what happens at the interface between silicon and its native oxide, silicon dioxide, when silicon has been cleaned in accordance with Muhammad Atalla's recipe — a fortunate and foundational circumstance in the history of modern electronics.

The absence of surface states — combined with the excellent insulating properties of silicon dioxide — finally allowed, in 1959, the practical realization of Lilienfeld's old idea: controlling a transistor through an electric field. The two individuals most responsible for this breakthrough were Bell Labs engineers Mohamed Atalla and Dawon Kahng (Fig. 27). Their invention became known as the MOSFET — the *Metal-Oxide-Semiconductor Field-Effect Transistor.*

Mohamed Atalla, an Egyptian-American engineer and materials scientist, joined Bell Labs in 1949. During the 1950s, he spearheaded the development of a technique that became fundamental to modern electronics: silicon surface passivation. By thermally growing a silicon dioxide layer on a silicon wafer and then removing it with hydrofluoric acid (HF), Atalla produced a clean, electrically "quiet" surface — one free from disruptive surface states. His original motivation was to suppress surface leakage currents that impaired the performance of diodes and transistors formed near the silicon surface.

But Atalla also had a broader vision. He was aware of the many unsuccessful attempts to build a functioning FET and recognized that the clean interface might be the missing piece. Acting on Atalla's insight — and with his encouragement — a young Bell Labs engineer, fresh out of graduate school, took up the challenge. In 1959, Dawon Kahng successfully fabricated the first MOSFET.

I saw Dawon Kahng once. My friend Serge Luryi knew him well — Kahng liked and respected Serge; they had worked together in the same laboratory at Bell Labs. One day, in 1982 or 1983, Serge needed to speak with Kahng, so he took me along to his office. I remember we were greeted by a confident man who chain-smoked big cigars and gave the impression of someone who had truly "seen a thing or two."

Afterward, as we walked along one of the endless corridors at Murray Hill, Serge explained who Kahng was. At the time, I didn't know the story of the MOSFET, which may be taken as a sign of my ignorance or perhaps as a testimony to the greatness of Bell Labs itself, where so many extraordinary things had happened that you could easily miss one or two, especially if they weren't within your particular field.

It took a few more years before I fully grasped the true significance of the MOSFET — and, with it, the greatness of Dawon Kahng.

The device was built in 1959 and demonstrated the following year, with Kahng filing for a MOSFET patent in 1960. As it happened, the first MOSFET was not a particularly fast device (speed would come later), and Bell Labs management received it coolly. But Kahng saw farther than his department head and lab director — perhaps because he was standing on the shoulders of these giants (just kidding). In a 1961 internal memo, he suggested that the MOSFET might be an ideal element for use in the emerging field of ICs. His foresight, however, failed to move his superiors.

When Bell Labs let an opportunity slip (which did not happen very often), others were quick to seize it. Researchers at Fairchild and RCA immediately recognized the advantages of the MOSFET. By the early 1960s, they had begun fabricating them, and by the mid-1960s, these devices were appearing in commercial ICs.

Today, the overwhelming majority of all transistors are MOSFETs. It is by far the most common electronic device, with billions embedded in every chip. In today's most advanced devices, MOSFETs have effective channel lengths approaching 3–5 nm. Their switching times are now measured in picoseconds — ten trillionths of a second — corresponding to frequencies around 100 gigahertz. This is not the clock frequency of a complex system, but it sets the baseline for the raw speed from which overall computational performance is ultimately built.

Between 1960 and 2026, it is estimated that approximately 50 sextillion (5×10^{22}) MOSFETs have been manufactured. Presently, they are being produced at a staggering rate of about 50 trillion per second. We have never seen such numbers applied to anything made by human hands. Nothing else is ever counted in sextillions.

CMOS

Most of those sextillions of MOSFETs found in ICs come in pairs — two transistors forming a structure called *complementary metal-oxide-semiconductor*, or CMOS. Following the invention of the MOSFET, Dawon Kahng demonstrated that it could be built in two versions: one with an n-type channel (NMOS) and another with a p-type channel (PMOS). Based on the NMOS configuration shown in Fig. 28, the reader can easily imagine its complement: In a PMOS, the silicon body is lightly doped n-type, the field-induced channel is p-type, and the voltage applied to the gate must be negative with respect to the body to induce conduction.

In 1963, two Fairchild engineers — Chih-Tang ("Tom") Sah and Frank Wanlass — realized that NMOS and PMOS transistors could be combined into a single device: CMOS. Their key insight was that this configuration could greatly reduce power consumption.

A single MOSFET operates in one of two states: an "on" state, where the gate voltage opens a conductive channel, and an "off" state, where the channel is closed. In the off state, the resistance is extremely high, and no current flows. In the on state, current flows through the channel, which causes power dissipation. If the transistor remains "on" for a prolonged period, the total energy dissipated can be significant.

Sah and Wanlass devised a clever workaround: They connected a pair of complementary MOSFETs — one NMOS and one PMOS — in such a way that the current does not flow in either static state; they found a way to achieve something very close to a "free lunch." Of course, there is no genuinely free lunch — but this one is very cheap. CMOS devices dissipate power only during the brief moments of switching between states. The only energy required is to charge or discharge the MOSFET's internal capacitance.

This remarkable efficiency is a testament to the unique capabilities of semiconductors — their ability to conduct via electrons or holes. Everything flows from that basic property: solar cells, diodes, bipolar transistors, FETs, and CMOS devices all depend on it.

CMOS can also be viewed as a fundamental logic element: an inverter. When operating between two voltage levels — say, 0 and 2 volts — CMOS reverses the logic state. A 0-volt input produces a 2-volt output, and vice versa. With this basic inverter, one can construct all the other logic gates required for digital computation. CMOS, in this sense, is the foundation of digital electronics.

FinFET

A planar MOSFET, with its gate sitting on top of the channel, is not the best design for exerting electrostatic control. As devices shrank, it became increasingly difficult for the top-mounted gate to maintain control over the channel. The solution was to change the geometry. In the *FinFET* — the modern evolution of the MOSFET — the silicon channel is shaped like a narrow vertical fin. The gate wraps around it on three sides: the two vertical sidewalls and the top (Fig. 29). This wrap-around structure facilitates gate control over the channel and reduces leakage currents. In effect, the gate "grabs" the channel more tightly, suppressing unwanted conduction and improving the device's behavior at nanometer scales.

The name "FinFET" comes from a raised source-drain ridge supposedly resembling a shark fin.

FinFETs — now ubiquitous, including in your smartphone — were invented in 1995 by UC Berkeley professor Chenming Hu (Fig. 30). FinFETs significantly improved upon planar MOSFETs, and Professor Hu has become a hero of IC technology. For this discovery, he received the National Medal of Technology and Innovation in 2016 — "the highest

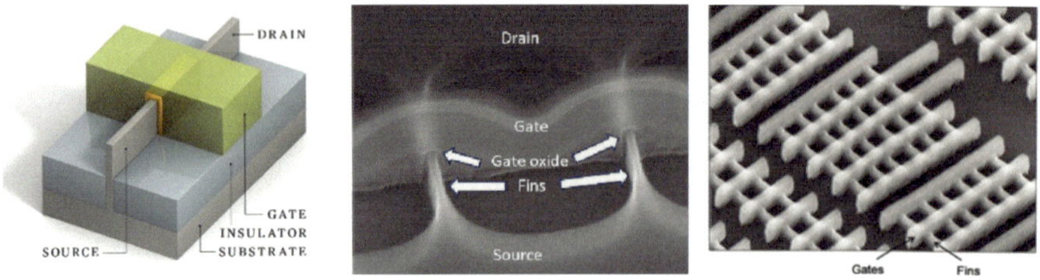

(a) (b) (c)

Fig. 29. (a) A 3D rendering of a single-channel FinFET structure. The thin, yellow frame around the raised source-drain fin represents the gate oxide; the gate, colored green, would be made of doped polysilicon. (b) Scanning electron microscope (SEM) photograph of a FinFET showing two out of ten parallel channels. One can make out the thin gate oxide separating the raised fins and the arching wrap-around gates. (c) SEM photograph of Intel's six-fin and two-fin FinFETs produced using 22 nm technology

Sources: (a) Illustration from a 2020 article about Chenming Hu in *IEEE Spectrum*, image by Emily Cooper; (b) Copyright of The Electrochemical Society. Reproduced by permission of IOP Publishing. All rights reserved; (c) image courtesy of Intel Corporation.

Fig. 30. Chenming Hu, inventor of the FinFET transistor, photographed in 1998.

Source: Image courtesy of Professor Chenming Hu and his administrative assistant, Ms. Charlotte Jones.

honor for achievement and leadership in advancing science and technology in the U.S." In 2020, Professor Hu was awarded the IEEE Medal of Honor.

Today, most FinFETs are made with several parallel channels; in Fig. 29(b), we see two fins (channels) out of a total of 10, while in Fig. 29(c), we see several multi-fin FinFETs. Such a device delivers a proportionally higher total current than a single-channel device. This higher current is beneficial because it provides a larger charge to the downstream transistor, driving it more decisively.

You might wonder why engineers did not build planar MOSFETs with multiple channels. The reason is that it would require increasing the device area (footprint) in proportion to the number of channels. In a FinFET, multiple channels increase the footprint more modestly because the FinFET is more three-dimensional than a planar MOSFET. A multi-channel FinFET consists of intersecting fins and gates, resembling a nanometer-scale sieve. Looking at Fig. 29(c), who would guess these interleaved micro-ridges represent the most sophisticated and efficient transistors ever made?

In addition to geometrical novelties, such as fins, wrap-around gates, and short channels, these new transistors also incorporate various material innovations, which we will not discuss here. The reader will find more on this in Gurvitch (2023), including the story of my own contribution to the so-called *high-k dielectrics*.

All in all, these new FETs are rather sophisticated creations — maybe a bit too sophisticated. At some point, the number of tricks used in their fabrication starts to feel excessive; they begin to resemble a sick old man kept alive by a battery of life-sustaining machines. One wonders: Is it worth the trouble? I imagine that semiconductor engineers today must feel nostalgic for the elegant simplicity of the original silicon MOSFET.

"Everything gets better and better": Moore's Law

In Lewis Carroll's *Alice in Wonderland* — a favorite source of quotes and epigraphs in books about science — Alice drinks from a bottle labeled "DRINK ME" and shrinks from her normal size to about ten inches tall, exclaiming, *"What a curious feeling! I must be shutting up like a telescope."* She waits a few minutes to see if she will shrink any further, but she does not.

Electronics seems to have taken a much deeper gulp from that bottle sometime around 1959, and it has been shrinking ever since. Unlike Alice, who merely shrank by a factor of five, electronic devices have shrunk by a factor of about a million between 1959 and today. And while Alice was content to let the shrinking stop, electronics is fighting tooth and claw to continue shrinking — even as it approaches fundamental physical limits.

This relentless trend of miniaturization is a direct consequence of the development of IC technology. It is known as *Moore's Law*, named after Gordon Moore, co-founder of Fairchild Semiconductor and later Intel, and a member of the prestigious "traitorous eight." Moore is seated against the wall in the upper-left corner of the group photo in Fig. 21.

Moore observed the trend that would bear his name in 1965, publishing his insights in a short article in the trade magazine *Electronics* (Moore, 1965, pp. 114–117). The piece addressed seemingly mundane issues, including the cost of electronic components, how that cost depends on manufacturing yield, and how it tends to decrease as more elements are packed onto a chip. Moore predicted that these trends would likely continue for at least 10 years. In 1965, chips contained roughly 50 components; Moore forecast that by 1975, that number would reach 65,000.

In that original paper, Moore did not explicitly describe the trend as exponential or specify a doubling time. He did not speak specifically about transistors but about "components." However, his numbers implied exponential growth, with a doubling time of roughly one year.

A decade later, in 1975, Moore revisited his now-famous "law" and refined it, predicting that the number of transistors on a chip would continue doubling approximately every two years. In practice, the industry outpaced even this revised forecast for a time: Throughout the 1980s and 1990s, the doubling time averaged about 18 months.

This exponential growth — astonishing in its longevity — has held for decades. Today, 66 years after the birth of the IC, Moore's Law continues to exert its influence. However, the pace has slowed in recent years, with the doubling time rising to about 2.5 years.

The growth is called *exponential* if a quantity doubles within a fixed period of time. The exponential function underlying Moore's Law can be written as $N(t) = N_0 \times 2^{t/T}$, where $N(t)$ is the number of transistors on a chip at time t, N_0 is the number at the starting point ($t = 0$), T is the doubling time, and t/T is the number of doubling intervals that have elapsed.

Let's examine the entire picture from 1959 to 2025. We take $N_0 = 2$ in 1959, the number of transistors in the earliest ICs made by Kilby and Noyce. In 2024–2025, Apple's M3 Max has approximately 92 billion transistors, NVIDIA's GPUs range between 80 and 100 billion, and some specialized chips (such as Google's TPU or high-density memory dies) may exceed even that, especially when using advanced packaging.

To construct a graph like the one shown in Fig. 31, we take the logarithm base 10 of both sides of the exponential formula: $\log[N(t)] = \log[N_0] + (t/T) \times \log(2)$.

Then we plot $\log N(t)$ versus t. If Moore's Law holds, the data points should fall along a straight line — and they do, approximately. The slope of this line is $\log(2)/T$, and by measuring that slope, we can estimate the doubling time T. The analysis of Fig. 31 gives $T = 1.88$ years, consistent with the historical average over much of the Moore's Law era.

In Fig. 31, we plotted 15 points representing the number of transistors on chips between 1959 and 2025. The points are labeled with numbers 1 through 15, and the legend specifying chip names corresponding to these numbers is given in the figure caption.

Of course, when dealing with exponentials — especially empirical ones — one should not expect perfect precision. The kind of agreement we

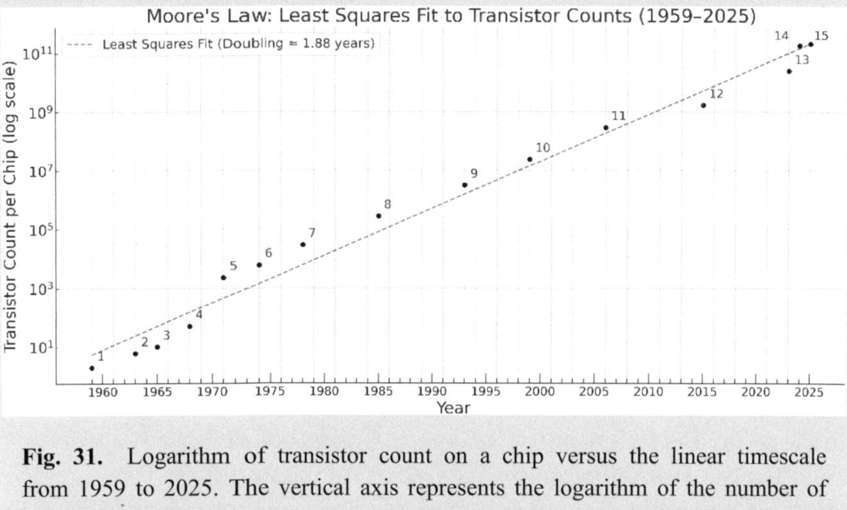

Fig. 31. Logarithm of transistor count on a chip versus the linear timescale from 1959 to 2025. The vertical axis represents the logarithm of the number of transistors, base 10.

Note: 1: Kilby and Noyce; 2: Fairchild Micrologic; 3: TI SN51x; 4: Signetics SE100; 5: Intel 4004; 6: Intel 8080; 7: Intel 8086; 8: Intel 80386; 9: Pentium; 10: Pentium III; 11: Core 2 Duo; 12: Skylake; 13: Apple M3; 14: Apple M3 Ultra; 15: NVIDIA Blackwell B200. The straight line represents the best linear fit to the data, corresponding to the average doubling time of 1.88 years.

Source: Image created by the author using ChatGPT-4o.

see here is impressive enough. It's also worth noting that while our plot tracks the number of transistors per chip, the area of a chip has not remained fixed — it has generally increased over time, especially as more cores were enclosed in a single package.

IC complexity — the level of integration — progressed steadily, from *small-scale integration* (SSI) in the early 1960s to *medium-scale integration* (MSI) in the late 1960s, then to *large-scale integration* (LSI) in the 1970s, to *very-large-scale integration* (VLSI) in the 1980s, and finally to *ultra-large-scale integration* (ULSI). We live in the ULSI era. My work at Bell Labs occurred during the VLSI era, while my adolescent years unfolded during the era of SSI.

Farewell to Moore's Law

Sometimes, we read that Moore's Law was a self-fulfilling prophecy because it was outlined in the *International Technology Roadmap for Semiconductors* (ITRS) reports and subsequently incorporated into the quarterly and annual plans of semiconductor companies. Some people say it was guaranteed to work; no mystery there. I think those who say that are confusing *cause* and *effect*: the issue is not whether Moore's Law was propelled by corporate decisions — it was — but that semiconductor companies *could* make such decisions. If they could make the doubling time equal to one year instead of two, they would do that, but they couldn't. And if one of the companies decided to slow down and take a "vacation" from Moore's Law, it would inevitably lose to its competitors and perish. As philosophers would say, Moore's Law reflects *objective reality*.

In my last book, I argued that electronics undergoes a true evolution, similar to the evolution in living nature. From that perspective, Moore's Law reflects a natural evolutionary rate of miniaturization. The objective factors have combined to give us Moore's Law, which cannot be sped up or slowed down at will. So, a question arises: When Moore's Law stops completely — perhaps in the late 2020-s — will that signify the end of e-evolution? No, evolution never stops; it is in its nature to continue indefinitely. The miniaturization may come to a halt, but other aspects will continue to evolve. Miniaturization is not necessary for evolution.

Biological evolution proceeded for three and a half billion years without scaling down basic biological structures such as DNA or proteins.

Moore's Law does not simply "happen," scaling is a formidable scientific and engineering challenge. It's not enough to shrink the minimum feature size — all other circuit parameters must be scaled simultaneously and correctly. Device dimensions, thicknesses, voltages, currents, and even doping levels must follow their own scaling rules, or nothing will work. Each parameter sips from its own bottle labeled "DRINK ME" — one for channel length, another for gate oxide thickness, a third for doping levels, and so on.

It took 13 years, from 1959 to 1972, before the need arose to construct a theory of IC scaling. Many researchers contributed to this effort, but Robert Dennard and his IBM colleagues provided the earliest and now classic treatment of this non-trivial problem (Dennard *et al.*, 1974).

The size scaling factor is denoted by κ (the Greek letter kappa — the grand, great-grandmother of the English *k*). All device sizes scale as $1/\kappa$. Typically, $\kappa = 1.41$ in one iteration of Moore's Law, ensuring that a device's area (its footprint) decreases by a factor of two: $1/\kappa^2 = 1/(1.41)^2 = 1/2$. The gate oxide thickness must also decrease as $1/\kappa$, but the doping level (the number of impurity atoms per unit volume) must *increase* as κ; the smaller the device, the higher its doping level. There are many other parameters whose correct scaling was dictated by Dennard's theory.

The capabilities of most digital electronic devices — processing speed, memory capacity, camera pixel count and so on — are tied to Moore's Law. Ray Kurzweil has produced particularly illuminating plots showing exponential growth of several computer parameters. The most informative of these graphs depicts the *computing speed (operations per second) per dollar* as a function of time from 1939 to 2004 (Kurzweil, 2005). The ancestor of all these graphs is still Moore's Law.

As we said, in 1965, Moore's primary concern was component prices. Moore's Law can be rephrased in those terms: the cost of making a transistor in an IC roughly halves every two years. In other words, while the price of a chip remains roughly constant, the chip does more for the same amount of money. Depending on the chip type, the cost per transistor ranges from a billionth to a hundred millionth of a dollar — a number almost too small to grasp.

But while transistors get cheaper, fabrication facilities become more expensive. As mentioned earlier, building a new semiconductor factory in the U.S. today costs tens of billions of dollars. This trend has its own law: *Rock's law*, named after Arthur Rock, Intel's first chairman, who stated that the cost of a semiconductor fabrication plant doubles every four years.

The resilience of Moore's Law has been extraordinary. No one, not even Moore himself, believed it would last this long. Predictions of its imminent demise have been made repeatedly, yet it has continued to operate. However, there is no doubt that it must eventually come to an end. Today, the smallest photolithographic features and MOSFET channels are just a few dozen atoms across. Shrinking further becomes not only difficult but also physically impossible. Alice was right to worry about endless shrinking: Nothing works when IC elements reach atomic scales. A metal made of a one-dimensional atomic chain cannot conduct electricity (there is a theorem in physics to this effect), and a gate oxide thinner than 1–2 nm cannot insulate.

Moore's Law has already slowed and is clearly at the end of its rope. Even the authoritative ITRS acknowledged this. Since 1992, the ITRS has published forward-looking reports every two years, projecting the future of semiconductor technology over the next 15 years. However, in 2015, it dramatically revised its previously optimistic 2013 forecasts, now predicting that Moore's Law would end by 2025, with some aspects of it ending even earlier. The ITRS took this forecast so seriously that it committed organizational *hara-kiri*: If there is no Moore's Law, there is no longer any need for the ITRS. Its last report was published in 2015.

Moore's Law — soon to be spoken of in the past tense — is unique. Nothing in human history, technological or otherwise, has developed so explosively for so long. There's simply nothing to compare it to. The observation Gordon Moore made 66 years ago became as iconic as some of the great laws of physics. I don't know which is more famous: Einstein's $E = mc^2$ or Moore's Law. Moore once joked, "Moore's Law is a violation of Murphy's law — everything gets better and better." Enjoy it while it lasts. As of 2026, it hasn't stopped completely, but its days are clearly numbered. The slope is flattening, the economic gains from further scaling are diminishing, and new tricks are running out.

Perhaps Murphy's law, in the end, triumphs over Moore's law and its extensions: the one law that continues to operate as reliably as ever is Rock's law — the cost of semiconductor fabrication facilities keeps doubling roughly every four years.

Chapter 2

From Abacus to the Internet

Because AI resides in computers, we are discussing their history, which, being the history of the highest technology developed by humankind, is fascinating. If man — *Homo sapiens* — is, as it were, the highest achievement of evolution in living nature, in the sense that he has the most developed brain among all animals, then evolution in electronics led to a comparable outcome with the creation of computers, and now also thinking computers endowed with artificial intelligence (AI).

So, let us start from the beginning. It all started innocently, with devices invented to assist people in calculations.

The Mighty Abacus

The oldest calculating device was the *abacus*, a system of movable beads (sometimes stones or twigs) arranged on grooves or rods (Fig. 1). It first appeared about 2500 BCE in ancient Mesopotamia (modern-day Iraq), then in Egypt and elsewhere. For arithmetic calculations, the abacus was as important as the wheel was for transportation. In other words, our civilization would not have developed without it. Let me explain why.

Expressing numbers in a *positional* system was a giant step forward. Some rudimentary positional systems appeared around 2000 BC, but the modern Hindu-Arabic decimal (base-ten) positional system, complete with zero, dates to the 7th century AD. We use it daily without consciously realizing its structure; we know that the number 23508 means $2 \times 10^4 + 3 \times 10^3 + 5 \times 10^2 + 0 \times 10^1 + 8 \times 10^0$, but unless we write a book like this one, we do not think much about it. By comparison, in the

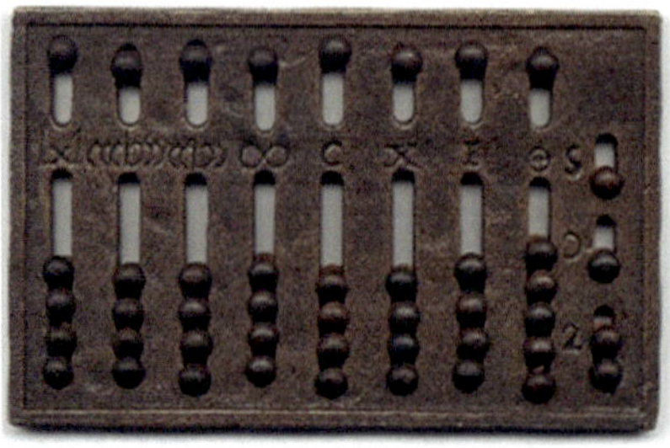

(a)

(b)

Fig. 1. (a): Roman stone abacus from the 1st century AD. **(b)** A Russian wooden abacus where each rod (except the one with four beads representing quarters, ¼ each) contains 10 beads, and the beads on the adjacent rods differ by a factor of 10. The beads on a rod above the short four-bead section have a value of 1 each; beads on the next higher rod are 10 each, etc. Each bead on the highest rod in this picture represents $10^7 = 10{,}000{,}000$, or ten million. The three rods below the short section contain fractions of 1/10, 1/100, and 1/1000. Beads become active when moved to the left side of a rod; the beads remaining on the right side are called residuals. No beads have been moved in this picture, so the number represented here is zero.

Sources: (a) Image from Creative Commons Lic. CC BY-SA 3.0; (b) image in the public domain.

Roman (non-positional) system, the same number would be written as XXIIIMMMDVIII — cumbersome and a little scary. However, cumbersome notation is only part of the problem. The actual disadvantages of the Roman system become apparent when we attempt to perform arithmetic with it. Manipulating Roman numerals is a nightmare; expressing fractions, as well as small and large numbers, is impossible. So, how, you might ask, did the ancient Romans survive? Presumably, they had to do some calculations. The answer is that they had an abacus!

The abacus, invented over 4.5 thousand years ago in ancient Sumeria and later either reinvented or adopted in Babylonia, Egypt, China, India, and Europe, was a material representation of the positional number system. In a decimal abacus (Fig. 1(b)), adjacent rods differ in value by a factor of 10: one bead moved to the left side on the next higher rod is equivalent to all 10 beads moved on the lower rod. A rod where no beads have been moved represents zero. With that, expressing and manipulating numbers becomes easy. In various designs and implementations, the abacus was continually used worldwide for over 4.5 thousand years, which may be a record for any human-made device, except for the wheel. In the first 3,000 years, it was indispensable, supplementing the nonexistent positional system and allowing arithmetic calculations with numbers large and small. Over the past 1,300 years, when it coexisted with the Hindu-Arabic number system, it became dispensable, as calculations could be performed equally well on paper. So little by little, the mighty abacus lost its significance. Yet, even today, it is used in everyday life in some countries.

Mechanical Calculators

An abacus is operated by hand, and doing arithmetic on it requires keeping simple sums and differences in one's head. Proficiency in using an abacus, especially when doing more than addition and subtraction, requires practice. Automated, faster mechanisms appeared as soon as they could be made; this happened at the beginning of the 17th century.[1]

The first mechanical calculator (Fig. 2) that could add and subtract (but not multiply and divide) was built in 1623 by a German professor of

Fig. 2. Replica of Schickard's mechanical calculator.

Source: Image by Herbert Klaeren, Lic. CC BY-SA 3.0.

[1]For a short time in the 20th century, it was believed that Leonardo da Vinci created the first mechanical calculator (adder). This belief was based on Leonardo's sketch, dated around 1500, showing a chain of 13 interlocking gears that would turn at 10 times larger angles in each successive gear. Leonardo's sketch was discovered in 1967, and a replica was built in 1968. A panel of experts met in 1969 or 1970 at the Massachusetts Institute of Technology to decide if it was a calculating device, and the panel concluded that it was not.

Fig. 3. Pascal's mechanical calculator (Pascaline) made in 1652.
Source: Image by Rama, Lic. CC BY-SA 3.0 fr.

mathematics (and of several other disciplines, including Hebrew, oriental languages, astronomy, and geography), Wilhelm Schickard (1592–1635) (Williams 1997, 119–124). The professor described his invention in a letter to his friend, Johannes Kepler, the famous scientist who figured out that the planets move around the Sun on elliptical orbits. Schickard's drawings and the mechanisms he built have been lost for over 300 years, and no one knew how they worked until a sketch was discovered in a letter between the pages of one of Kepler's books. This book was kept in the Pulkovo Observatory near Leningrad (St. Petersburg), where the sketch (made, I assume, either by Kepler or by Schickard) was discovered in the 1950s.

The second mechanical calculating machine was constructed by the great French physicist, mathematician, and philosopher Blaise Pascal (1623–1662). Pascal built his first calculator in 1642 when he was only 19, and then he made about 50 versions in later years; at one point, he tried to make a business out of it. Several of these *Pascaline* calculators (Fig. 3) have survived.

Like Schickard's machine (of which Pascal was unaware), the Pascaline could add and subtract but not multiply or divide. Being a philosopher, Blaise Pascal took a broad view of his invention. He wrote in 1642, "The arithmetical machine produces effects which appear nearer to thought than all the actions of animals." This may have been the first time in history when *machines* and *thought* were connected in one sentence.

The great mathematician (and philosopher, linguist, and notable German diplomat) Gottfried Leibniz (1646–1716; Fig. 4) saw Pascal's design and decided to find a way of adding multiplication to the first two

Fig. 4. Gottfried Leibniz (1646–1716) in a 1703 portrait. His invention of the mechanical calculator pales in comparison to his scientific credits: He is the founder of mathematical logic, the inventor of the binary number system, and an independent co-inventor (with Isaac Newton) of calculus. A true polymath — perhaps we can say "*polygenius*" — he also made significant contributions to linguistics.

Source: Image in the public domain.

arithmetic operations. His machine, called *Leibniz's wheels*, was built in 1674. It was lost for about 200 years until it was rediscovered in 1879 in an attic of the University of Göttingen building.

Leibniz did far more for computers than the design of a mechanical calculator, as we will see shortly.

There were others: Englishman Samuel Morland (1625–1695) invented several calculators and published a book about them in 1673. A clockmaker to King Louis XIV, René Grillet, made a calculator that did not survive until our time but which he described in print in 1678. The best calculating machine was invented in the early 1820s in France by Thomas de Colmar. It was a reliable, commercially produced calculator based on the earlier design by Leibniz. De Colmar's *arithmometer*, as it was known, was produced with modifications and improvements until the start of the First World War.

Meanwhile, several other mechanical and electromechanical calculators were invented, including machines where a user had to turn a crank to perform arithmetic operations — a design commonly seen in movies and occasionally in real life. They were also called "arithmometers," although they differed from De Colmar's design. They survived as cash registers for a long time, at least until the 1960s. I remember seeing such cash registers in my youth.

Mechanical calculators are not *computers* because they are not *programmable*. A computer is a machine that can be instructed to automatically perform an arbitrary sequence of operations, either arithmetic or logical. It differs from a calculator in that it can be programmed to solve any solvable problem; it is a *universal* machine. Yet, computers would not have appeared without those mechanical predecessors we described. The mathematical logic and the binary (as opposed to decimal) number system were even more significant ingredients.

The Binary Number System and Mathematical Logic

Binary number systems have been toyed with since ancient times. In 1703, Leibniz published a paper entitled *"Explanation of Binary Arithmetic, which uses only the characters 1 and 0, with some remarks on its usefulness, and on the light it throws on the ancient Chinese figures of Fu Xi."* Today, we would not give a paper such a long title, but in 1703, it was acceptable. Leibniz was interested in a binary number system for a profound reason: he realized that it could be used in mathematical logic, which he had founded 30 years earlier.

> Binary numbers are formed like decimals (i.e., our usual numbers expressed with 10 symbols: 0, 1, 2, 3, 4, 5, 6, 7, 8, 9) but with only two symbols: 0 and 1. A digit (0 or 1) in a particular position within the number is multiplied by a power of 2 (as opposed to a power of 10 in a digital system) determined by that digit's position. The utmost right is designated as a zero position, so the multiplier of the last digit is $2^0 = 1$ (the same as $10^0 = 1$ in a decimal system); the next multiplier is $2^1 = 2$, and so on. Let us take the binary number 100101 and see what it corresponds to in the decimal system. Following positional rules, $(100101)_2 = [(1) \times 2^5] + [(0) \times 2^4] + [(0) \times 2^3] + [(1) \times 2^2] + [(0) \times 2^1] + [(1) \times 2^0] = (137)_{10}$; So, 100101 in a binary system corresponds to 137 in a decimal system. The inverse conversion, from a decimal number to a binary number, is only slightly more involved.

And now about logic. Classical *syllogistic logic* originated around 350 BCE in Greece with the work of Aristotle, who gave the following definition:

> A syllogism is a speech (or discourse) in which certain things being stated, something other than what is stated, follows of necessity from their being so. (Aristotle 1984, 24)

Syllogistic logic is a form of deductive reasoning that evaluates arguments based on their structure, aiming to derive a conclusion from two given premises. A syllogism consists of three parts:

1. Major Premise: A general statement (e.g., "All humans are mortal").
2. Minor Premise: A statement related to the major premise (e.g., "Socrates is a human").
3. Conclusion: A logical deduction based on the two premises (e.g., "Therefore, Socrates is mortal").

More generally, logic is the study of a thinking process that is supposed to organize our thoughts and make them mistake-free. Logic teaches its students how to arrive at correct conclusions once the chain of arguments starts from meaningful premises (if premises are not meaningful, being logical isn't going to help).

Nobody seems to have benefited in this way from the study of logic. Case in point: Until the beginning of the 20th century, classical logic was part of a school curriculum, and later logic as a discipline was eliminated from school programs. Have the graduates who studied logic been better equipped to think and argue than those who didn't? Probably not, or people would notice the difference, and the study of logic would have been reinstated. Is classical logic completely useless? Perhaps not. The study of logic dissects human thought and exposes its inner structure, which may interest linguists, mathematicians, or computer designers. For the latter, it is crucial to examine and formalize the steps that must be taken to achieve the goal, whether that goal is calculational or logical.

Gottfried Leibniz was the first to see that classical logic could be beneficially combined with mathematics. He started working on the foundations of mathematical logic in 1670 and worked on it on and off for 20 years. Being a polymath has its disadvantages: Leibniz's work on this subject, scattered in his letters and papers (he wrote tens of thousands of letters), was little known until 1895, when the great admirer of Leibniz, French mathematician Louis Couturat, dug them up in the Hanover Royal Library, publishing his findings in 1903. When Couturat and others examined Leibniz's archives, they discovered that he attempted to create an algebra of logical concepts roughly equivalent to what George Boole, Augustus De Morgan, and others had developed 150 years later. The opinions concerning the completeness of Leibniz's scheme varied, but nobody can deny that he was the forerunner and founder of mathematical logic. And, as we said above, he was the first to point out the benefit of using a binary number system, where 1s and 0s can represent the duality of logical outcomes — statements being either *true* or *false*.

We will now jump forward approximately 150 years to a time when this subject became part of mathematics. Unknowingly following Leibniz and going further than he could in mathematical rigor, British mathematician George Boole (Fig. 5), working in the mid-19th century, also realized that logic could be formalized and presented as a kind of algebra. What does this mean?

Ordinary algebra operates on symbols that may, in principle, take numerical values; algebra treats them as labels for these values, which

Fig. 5. George Boole (1815–1864), creator of mathematical logic.

Source: Image in the public domain.

gives algebra its generality. When we write $a + b = c$, we allow for an infinite number of possibilities for the numerical values of a and b, as long as c is the sum of these values.

In Boolean algebra, the values of algebraic symbols are not numbers but truth values: either “true,” represented by 1, or “false,” represented by 0. One assumes nothing in between — a meaningful statement can be either true or false. In Boolean algebra, as in Aristotelian logic, there are no half-truths.

These truth values can be manipulated using logical operations; the three basic ones are called AND, OR, and NOT. Fearing the decline in my book sales, I will refrain from defining them here. What matters is that all other logical operations can be constructed from these three; AND, OR, and NOT are sufficient to build a complete logic system. There are even more powerful operations, such as NAND, which alone can generate all of logic.

Once the basic operations have been defined, one can begin formulating mathematical theorems that can be proven, each representing a logical conclusion derived from certain premises. The result, akin to a syllogistic conclusion in Aristotelian logic, is obtained not verbally but mathematically, using Boolean algebra.

I won’t go further into this here. An interested reader can find more information about Boolean logic in my previous book (Gurvitch, 2023) or specialized books on the subject. It is a beautiful and deceptively simple framework, at least at first. Then, as with all things mathematical, it quickly gets complicated.

Mathematical logicians were standing over AI’s cradle in the 1940s and played a significant role throughout its history. We will meet a few of them in the pages that follow.

Claude Shannon and the implementation of mathematical logic in hardware

In 1936–1937, Claude Shannon (Fig. 6), a graduate student at the Massachusetts Institute of Technology (MIT), showed that Boolean logic is essential to constructing sophisticated calculating machines. As part of his master’s thesis, Shannon mathematically proved the equivalence of Boolean logic and electrical circuits containing two-state switching elements. In other words, a 21-year-old genius (for he was a true genius) demonstrated that all conceivable logical operations can be implemented

Fig. 6. Claude Shannon (1916–2001), photographed in 1952 with his "smart mouse," Theseus, which was used in the first AI experiments; Theseus learned how to escape from a maze (in Chapter 4).

Source: Image courtesy of AT&T Archives and History Center.

in hardware, and the only ingredients required for this are wires (more generally, electrical interconnects) and two-state (binary) switching elements. The nature of these switching elements does not matter.

This spectacular result was only the beginning for Claude Shannon. Ten years later, while working at Bell Labs, Shannon developed Information Theory, a cornerstone of all information-related fields. He defined the basic unit of information, the *bit* (see the following section), which allowed to treat information mathematically. He also founded the theory of secret codes, known as mathematical cryptography. Shannon was also one of the founding fathers of AI and the first to conduct experiments in this field. In the late 1940s, he discussed a computer program for playing chess. We will detail his contributions to AI in Chapter 4.

Following Leibniz, Boole, and Shannon, all computers use binary codes in which high and low voltage levels correspond to 1s and 0s. When a computer calculates, it interprets a sequence of 1s and 0s as a number written in the binary (base 2) representation. However, when it performs a logic operation, the same 1s and 0s are interpreted as Boolean symbols, signifying "true" and "false" values. This duality of 1s and 0s constitutes the innermost source of computer power.

Bits and bytes

Each symbol, 1 or 0, carries one *bit* of information. Shannon introduced "bit" as a shorthand for the "binary digit." One bit is the amount of information we learn when we receive a "yes" or "no" answer to a question that is appropriately formulated. Mathematically, "yes" is represented by "1" and "no" by "0." The bit is the *minimum* unit of information. For example, if I ask you, "Is your name John?" your answer, if it is given as a minimal response to my question, can be "yes" or "no." You may go beyond that minimum and say, "My name is not John; it is Bill." If you do that, I will receive more than one bit of information, which is more than requested.

A *byte* is defined as an 8-bit symbol, such as 10011011; it contains eight answers to eight "yes" or "no" questions.

Logic gates

Logic operations and numerical computations are implemented in computers using *logic gates*. A logic gate is a small device delivering a logical output of either "true" (1) or "false" (0), based on 1s and 0s received at its inputs. Three standard gates correspond to the basic Boolean operations: AND, OR, and NOT. There are also more sophisticated gates, such as NAND and NOR. These more advanced gates are self-sufficient; a complete logic family can be constructed using only NANDs or only NORs.

Logic gates can be physically constructed in a variety of ways. In 1937, when Claude Shannon wrote his famous master's thesis, they could be built using electromechanical relays or vacuum tubes. After 1950, logic gates were constructed using transistors and CMOS (as mentioned in Chapter 1). More exotic forms have also been developed: experimental photonic gates that manipulate light, molecular gates based on single molecules, and even biological gates, such as the neurons in our brains, which we will discuss in Chapter 4.

Quantum computing introduces yet another kind of gate — quantum gates — which operate according to the principles of quantum mechanics rather than Boolean logic. We will touch on these briefly in the following chapter.

Today's computer processors are built on transistors and contain a vast number of logic gates — not just thousands or millions but tens of billions. For example, NVIDIA's Blackwell B200 processor, released in 2024, contains over 200 billion transistors. Since a typical logic gate uses about four to six transistors, this chip likely implements on the order of 30–50 billion logic gates.

Computers' Romantic Pre-history: Charles Babbage and Ada Lovelace

Charles Babbage: the most original mind of the 19th century

The first *programmable* calculating machines were designed and partially built by British polymath (mathematician, engineer, economist, physicist, geologist, philosopher, and theologian) Charles Babbage (left image in Fig. 7), who worked on his calculational projects from 1819, when he was 22, to his death in 1871, when he was 80. He was obsessed with

Fig.7. "Grandfather of the Computer" Charles Babbage (1791–1871) and Ada Lovelace (nèe Byron) (1815–1852), Babbage's collaborator and the first computer programmer in history

Sources: Left image from Wellcome Collection, London, UK, Lic. CC BY 4.0. Right Image in the public domain; this portrait was painted in 1836 by Margaret Sarah Carpenter and is now in the Government Art Collection at 10 Downing Street in London.

mathematical accuracy all his life. Initially, Babbage wanted to improve non-programmable mechanical calculators, and to that end, he embarked on designing a machine to automate tedious calculations required to compose mathematical tables. Volumes of published tables existed at the beginning of the 19th century to benefit engineers, astronomers, and anyone who needed them. These tables were painstakingly produced by human "calculators" who spent years doing this tedious work. Babbage was particularly unhappy that all these tables were full of mistakes that were very difficult to find and correct (Williams 1997, 160, 163). While a student at Cambridge, he began thinking about a machine that would compute values for such tables with much greater precision than that available in Pascal's and Leibniz's calculators. And Babbage wanted his device to print the results. In 1819, he started working on designing such a machine in earnest.

The *Difference Engine* was supposed to perform calculations based on Isaac Newton's mathematical method of *finite differences* (hence the word "difference" in the name), which required only repeated additions.[2] It was not what we would call a *computer* but rather a specialized *calculator*. The machine was purely mechanical; no electricity was involved. (Naturally, in 1819, there was no power grid, only Volta's batteries, but they could not power Babbage's machine.) By 1822, Babbage had built a small prototype, known as the *Difference Engine 0*, and sought support for further work through the Royal Society, of which he was a member, envisioning that the project would take three years. The Royal Society recommended the project to Parliament, and in 1823 Babbage received some government support. However, the undertaking proved to be more complex than imagined. Over the years, Babbage contributed a substantial amount of his own money to the funds he received from the government. The British government supported Babbage's project from 1823 to 1842, although it did so reluctantly, advancing the promised sums with delays that lasted years. At one point, a member of Parliament suggested, with an undeniable sense of humor, that Babbage's machine should be set to work calculating how much it had already cost the British government (in Williams 1997, 169).

Plagued by numerous technical and personal problems, Babbage managed to build the second prototype, the *Difference Engine 1*, by 1832.

[2]It may be confusing that finding *differences* requires *additions*, but this is how this Newtonian method works.

This prototype was significantly smaller than the complete machine and operated with limited precision, processing 6-digit numbers instead of the planned 20-digit numbers. After 1842, Babbage shifted his attention to a new project, which we will discuss shortly. Nonetheless, in 1846–1849, he returned to the Difference Engine and made detailed plans for the improved *Difference Engine 2*. That machine was intended to manipulate 31-digit numbers — unheard-of precision at that time and for a long time. This project was not completed in Babbage's lifetime. However, a century and a half later, two working models of the Difference Engine 2 were built according to Babbage's plans: one was completed in 1991 in the U.K., and another, in 2008, in the U.S.

These vast machines, built in the 20th century according to Babbage's plans, consisted of thousands of parts and weighed several tons, yet worked flawlessly, calculating to the planned precision of 31 digits. If the Difference Engine 2 calculated the famous Pi (π), it would get 3.14159265358979323846 2643383279[3] — a most impressive accuracy, considering that this mechanical machine was designed in the first half of the 19th century.

If the Difference Engine was impressive, the second Babbage machine, the *Analytical Engine*, was even more impressive, notwithstanding the fact that it was never built. The Analytical Engine would perform all arithmetic operations (not just addition) with the same precision as the Difference Engine 2, and, most significantly, it would be *programmable*. If such a machine were indeed built in the mid-19th century, as Babbage hoped, it would surely have changed the course of history. Alas, it was not to be. The Analytical Engine was not built in the 19th century, nor ever. The first general-purpose computer — an electronic one — appeared a hundred years after Babbage started working on his mechanical dream machine. And we cannot even claim that it arose from that early work by Babbage. But despite remaining a dream (a rather detailed one), the very

[3] Pi (designated by the Greek letter π) is the ratio of a circle's circumference to its diameter. Pi has always fascinated people. The first man who developed a method of calculating π with arbitrary precision was the great Greek mathematician Archimedes. In 2016, one of Google's employees, Emma Haruka Iwao, computed over 31 trillion (31×10^{12}) digits of π, which is a *trillion* times (1 with 12 zeros) more than what could be calculated on the Difference Engine 2. And that does not mean that the job is finished: π has an infinite number of digits, and being an irrational number, these digits cannot, in principle, form repeating patterns.

concept of the Analytical Engine and all that surrounded its planning and unfinished construction made Charles Babbage's name immortal.

Babbage started to think about this second machine in 1834, and in 1835 he wrote in a letter to his correspondent:

> ...for six months been engaged in making the drawings of a new calculating engine of far greater power than the first. I am myself astonished at the power I have been enabled to give to this machine; a year ago I should not have believed this result possible. (quoted in Williams 1997, 178)

He worked on the Analytical Engine from 1834 to 1849 and, after a seven-year break, from 1856 until his death in 1871. Seeing all the difficulties ahead, he endlessly redesigned his dream machine and occasionally constructed small sections of it. Although Babbage published several books, he never published the plans for his calculating engines. He left them in his private papers, on thousands of pages in his notebooks, and in a set of 300 large, professionally drawn plans. In 1837, he produced a description of the Analytical Engine, but that description was not published during his lifetime (it was published in 1973).

The Analytical Engine was envisioned as a general-purpose automated calculating machine capable of performing a wide variety of tasks. It had the essential structure of a modern computer, including the *input* for data and programs, which Babbage called the *Reader*; the *output*, which was a printer; the *central processing unit* (CPU) (the *Mill*); and the *memory* (the *Store*). It could be classified as a *Universal Turing Machine*, the concept introduced by Alan Turing in 1937. Programs and data were entered using *punch cards* (cardboard cards perforated with holes), which Babbage adapted from the automated textile-weaving machines (looms).

Punch cards as a medium for computer input survived into the 1970s and even longer in various non-computer applications, such as voting ballots. I used them in the mid-1960s in the Soviet Union, where computer technology was lagging behind the West for at least a decade, thanks to the Stalinist era's hostile attitude toward computer science (cybernetics). Students at the Leningrad Polytechnic Institute were required to write a small program, punch it into the cards, and run it on a computer (Ural or

Minsk, one of the two domestic models at the time). Vacuum tubes in that room-sized machine burned with depressing regularity, leaving it offline for hours at a time, which did not unduly upset us, the students. The program was punched into a stack of cards using a device that translated program commands into the positions of punched holes. I had not realized then that I was using late 18th-century technology.

Babbage planned the Analytical Engine's memory capacity to comprise a thousand 50-digit numbers (or *words* in computer terminology), which exceeded the memory capacity of any 20th-century electronic computer made before 1960. The entire machine, if completed, would be approximately 15 ft. high. The mill (processor) would be five ft. in diameter, and the store (memory) would project 10 or 20 ft. from the side of the machine [Williams 1997, 183]. The construction would be so large and heavy that its creator envisioned powering it with a steam engine!

The first software written by Lord Byron's daughter, Ada Lovelace

And then there was software. In his private notes, Babbage outlined the programming principles he envisioned for his Analytical Engine. He had rather definite ideas about these programming principles, which historians later extracted from his writings. These principles encompass essential features widely used in computer programming today, including *loops*, *subroutines,* and *conditional branching.*[4]

Preparations for the Analytical Engine's programming were performed in the early 1840s by an unlikely yet colorful person who has fascinated historians, writers, playwrights, and feminists ever since. Augusta "Ada" Byron (in marriage, Countess of Lovelace; right image in Fig. 7) was the only legitimate child of the great English poet George Gordon Noel, sixth Baron Byron, known to the whole world as Lord Byron. The nickname "Ada" was given to her by her famous father. Ada

[4]A *loop* is a part of a program that repeats a set of operations over and over for a specified number of times; a *subroutine* is an autonomous part of a big program; and *conditional branching* refers to jumping from one part of a program to another, the jump conditioned on a specific quantity reaching a pre-set value.

could not remember him, as her mother, Anne Isabella Noel-Byron, 11th Baroness Wentworth and Baroness Byron (*née* Milbanke) — all referring to the same person, mind you — had separated from Lord Byron soon after Ada was born. Lord Byron only saw his daughter in the first month of her life, but he always kept her portrait, in which she was depicted when she was four. Soon after she was born, he left England and died in Greece in 1824, when she was eight years old. Lord Byron died at 36, and his daughter Ada was destined to die at the same young age.[5] At her request, she was buried next to her father.

When Ada was 17, she met Charles Babbage. She became fascinated with him and later with his Analytical Engine, offering her considerable mathematical talent and staunch support for his projects, which could be valuable given her social standing and connections. Babbage, in turn, came to appreciate her intellect and mathematical abilities; he even introduced her to his friend Augustus De Morgan, one of the early British mathematical logicians, who gave Ada private lessons.

In 1843, Ada Lovelace published a paper known as "Notes," which became an essential resource for all future studies of the Analytical Engine. The genesis of these Notes is somewhat convoluted. In 1843, Ada translated a French paper into English and supplemented the text with her extensive notes. The original article was published in 1842 in a Swiss journal by Italian military engineer Luigi Manabrea,[6] who in 1840 attended a series of seminars on the Analytical Engine given by Charles Babbage at Turin University in Italy. Manabrea published his notes under the title "Sketch of the Analytical Engine Invented by Charles Babbage, Esq." Ada Lovelace published her translation of these notes under Manabrea's name, modestly signing only her initials, "A. A. L." It is now traditional to list them both as authors: Manabrea and Lovelace (1843). Manabrea's text constitutes about a quarter of the document, while Lady Lovelace's notes comprise three-quarters. Much of what Ada has

[5] In the 18th and 19th centuries, many geniuses died at age 36–37: Henry Purcell, Wolfgang Mozart, George Byron, Alexander Pushkin, Robert Burns, Sadi Carnot, Vincent Van Gogh, Arthur Rimbaud, and Heinrich Hertz, to name a few. Of course, some died even younger, but the ages of 36–37 seemed to be especially dangerous for true geniuses.

[6] Luigi Manabrea had a distinguished career. He reached the military rank of lieutenant general, was appointed a senator, and received the title of count. At the height of his career (in 1867–1869), he was Prime Minister of Italy under Victor Emmanuel II. Yet, his true legacy rests with his fleeting involvement with Babbage and Lovelace.

written was the result of daily consultations with Charles Babbage, who was pleased to see his methods described and publicized; apparently, it was difficult for him to do it alone. The Notes described how mathematical procedures could be broken down into four arithmetic operations and implemented on the Analytical Engine. They also included Babbage's ideas about programming and some of Ada's thoughts and algorithms.[7]

There are two reasons why the Notes made Ada Lovelace's name famous, albeit many years later: She was the first person in history to write a computer program, and she expressed some general ideas about computers that were ahead of their time by a good century. One view expressed by her was that a machine such as the Analytical Engine is capable of any conceivable calculation. She wrote in Note A:

> The Analytical Engine…is not merely adapted for *tabulating* the results of one particular function and of no other, but for *developing and tabulating* any function whatever. In fact the Engine may be described as being the material expression of any indefinite function of any degree of generality and complexity…[italics kept as in the original]. (Manabrea and Lovelace 1843, Note A).

In other words, she characterized the Analytical Engine as a *general-purpose computer*, something that Babbage certainly had in mind but did not bother to put in writing, at least not in this straightforward way. We will shortly see that Alan Turing defined the same concept, his Universal Machine, albeit with greater mathematical rigor, in 1937.

Moreover, Ada Lovelace was the first to muse about the possible uses of computing machines beyond numerical calculations. In that, she seemed to be ahead of her mentor, not to speak of her other contemporaries. Here I present (somewhat shortened and slightly corrected — forgive me, Countess[8]) passages from her Note A:

> The Analytical Engine is an *embodying of the science of operations*… figures, the symbols of *numerical magnitude*, are frequently *also* the

[7]An *algorithm* is a set of operations required to perform a task or solve a problem. A *program* is a sequential list of commands written in a specific computer language that expresses an algorithm in a way that will be understood by the computing machine.

[8]While expressing profound thoughts, Ada sometimes used impossibly long and cumbersome expressions in her writing. Nonetheless, with a bit of persistence and patience, we can understand what she was trying to say. Her writing style can be forgiven in view of her content.

symbols of *operations*...The operating mechanism...might act upon other things besides *number*, were objects found [whenever objects are found] whose mutual fundamental relations could be expressed by those of the abstract science of operations...Supposing, for instance, that the fundamental relations of pitched sounds in the science of harmony and of musical composition were susceptible of such expression and adaptations, the [Analytical] Engine might compose elaborate and scientific pieces of music of any degree of complexity or extent [italics kept as in the original; text in square brackets — ours]. (Manabrea and Lovelace 1843, Note A)

Although somewhat vague and convoluted, these passages can be interpreted as expressing the idea of what was in the 20th century called a *general-purpose computer with stored memory*, which treats numbers, logical values of *true* and *false*, and even programming instructions on equal footing. This text about manipulating symbols (operations) instead of or alongside numbers constitutes a remarkable insight into the nature of computers. As one historian of science has put it, Ada's idea marks a transition from *calculation* to *computation*. No one else articulated such thoughts for the next hundred years. And all of that while being a well-dressed beauty, a countess, and the only daughter of Lord Byron!

Countess Lovelace commented on the use of *programming cycles* in Note F. In the context of the Analytical Engine, this meant repeatedly using the same cards. She showed how this could drastically reduce the number of cards required.

Finally, at the end of the document, in Note G, Ada Lovelace says:

We will terminate these Notes by following up in detail the steps through which the Engine could compute the Numbers of Bernoulli, this being (in the form in which we shall deduce it) a rather complicated example of its powers. (Manabrea and Lovelace 1843, Note G)

Then, she proceeded to present a complete computer program, specifying the punch cards that had to be prepared to calculate an infinite series of Bernoulli numbers.[9] Interestingly, Babbage assisted her in that task

[9]Bernoulli numbers form an infinite series obtained via a specific set of mathematical operations; they are used in number theory. The first six Bernoulli numbers are 1, ±1/2, 1/6, 0, −1/30, 0. As you can see, the next Bernoulli number cannot be predicted; it can only be calculated if one knows the algorithm.

"to save Lady Lovelace the trouble," but he made a mistake, which she noticed and which was promptly corrected (Isaacson, 2014, p. 21).

In that Bernoulli numbers exercise, Ada prepared a list of coding instructions and a carefully drawn diagram showing the flow of operations, which she proudly presented to Babbage. She made her husband, William, Earl of Lovelace, make the ink drawings from her pencil sketches; little did he know that this would be the *only* thing in his life (apart from being married to Ada) for which posterity would remember him. This presentation on Bernoulli numbers is the chief reason she is considered the world's first computer programmer. In the 20th century, the Babbage–Lovelace algorithm was checked and found to be mistake-free.

Neither Charles Babbage in his Analytical Engine endeavor nor Ada Lovelace in her programming had a serious following. The first machine, the Difference Engine, had some followers in the 19th and 20th centuries (Williams 1997, 170–177), and some clones were built by Babbage's son, Henry. One American computer, the *Harvard Mark I*, built in the 1940s, was similar to the Difference Engine. However, the absolute marvel, the programmable Analytical Engine, did not have a following, nor was it ever built, except for the part of the Mill built by Henry in 1910, when he was 86 years old. The Analytical Engine's Mill can be seen in the Science Museum in London, alongside the working model of the Difference Engine. The museum also holds thousands of pages of the original documentation for the Analytical Engine. The unfinished project remained a historical curiosity, which was forgotten by the mid-20th century. The creators of modern computers — Turing, Zuse, von Neumann, Mauchly, Eckert, and others — were either unaware of the Analytical Engine or did not give it much thought. We cannot blame them since it was never built. Turing once referred to Ada Lovelace when considering thinking machines (artificial intelligence), noting that she did not believe in such a possibility (as evidenced by a passage in Notes to that end); at least he was aware of her existence. Yet, Alan Turing had not noticed that Ada had realized and stated in her Notes that the Analytical Engine would be a universal machine.

The Analytical Engine was studied only toward the end of the 20th century, particularly in the 1980s and later. These studies revived interest and sparked new admiration for these computer pioneers. Charles Babbage and Ada Lovelace were ahead of their contemporaries and several generations of their descendants. It is admirable, but it often leaves such people feeling lonely and misunderstood in their time. In our time, there are proposals to build a model of the Analytical Engine, just as it was done for the Difference Engine. Alleged uncertainties in Babbage's Analytical Engine plans and the lack of funding have so far prevented this from being done.

If one day the Analytical Engine is built according to Babbage's plans, I would like to see it powered by a steam engine — that would be a sight to behold!

After Babbage's hopelessly early, heroic, and unfinished attempt to create a programmable machine, little progress was made in the digital computer field until the late 1930s. Various digital calculators were developed, initially mechanical and later electromechanical, but not a *computer*.

Analog Calculating Machines

We call machines that treat information discreetly, breaking it into small chunks (bits) and manipulating these chunks — *digital* machines. If a digital machine were building a castle, that castle would be made of bricks. A digital machine may use binary (base 2) or decimal (base 10) notation. The latter is less convenient for performing logic operations but can still be used. Despite its mechanical construction, the abacus should be considered a digital machine. In it, the numbers are represented discreetly: The bead is either on one side of a rod, engaged, or on the other side, dormant, with no meaningful positions in between. All the mechanical calculators and Babbage's computers described above were also digital. Therefore, it would be incorrect to assume that digital calculations are a modern invention. What is modern is Shannon's scheme: two-state switching elements connected by wires.

Digital methods have won; there are hardly any analog calculating machines left. Yet, historically, from ancient times to the mid-20th century, analog devices held a place in the computing world, and their existence was justified by the fact that they were faster than their digital counterparts for a long time. Let us now say a few words about these wonderful devices.

Analog machines, mechanical or electrical, deal with continually changing quantities. The castle built by an analog machine would be made from clay. Analog computational aids have been known since antiquity.

Fig. 8. Part of the Antikythera Mechanism in the National Archeological Museum in Athens, and one of the many proposed models of how it may have worked.

Source: On the left: The photo from the Museum is by Francesco Bini, Lic. CC BY-SA 4.0. On the right: Exploded model of Cosmos gearing, by T. Freeth *et al.* Creative Commons Attribution 4.0 International license.

A device called an *astrolabe*, which helps with astronomical observations and calculations, is at least 2,000 years old.

The ancient Greeks used rather sophisticated analog devices; one was discovered in a shipwreck in the Aegean Sea in 1901. It is called the *Antikythera Mechanism* (Fig. 8), named after the small Greek island of Antikythera where it was recovered. Although badly damaged by water, this device reveals an impressive construction of 30 interconnected bronze gears and asymmetrical circles — *epicycles* used in astronomy to describe the apparent motions of heavenly bodies. It was determined that the ship was traveling from Rhodes to Rome. Antikythera may have been the very marvel described by Cicero, who wrote in 54–51 BCE that when he visited Rhodes, he was shown a device that could accurately reproduce the motions of the Sun, the Moon, and the five planets (Williams 1997, 195–197). Nothing approaching the complexity of the Antikythera mechanism was constructed in Europe before the end of the 16th century.

By far, the most prevalent analog instrument was the *slide rule*. It was invented in the 17th century following the introduction of a *logarithmic function* in 1614 by the Scottish mathematician John Napier. The slide rule emerged as a practical application of logarithms. The idea originated with Edmund Gunter, an English mathematician, who in 1620 devised a logarithmic scale — a number line spaced according to logarithmic values, allowing multiplication and division to be performed by measuring distances. The slide rule proper, however, was invented by William Oughtred around 1622. Oughtred placed two sliding logarithmic scales next to each other, enabling users to mechanically add and subtract distances, which is equivalent to multiplying and dividing numbers. The relative position of the sliding parts changes continually, making the slide rule an analog device. This innovation facilitated rapid computations and became an essential tool for engineers, scientists, and anyone who needed to multiply and divide beyond the multiplication tables of small integers. Suppose you want to multiply 4.35 by 15.70; on paper, it would take you a few minutes, but on a slide rule, only a few seconds. The slide rule lasted for three and a half centuries, from 1622 to about 1970. During that period, it could be found on every scientist's desk and in every engineer's pocket.

The logarithmic function can replace multiplication and division with addition and subtraction. Mathematically, this is expressed through the well-known formulas $\log(A \times B) = \log(A) + \log(B)$ and $\log(A/B) = \log(A - \log(B)$. Thus, if we mark logarithmic values on adjacent sliding scales and mechanically add distances corresponding to log(A) and log(B), the point corresponding to $\log(A) + \log(B)$ will give us $\log(A \times B)$, and we will be able to read the value $A \times B$ from the scale. Reducing the distance log(A) by log(B) gives log(A/B) and thus A/B.

The slide rule remained in widespread use until the rise of electronic calculators in the 1970s, when digital computation largely replaced manual logarithmic tools. There were many versions of a slide rule; one, made in the 1950s, is shown in Fig. 9.

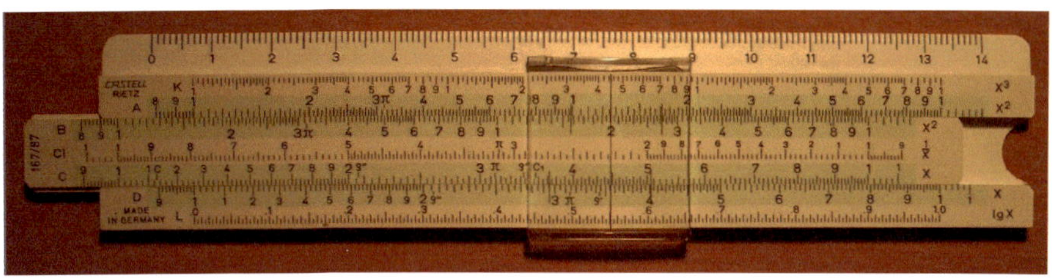

Fig. 9. This slide rule was made by the German company Faber-Castell in 1953; the arrangement of scales on it, known as "System Reitz," was developed in 1902 by Max Reitz. It has six or seven scales. Slide rules with up to 31 scales could facilitate calculations involving logarithmic, trigonometric, hyperbolic, and specialized engineering functions, as well as finding square and cubic roots of numbers.

Source: Image by Retoa, Lic. CC BY-SA 3.0.

Numerous other analog devices replaced lengthy calculations with a quick mechanical procedure. There were, for example, *tide predictors* that could accurately tell the times of high and low tides for various seaports. Lord Kelvin (William Thomson), a famous British scientist, was actively interested in analog calculating devices. Between 1872 and 1881, he designed and built several complex mechanical tide predictors. In the 20th century, such machines were perfected to such an extent that they could predict water levels in a given location with a precision of 0.1 ft (approximately 3 cm) for each minute of the year (Williams, 1997, 201). They were so valuable for naval operations that they were classified as military secrets during both the First and Second World Wars. Following the Second World War, such analog devices were gradually replaced by digital electronic computers. The last tide-predicting machine was built in the mid-1950s in East Germany.

Lord Kelvin and his elder brother James,[10] in the 1870s, constructed a mechanical device capable of numerically solving (integrating) differential equations. This problem was considerably more general than predicting the tides.

A mechanical device, designed by the French mathematician and physicist Gaspard-Gustave Coriolis in 1836, could solve differential equations of the first order, i.e., find a function $f(x)$ from an equation containing an unknown function's derivative $df(x)/dx$. The Thomson brothers addressed a more challenging problem: finding a function $f(x)$ from an equation containing both the first and second derivatives, $df(x)/dx$ and $d^2f(x)/dx^2$. Later, they worked on a device capable of solving equations containing even higher derivatives.

The numerical integration device designed by the Thomson brothers was never fully completed in its hardware glory, but it was not forgotten. In 1931, MIT professor Vannevar Bush took up the challenge where it had been left off 60 years prior.

[10] James Thomson (1822–1892), although not as famous as his brother William, was himself a talented physicist and engineer. In addition to the mechanical integrator of differential equations, on which he collaborated with William, James invented a mechanical spectrum analyzer in 1873 — a device designed for the analysis of periodic functions. One curious note: If you look at James Thomson's photograph, you will notice an uncanny likeness to his great Russian contemporary, the writer Leo Tolstoy!

Bush's *Differential Analyzer* was a sophisticated electromechanical analog machine capable of integrating a system of differential equations with up to 18 variables. It was a remarkable achievement but without a follow-up. The future belonged to digital machines, and even the best analog ones had almost no impact on that digital future. It is as if a physicist, aspiring to understand how the world works, would refuse to invoke the idea of discrete atoms and elementary particles. The *continuous world* program could work for a while, but eventually, it would reach its limits simply because it clashed with how the real world is put together. Computer science's underlying reality is also fundamentally discrete rather than continuous. So, Vannevar Bush's analog computing machine, valuable as it was, was historically a dead end — the last surviving *dinosaur* of its kind.

Vannevar Bush was a talented and influential man. He became MIT's vice president and dean of the MIT School of Engineering in 1932 and later served as president of the Carnegie Institution. During the war, Bush headed the powerful Office of Scientific Research and Development (OSRD), which was responsible for the Manhattan Project, radar, and all government-funded research and development. After the war, Bush played a crucial role in convincing Congress to establish the National Science Foundation. In 1945, Bush wrote an essay called "As We May Think," predicting future computer trends and products.

Electromechanical Digital Dinosaurs

In the 1930s and 1940s, several calculators were still built using mechanisms developed a hundred years earlier by Charles Babbage — except that, in the 20th century, the motions of shafts, gears, drums, and wheels were powered by electricity. Electricity was also used for control functions and to punch holes in the input and output cards. If electromechanical relays were present in these machines, they wouldn't have been used for logic operations but simply as electrically triggered switches. Despite this distinctly retro flavor, some of these machines were impressive, and all were useful.

In 1931, IBM introduced the *IBM 601*, a multiplying calculator. The user would input two eight-digit numbers to be multiplied, which were then punched onto a cardboard punch card. The machine computed the product in about one second and punched the resulting eight-digit answer onto the same card. The IBM 601 was followed in 1939 by a slightly faster *IBM 602*, which could handle 12-digit numbers and perform all four arithmetic operations. An improved *602-A* version appeared in 1948.

If not for the dates, one might think we were describing the early efforts of Charles Babbage before he brought his Difference Engine to full precision. But such was the state of mechanical calculation at the time — a situation that would soon change.

A far more sophisticated digital dinosaur, the *Harvard Mark I* (Fig. 10), was built as a collaboration between Harvard University and IBM. It was conceived by Howard Aiken between 1935 and 1937, approved by IBM in 1939, and completed at an IBM plant in January 1943, with Aiken overseeing its construction.

Aiken, the principal creator of the machine, was inspired by Babbage's Difference Engine, borrowing some key ideas, including the use of punched cards for programming (later replaced by continuous punched paper tape). In a poetic gesture, Aiken even incorporated into his machine two 10-position wheels originally designed by Babbage's son, Henry, for

Fig. 10. Harvard Mark I was 51 feet (16 m) in length, 8 feet (2.4 m) in height, and 2 feet (0.61 m) deep. This gigantic electromechanical calculator contained almost a million parts and weighed over 4 tons.

Source: Image courtesy of the Collection of Historical Scientific Instruments, Harvard University.

one of several demonstration models of the Difference Engine. That model had made its way from England to America in the early 20th century. The wheels had been quietly collecting dust in the attic of Harvard's Science Center when Aiken discovered them and decided to include them in Mark I as a tribute to Babbage.

By the end of its construction, the Harvard Mark I was an electromechanical behemoth: It weighed 4.3 tons and contained thousands of switches, relays, rotating shafts, wheels, and clutches — altogether over 765,000 electromechanical components within its sleek casing. Although not a general-purpose computer, Mark I represents a milestone in computing history. It was designed to solve systems of *nonlinear differential equations*, requiring intensive calculations. Aiken had always been more interested in calculation than in any specific calculational technology. His decision to focus the Mark I on nonlinear problems reflected a desire to tackle real-world systems that could not be modeled with simpler linear equations.

> The machine was known as Harvard Mark I at Harvard, but it was called the *Automatic Sequence Controlled Calculator* (ASCC) at IBM. The president of IBM, Thomas Watson Sr., who initially decided to help Aiken despite seeing no monetary reward for IBM, grew unhappy with him upon completing the project. He felt that the press release at the opening ceremony in August 1944 at Harvard University was unfair to IBM. That press release did not give sufficient credit to the three IBM engineers who worked with Aiken on the project; two were not even mentioned. Aiken and Harvard's officials later apologized to Watson and the engineers.

IBM funded this work, assisted by a military grant obtained by Aiken, who held dual roles as an assistant professor at Harvard and a commander in the U.S. Naval Reserve. Mark I, billed as a gift from IBM to Harvard, was relocated there and began operation in 1944.

It may be tempting to look down on this 4.3-ton electromechanical monster built 80 years ago; after all, the future belonged to electronic machines. If we do so, however, we would be unfair. Harvard Mark I worked very well, performing much-needed wartime calculations that no other machine could make, and it was decommissioned only in 1959. One of Mark I's users in 1944 was the ubiquitous John von Neumann, who desperately needed help performing massive calculations in the

Manhattan Project. Von Neumann pulled some strings and secured time on the newly constructed Mark I. No one knew what he was calculating; the atomic bomb was a highly classified secret in 1944.

After the war, Howard Aiken built three successors to the Mark I: the Mark II and Mark III were electromechanical calculators, but the final machine in the series, the Mark IV (completed in 1952), was entirely electronic and featured an internally stored program. It remained in use until 1962.

Another reason Howard Aiken is fondly remembered in the history of computing is his significant contributions to education. He established one of the earliest academic programs for computing at Harvard, laying the groundwork for what would become *computer science*. His students, earning M.S. and Ph.D. degrees under his guidance, were among the world's first formally certified "computer scientists."

Alan Turing and the Beginning of Computer Science

The brilliant British mathematician and most celebrated codebreaker in history and the man who arguably contributed more than anyone else to the Allied victory in the Second World War, Alan Turing (Fig. 11), formulated the scientific definition of the general-purpose computer. He also outlined the limits of algorithmic computations and was one of the founders of AI (see Chapter 4). Here, we recount his brilliant and tragic life

Fig. 11. Alan Turing (1912–1954), photographed in 1936.

Source: Image from Princeton University Archives; photographer unknown.

story and discuss his significant contributions to mathematics and computer science.

Like most mathematicians, Alan showed his talents early. His math teacher at school, observing what he could do at 15, proclaimed him a "boy genius." In 1928, at the age of 16, Alan read and allegedly understood Einstein's book on Special and General Relativity. In 1929, three years after the appearance of quantum mechanics, he read its foundational papers, such as Schrödinger's paper. A little later, he studied a book titled *Mathematical Foundations of Quantum Mechanics* by John von Neumann. We discuss von Neumann in greater detail later in this chapter; the two men strongly influenced each other. Upon entering King's College, Cambridge, Turing studied mathematics with professors G. H. Hardy and Max Newman, graduating in 1934 with first-class honors in the subject.

Turing and Hilbert's challenge

Mathematics developed in many directions during the first half of the 20th century. One of them was the work on the *foundations of mathematics*, addressing questions of rigor and consistency in mathematical theories. Mathematicians were ever concerned with the correctness and logical consistency of their work. Still, in the 20th century, their attention to these foundational problems increased, thanks to the influential views of the mathematical genius David Hilbert (1862–1943). Hilbert, one of the most prolific mathematicians of all time, left a vast mathematical legacy. He held court at Göttingen from 1895 to 1930, where he had 69 Ph.D. students, many of whom became the leading mathematicians of the 20th century. One of his legacies is known as the "Hilbert problems."

In 1900–1902, Hilbert published 23 unsolved mathematical problems. The choice of subjects in these 23 problems influenced mathematics for the next 100 years and beyond. Several Hilbert problems remain unsolved today, despite the extensive work that has been done on them over the intervening 125 years (some having been shown to be unresolvable in their original formulation). However, some of Hilbert's problems have been solved, and those who were talented and lucky enough to solve them became the leading mathematicians of their time. There is no higher honor in math than to solve one of Hilbert's problems.

In 1900, Hilbert's problem #2 was formulated as follows: *Prove that the axioms*[11] *of arithmetic are consistent.* This question remained unanswered in 1928 when Hilbert reformulated and somewhat detailed it, posing it as the following three questions (here slightly simplified):

(1) *Is a mathematical theory based on axioms and the rules of logic* ***complete***?
Here, the word "*complete*" means that any statement can be proven or disproven.
(2) *Is such a mathematical theory* ***consistent***?
Here, the word "*consistent*" means that no statement can be shown to be simultaneously true and false.
(3) *Given a statement formulated within a system of axioms, is there an algorithm (a set of well-defined mathematical and logical operations) that can tell whether this statement is, in principle, provable?*

The word "*provable*" means the statement can be logically deduced from the system's axioms. This is the last of Hilbert's three 1928 challenges, bearing a daunting German name, *Entscheidungsproblem*, which translates as "decision problem."

While still in Cambridge, Turing started thinking about a solution to the *Entscheidungsproblem* in 1935. He continued working on it in America at Princeton University, where he was invited in 1936 by an American mathematician and logician, Alonzo Church. Turing's and Church's parallel work resulted in solving Hilbert's *Entscheidungsproblem* using two very different methods. Church solved Hilbert's problem with a technique called "Lambda-analysis" and published his solution in 1936. Turing also solved it, publishing "On Computable Numbers, with an Application to the *Entscheidungsproblem*" in January 1937. Both Church and Turing have shown that there are "undecidable" problems for which an unambiguous "yes, it can be proven" or "no, it cannot be proven" cannot be given.

It should be noted that Church's and Turing's results were in the same vein as those obtained by the Czech–Austrian–American mathematician

[11] An *axiom* is a premise or a statement taken to be obviously true. Axioms serve as building blocks of mathematical constructions. For example, Euclidian geometry is based on five axioms from which all the theorems of geometry can be derived.

and logician Kurt Gödel. The latter, in 1931, published two *incompleteness theorems*, answering the first two of Hilbert's questions in the negative. To avoid using exact but demanding language typically employed by mathematicians, let me provide simplified statements of Gödel's incompleteness theorems:

(1) *Given the system of axioms on which mathematical theory rests, there will always be statements that are true but cannot be proven.*
(2) *One cannot prove that a given mathematical (axiomatic) system is consistent* (as we have mentioned above, the term "*consistent*" means that no statement can be shown to be simultaneously true and false within that system).

These results shocked the mathematical community. Before Gödel's results, Hilbert himself was optimistic about his questions, stating in 1930, "There is no such thing as an unsolvable problem." Gödel's negative results pointed to the dark side of mathematics, showing its limits and warning about the possibility — perhaps even the inevitability — of making grave mistakes. I should add, however, that mathematicians I queried on that subject suggested (as best I could understand them) that Gödel's theorems do not interfere much with the mathematical mainstream. So far, the examples of unresolvable problems people have come up with have tended to be trivial and mathematically uninteresting. There is no guarantee that it will always be this way, but the initial shock mathematicians experienced in 1931 has been softened; they recovered.

It is also interesting that John von Neumann, whom we have already mentioned and will discuss further, came up with a resolution to the second incompleteness theorem simultaneously with Gödel. He communicated his result to Gödel in a letter; in his reply, Gödel sent von Neumann a preprint of his paper, which contained both theorems.

Why do we care about Gödel's theorems and Turing's proof? We care because the novel mathematical language in which Turing presented his proof was the language of the discipline that did not exist in 1937 — it was the language of *computer science*. The best I can do is present how Turing approached the problem and what he concluded, rather than his solution. At least the reader will see how Turing's thinking differed from what one expects from a mathematician and how it could inspire the new field we now call computer science. For the first time in the history of mathematics, as far as I know, a mathematician solved a problem by analyzing a *thought experiment* involving an imaginary computing machine.

> Turing imagined a computing machine fed by a *tape* divided into *cells*, much like a film divided into frames, with each cell containing a single symbol. These symbols (commands or instructions) come from a finite (non-infinite) list. The tape can be of any arbitrary length, as necessary. The tape can move forward or backward through the machine, advancing one cell at a time. The machine has a central device, which today we would call a *central processing unit* (CPU), and Turing refers to as a "head." The head examines a sequence of symbols on a tape, one by one. It can also manipulate the content of a cell under examination: It can overwrite the symbol found in a cell with a new one taken from a list, or it can leave that original symbol intact, and it can erase the original symbol, leaving the cell empty. There is a register (memory) keeping track of the *states*. It records everything that has been read and written so far, with a new state arising at the end of each step. The total number of states equals the number of steps taken so far, and all of these are preserved in memory. The machine reacts to a *combination* of its present state and the symbol it encounters on a tape, acting according to a finite table of rules. The machine's reaction concerning the tape's motion can take three forms: It may advance the tape forward or backward by one cell, or stay on the same cell and reread it. Because the tape can move back and forth, the newly written symbol has a chance of being eventually read by the machine.

Turing has shown that the machine he imagined is the simplest possible machine capable, when supplied with the list of symbols and the table of rules, of solving any mathematical problem. It is not the optimal device, nor the fastest (in fact, it would be horribly slow) — but it is the simplest.

Turing devised a minimal set of rules for his machine to be *complete*. Deny Turing's machine any of its features — for example, deny it the ability to leave a cell unchanged — and it will lose its universality, the ability to solve any solvable problem.

Having defined the machine, Turing proceeded to rigorously prove that, given any algorithm, such a device could implement it. That was the "good news." However, the "bad news" soon followed: Turing had shown that, for a given problem, it is *impossible to determine* whether his machine will successfully solve that problem. He demonstrated that it might stop working altogether (halt) or repeat the same step, like a vinyl record getting stuck in the same groove and repeatedly playing a musical phrase. In other words, there will be instances where a machine is unsure of its next step. Turing proved that it is impossible to predict ahead of time whether a given problem will lead to such an impasse.

So, Turing and Church answered the third part of Hilbert's challenge in the negative — one cannot determine whether all problems can be solved. There exist *undecided* problems, and it is impossible to predict whether a given problem belongs to the "decided" (solvable) or the "undecided" (unsolvable) type. This conclusion was, it seems to me, much softer than Gödel's alarming incompleteness theorems. I must admit that I do not find it disturbing at all. Suppose I am shown a complex mathematical problem. All I know for certain is that I cannot solve it, even if I work hard on it for the rest of my life. However, whether it can be solved by someone else or even be solved in principle, I do not know. So what? It seems natural to me. That is the signature of a complex problem. Why would I expect to be able to predict whether it could be solved? I do not see it as a meaningful concern. But Hilbert, Turing, Church, and all the other mathematicians saw it differently. So, let me leave it at that before my mathematical readers (if there are any) throw this book out of the window in disgust.

While mathematicians were impressed with the resolution of Hilbert's problem, future computer scientists took note of the first part of Turing's proof: For any algorithm, it is possible to construct a machine to implement it. That is, one can program a machine to solve any mathematical or logical problem that, in principle, can be solved; the general-purpose computer can be made.

Turing further demonstrated that his device could simulate any general-purpose machine, regardless of its complexity. All such equivalent machines are called *Turing-equivalent* or *Universal Turing Machines* (UTMs). When we talked about Babbage's Analytical Engine, we, following Ada Lovelace, noted that, if completed, it would be a UTM. In principle, any question that can be answered with a computation can be answered using any UTM — past, present, or future. All the UTMs (general-purpose computers) are, in this sense, equivalent to Turing's one-dimensional tape machine.

To summarize, the conclusions of interest to computer designers that could be derived from Turing's 1937 paper were that a general-purpose computer is possible and can be implemented in various ways. These conclusions inspired John von Neumann and others a decade later to create the *stored-program* electronic computer, of which we will speak later. A well-known American computer scientist, Marvin Minsky, wrote in 1967:

> Turing's paper...contains, in essence, the invention of the modern computer and some of the programming techniques that accompanied it. (Minsky 1967)

Turing's unconventional methods have found new life in the 21st century, influencing number theory and other areas of mathematics when combined with more traditional techniques. I recently listened to a talk titled "Why Philosophers Should Care About Computational Complexity," delivered by Scott Aaronson, an American theoretical computer scientist and Chair of Computer Science at the University of Texas (Aaronson 2025). In his presentation, Professor Aaronson explained how certain deep mathematical and quantum-computing problems can be rephrased in the language of Turing machines. In particular, one can define a number — a super-astronomically large one — that represents the number of steps a Turing machine would take before halting. In this way, Alan Turing provided mathematicians with yet another tool to approach the most mysterious of concepts: *infinity*.

Enigma and the Bombe

After Princeton, in 1938, Turing returned to England. The war with Hitler's Germany was inevitable, and from September 1938, he occupied himself with questions of encryption and code-breaking. What he did during the war is well known — breaking the "unbreakable" German Naval code, which was newly encrypted daily using a sophisticated electromechanical *Enigma*[12] machine. This feat was achieved thanks to Turing's and

[12]Enigma was developed by German engineer Arthur Scherbius at the end of the First World War. The company, co-founded by Scherbius, began marketing the machine in 1923, initially targeting commercial markets. It was adopted by the German Navy in 1926 and later by the German Army and Air Force.

his fellow codebreakers' talents and because they constructed an electromechanical device (called the *Bombe*), which Turing personally designed. The Bombe allowed the Bletchley Park team to find daily settings on the Enigma. German operators randomly changed these settings daily, significantly complicating the deciphering of the encrypted German messages. The new Enigma settings were determined based on freshly intercepted messages that had been passed through the Bombe "grinder." This remarkable story is described in the 1983 book *Alan Turing: The Enigma* by Andrew Hodges. The 2014 movie *The Imitation Game* is loosely based on that book.

Turing performed the highest-caliber cryptography work at Bletchley Park. While working there, he wrote two papers on cryptography, deemed so valuable that they were unclassified and published only 70 years later, in 2012 — the Brits are good at keeping their secrets!

Knowing the Enigma code and keeping this knowledge secret from the enemy was so crucial for the Allies that they allowed certain German moves to proceed unopposed so that the Germans would not suspect that all their messages were being read. This required sacrificing some ships and the lives of British sailors, but it was done to safeguard the breaking of the Enigma.

Bletchley Park was the location of the *Government Code and Cypher School* (GC&CS), a vast operation that employed approximately 10,000 people toward the end of the war. They worked on deciphering German, Italian, Japanese, and Soviet communications. The expertise of those hired to work at Bletchley Park ranged from champion chess players and crossword puzzle competition winners to historians and mathematicians. However, it is universally agreed that, in retrospect, Alan Turing played the leading role in that great effort. Winston Churchill decisively assisted Turing and his fellow codebreakers after they wrote to him in 1941, asking for additional support. Churchill reacted to that letter by giving his chief military assistant, General Ismay, the following directive: "*ACTION THIS DAY. Make sure they have all they want on extreme priority and report to me that this has been done.*"

After the war, between 1945 and 1947, Turing worked on designing the *Automatic Computing Engine* (ACE) at the National Physical Laboratory (NPL) located in Teddington, Middlesex. In 1946, he presented a paper detailing the design of this stored-program general-purpose machine. In February of the same year, the first American computer, ENIAC, began operating at the University of Pennsylvania in

Philadelphia. Turing's work on the ACE and his description of its design were well known to John von Neumann, and through him it influenced subsequent American work on the ENIAC and other computers.

In 1948–1950, Turing wrote about "thinking machines," what we now call *artificial intelligence*. This theme is central to our book; we will postpone discussing Turing's AI contributions until Chapter 4, and of his views on this subject till Chapter 5.

There exists an opinion that Turing and his team at Bletchley Park shortened the war by two to four years, saving millions of lives. I cannot think of any other person in history (except possibly for Alexander Fleming, the inventor of penicillin) who could claim to have saved millions of lives. It is easy to think of those who were instrumental in *killing* millions of people, but not the other way around. It arguably qualifies Alan Turing as one of the greatest heroes of all time. Initially, the British government appeared to understand this. In 1946, after the war, Turing was appointed by King George VI, father of Queen Elizabeth II, an *Officer of the Order of the British Empire* (OBE), one of the highest honors possible in Great Britain. On the scientific side, in 1951, he was elected a *Fellow of the Royal Society* (FRS).

Unfortunately, it is also well known that shortly after, in 1952, Great Britain initiated Turing's legal prosecution for homosexual conduct. He was convicted based on a cruel and outdated 1885 law, and he was given a choice of either imprisonment or chemical castration. He chose the latter. His employment at the government encryption center was terminated, his security clearance was revoked, and he was denied entry into the U.S., a country where he could have continued working productively. Alan Turing died two years later, on June 8, 1954, at the age of 41, of cyanide poisoning. An inquest determined that he had committed suicide. Beside his bed, there was a partly eaten apple. Although it was not tested for potassium cyanide,[13] many believed he committed suicide by injecting an apple with that quick-killing poison and then biting it. Some say that his death could have been accidental, caused by inhaling vapors of cyanide-containing chemicals that Turing used in his experiments (he was fond of chemistry and electrochemistry and conducted experiments at home). It is also said that in 1976, Steve Jobs chose the bitten-apple logo for his company in honor of Alan Turing — if true, it was a thoughtful and honorable move.

---

[13] It seems obvious that it should have been tested.

James Barrat writes in his book that one of the first to speak out against the government's treatment of Alan Turing was his Bletchley Park colleague and friend, Irving J. Good, whom we will discuss in Chapter 5. Good said, "I won't say that what Turing did made us win the war. But I daresay we might have lost it without him." (Barrat, 2013, p. 110).

In 2009, the British Prime Minister, Gordon Brown, acknowledging the petition signed by some 30,000 individuals, issued an apology in which he described the treatment of Turing as "appalling," saying, "We're sorry, you deserved so much better." We cannot agree more. However, Brown could not lift the conviction, which was removed in 2014 by Queen Elizabeth II under the *royal prerogative of mercy* after a lengthy debate and several more petitions.

Relay-based Digital Machines

In Chapter 1, we stated that an on-off electromechanical relay is ideally suited for implementing Boolean logic, providing a physical embodiment for truth values, namely ones and zeros. We also referred to the relay as a bridge between mechanical and electronic computing machines. It was a short bridge. By the time this idea was implemented, the remaining lifespan of relay-based machines was brief; the era of electronic computers was dawning, and relay computers could not compete with machines that were thousands of times faster. You might think that someone could have pioneered computers based on electromechanical relays in the 19th century, but that did not happen. Perhaps the realization of this sound idea had to wait for the appearance of the theoretical work of Claude Shannon, who, as we said, proved in 1936–1937 the universality of calculational circuits containing elements capable of switching into two binary states.

George Stibitz

It is said that George Stibitz (1904–1995; Fig. 12) conceived the idea of using electromagnetic relays in digital computers independently of Claude Shannon, and at the same time, in 1937.

Stibitz received a Ph.D. in mathematical physics from Cornell University in 1930 and was hired as an electrical engineer at Bell Labs in the same year. In 1937–1940, while at Bell Labs, he constructed the first

Fig. 12. George Stibitz (1904–1995), the creator of the first relay-based digital calculators.

Source: Image taken from the site associated with the Department of Mathematics and Computer Science archives at Denison University.

binary adder (a device that adds numbers expressed in the binary system). It was based entirely on electrical relay circuits, which secured Stibitz's place in computer history. Subsequently, Stibitz oversaw the work on a more ambitious digital calculating machine, the *Complex Number Calculator*, completed at Bell Labs in 1939. This sophisticated relay-based machine could perform calculations with complex numbers.[14] Moreover, Stibitz's machine could be operated from a distance of hundreds of miles, with commands sent through telephone lines. Stibitz's 1940 demonstration of the remotely controlled Complex Number Calculator was witnessed by the future computer pioneers John Mauchly, John von Neumann, and Norbert Wiener (Isaacson 2014, p. 50).

[14] Complex numbers are widely used in mathematics, physics, and engineering. The mathematics of quantum mechanics is based on them. Dealing with them is more involved than with normal numbers.

Fig. 13. This is an early photograph of Konrad Zuse (1910–1995) with one of his Z-computers.

Source: Photographer unknown. Rights holder untraced despite good-faith efforts. If you hold rights to this image, please contact the publisher.

Konrad Zuse

Konrad Zuse (Fig. 13) was a talented and prolific computer designer who independently developed key computer concepts while isolated from European and American developments during the Second World War. He was also the first to design and build a Turing-complete (universal) computer based entirely on relay switches.

Unfortunately, Zuse began his computer work in 1935, in Germany, soon after Hitler became the *Reichskanzler*. Zuse was completely apolitical (or so it was presented after the war), and he never joined the Nazi Party. Still, he could not escape the conditions in which he lived. Although his initial work went unnoticed by the Nazi authorities, eventually, after he created his most famous machine, Z3, they caught up with him. In 1941–1942, his work on Z4 received funding from the German Aircraft Research Institute, where Zuse's machines were used to improve the aerodynamics of gliding bombs — not an admirable part of his biography.

Over the years, Zuse created seven computer models: Z1–Z5, Z11, and Z22. All except Z22 were electromechanical machines based on relays. Zuse began thinking about computers in 1935; his Z1 and Z2 models were constructed between 1936 and 1939. Model Z3, a technical marvel for its time and a milestone in computer history, was built from 1939 to 1941 — what a remarkable period to achieve something noteworthy in Germany! Zuse was the first to develop the *stored-program* concept, despite the common attribution of this idea to von Neumann (see the following). His priority is reflected in the two 1936 patent applications.

Z3 was the world's first programmable, fully automatic binary computer with memory — the first general-purpose computer. It contained 2,000 relays and operated at a clock frequency of 5.3 Hz (5.3 cycles per second). Program code and data were stored on punched tape, derived from a discarded 35 mm movie film, reminiscent of Turing's abstract universal machine of 1937. Zuse wanted to replace relays with electronic vacuum tubes but could not secure the funding. Germany had other priorities and did not have someone like Winston Churchill to appreciate and support the novel computing initiative. These disastrous priorities led to Allied bombing raids on German cities, and in 1943, Z3 was destroyed during a bombing raid on Berlin. Z3 was *almost* Turing-complete; it lacked *conditional branching*, which is essential for sophisticated programs (something that Babbage realized a hundred years earlier).

In 1941, Zuse and his engineers established a company to manufacture their computers commercially. Z4 was completed in 1945, and after the war, Z5, Z11, and Z22 followed. Starting from Z3, all of Zuse's computers were programmable. Despite being slow, Z4 was good at numerically solving differential equations. As we said, only the seventh Zuse model, Z22, built in 1955, was an electronic machine based on vacuum tubes.

The Dawn of Electronic Computers

John Atanasoff and Clifford Berry

Professor John Atanasoff (Fig. 14, left) and his graduate student, Clifford Berry, at Iowa State College, designed and built America's first electronic digital calculator. They called it ABC, which conveniently stood for *Atanasoff–Berry Calculator*. Atanasoff, like Aiken, was driven by the desire to automate demanding calculations of high practical value in

Fig. 14. John Atanasoff (1903–1995) designed and built (with his graduate student Clifford Berry) the first (almost) electronic computer, the Atanasoff–Berry Calculator (ABC), in 1937–1942. In the photo of ABC, we see some of the vacuum tubes, the control panel on the right, and, in the center back, a rotating cylinder containing a capacitive memory.

Source: Images courtesy of Iowa State University Special Collections and Archives.

physics and engineering. For Aiken, it was a nonlinear differential equation; for Atanasoff, it was a system of linear algebraic equations. The calculations required in both cases were prohibitively complex if done "by hand."

ABC was designed to solve up to 29 linear equations with 29 unknowns. It was not reprogrammable; its program was hard-wired for this one task. Therefore, like Mark I, it was not Turing-complete. It was not a computer but rather — what we would call today — an *arithmetic logic unit* (ALU), which is a part of a modern CPU.

The project began in 1937; a small, 11-tube prototype was constructed in 1939, and the complete machine, comprising over 300 vacuum tubes, was completed and tested in 1942. The ABC was essentially electronic and used binary arithmetic. The alternating current power-line frequency of 60 Hz set the primary clock rate.

The machine worked well but was abandoned soon after testing. Atanasoff had to leave college for war assignments.

Atanasoff and Berry's design included remarkable novelties, such as using *vacuum tubes as switching elements* and *a regenerative capacitor memory* separate from computing functions (i.e., memory separate from the CPU or, in this case, the ALU).

The memory in ABC was slow and small but clever. Capacitors are natural for building memory: A charged capacitor develops a voltage, signifying a logical "1." A discharged (or empty) capacitor has zero voltage, indicating a logical "0." The problem is that capacitors do not stay charged forever; the charge tends to leak from the capacitor plates. To keep their capacitors charged, Atanasoff and Berry placed them on rotating cylindrical drums, with electrical contacts (metal brushes) touching capacitor terminals each time the drum rotated. We can see such a drum with contacts in Fig. 14 (right). These intermittent electrical contacts read the capacitor states (1 or 0), refreshing those that require recharging.

ABC had two capacitor drums; together, they stored 60 binary 50-digit numbers (or "words" in computer terminology). The total memory in ABC was $60 \times 50 = 3000$ bits, or 3 kb (kilobits). The drums made one revolution per second, performing reading and writing at a frequency of 1 Hz. The machine could add or subtract thirty 50-digit binary numbers per second.

Capacitors, the basis of computer memory, were revived years later when the semiconductor *Dynamic Random Access Memory* (DRAM) was invented in the 1960s. We will discuss computer memory later in the chapter.

In the 1970s, Honeywell and Sperry Rand were involved in a lengthy patent dispute. The latter held rights to the ENIAC 1964 patent for the first fully electronic computer. The former attacked these rights because Atanasoff and Berry had priority. It also played a role that John Mauchly, one of the ENIAC inventors, visited Atanasoff at Iowa State in 1941, where he learned all he could about his machine. The original ENIAC patent was filed in 1947 but was not granted until 1964. Under the patent law, a patent is valid for 17 years from the date of award. For example, a patent granted in 1964 would remain valid until 1981. The U.S. District Court sided with Honeywell, agreeing that Atanasoff had priority, and the ENIAC 1964 patent was invalidated in 1973,[15] placing electronic digital computers into the public domain. The court also stated that the ENIAC inventors had derived the subject matter of the electronic digital computer from Atanasoff, thus reaffirming the significance and originality of Atanasoff and Berry's design.

[15] I noticed a striking coincidence between the ENIAC patent dates and my own life: Eckert and Mauchly filed the patent on June 26, 1947, and it was invalidated on October 19, 1973. I was born on June 25, 1947, and I entered the United States as an immigrant in October 1973.

British Colossus computers

ABC, which was developed between 1937 and 1942, was not programmable; it was not a UTM; it was not even a *computer*, but an electronic *automatic calculator* — the name given to it by the modest and unassuming Atanasoff was correct. The first *electronic* and *programmable* (i.e., general-purpose, Turing-complete) computers appeared in Great Britain in 1944 and the U.S. in 1946.

The British *Colossus*, or rather a series of *Colossuses*, was built in Bletchley Park by Thomas Flowers and co-workers, with input from Max Newman, Turing's math professor from Cambridge. These machines were electronic, based on vacuum tubes. The first Colossus began operation at the start of 1944. Its programming was external, utilizing switches and plugs; it did not have a stored program loaded into memory. The fact that Britain was the first to develop general-purpose electronic computers was not known until the mid-1970s when the Colossus program was declassified. It's a shame; they were remarkable machines, the first of their kind, satisfying all UTM requirements. By the war's end, there were 10 such machines at Bletchley Park. Like Turing's *Bombe*, all the Colossuses were decoding intercepted enemy messages. While the Bombe specialized in decoding naval messages, the Colossuses focused on deciphering army messages. Thomas Flowers worked with Turing on the Bombe, and impressed with his work, Turing recommended him to the division dealing with the German army's cipher machine, the *Lorenz*.

Colossus machines were shrouded in secrecy, and few people outside British cryptography headquarters were aware of their existence. Consequently, they had no followers and did not influence the history of computers. Even now, when we are aware of them, the British Colossus is rarely referred to as the first electronic computer. Clio, the muse of history, is a capricious creature. She dislikes rewriting history, even when new information becomes available.

John "Pres" Eckert and John Mauchly are building ENIAC, EDVAC, and UNIVAC I

Meanwhile, in America, another general-purpose machine — by most accounts, the ancestor of all electronic computers — called the Electronic Numerical Integrator and Computer (ENIAC), which we have already mentioned, was built. The principal designers of ENIAC were John Presper "Pres" Eckert and John Mauchly (Fig. 15). In 1942, Mauchly

Fig. 15. John Mauchly (1907–1980; left) and Pres Eckert (1919–1995; right), with Major General Gladeon M. Barnes (1887–1961; center). The general, chief of research and engineering in the U.S. Army Ordnance Department, financed and oversaw the ENIAC project. The photo shows the three men standing at the University of Pennsylvania in front of ENIAC, pretending to read a sizeable government-issued notebook.[16] This photo may have been taken on February 15, 1946, when ENIAC was officially dedicated.

Source: Image courtesy of the University of Pennsylvania Archives; in the public domain.

proposed to the U.S. Army that he and Eckert would build a programmable machine that was a thousand times faster than existing electromechanical calculators. The Army's Office of Scientific Research agreed, and the brilliant duo delivered on their promises.

ENIAC was launched in 1946 at Moore's School of Electrical Engineering at the University of Pennsylvania. The U.S. Army funded the

[16] I recognize the notebook! When I started as a graduate student from Stony Brook University at Brookhaven National Laboratory in 1974, I signed out several notebooks like that one from the stockroom. They had the same luxurious leather (perhaps imitation) binding, and the pages were numbered. They were so beautiful that I wouldn't dare write anything unimportant in them. I used some of these books for my lecture notes at Stony Brook, where I studied. I still have them, along with my old lecture notes — technically, stolen government property.

project for $487,000 (over $8 million in today's dollars). Like the British Colossus, the machine was initially developed in secrecy, but the secrecy was lifted when the war ended. ENIAC ran from the end of 1946 to 1955, solving numerous time-consuming computational problems, including the composition of artillery firing tables and the feasibility of a thermonuclear (hydrogen) bomb. ENIAC was the first Turing-equivalent electronic computer in the U.S.

Similar to the British Colossus, programming was external to the machine, utilizing thousands of switches and cables. Each time the program was changed, all these switches and cables had to be rearranged, which could take weeks. In 1948, however, ENIAC was retrofitted with the ability to store programs in its memory. That memory was initially made with *mercury delay lines* (Eckert's idea), but in 1953, delay line memory was replaced with *magnetic core memory*. ENIAC was a *decimal* machine that worked with numbers 0–9 rather than 0 and 1. Its clock cycle was 100,000 Hz, and it could make 5,000 additions of ten-digit numbers in one second — a blinding speed compared to all the previous electromechanical machines. *Conditional branching* and numerous *subroutines* (standard components of programs that perform a specific task) were used more frequently than in earlier calculating devices. Several of the 17,500 vacuum tubes burned out daily, making the machine functional only about half the time. In later years, however, the downtime was significantly reduced.

Before ENIAC was finished, Mauchly and Eckert proposed another computer, the EDVAC. This one had *stored program* capability from the outset and included other enhancements. Von Neumann provided some input into the design of EDVAC as a consultant to Mauchly and Eckert. EDVAC was completed in 1949 and continued to operate, with modifications, until 1962.

These were all general-purpose machines, but as George Orwell would say, *some were more general-purpose than others*. It is especially applicable to the next one, the *Universal Automatic Computer*, UNIVAC I. It was also designed by Eckert and Mauchly, this time in their own company, *Eckert–Mauchly Computer Corporation* (EMCC). Upon completing EDVAC, the two brilliant designers resigned from the University of Pennsylvania, partly due to a disagreement over patent rights, and went into business for themselves, designing *commercial* computers. At the time, the idea of a computer used in business seemed radical and fantastic. They founded EMCC in 1947, which, because of financial difficulties, became part of Remington Rand in 1950, subsequently becoming Sperry

Fig. 16. UNIVAC I in an office installation and the insides of the machine.

Source: Images in the public domain.

Rand (resulting in the litigation several years later between Honeywell, which held rights to ABC, and Sperry Rand, which held rights to ENIAC).

The UNIVAC I was designed for commercial and civilian applications rather than for military use. It utilized 5,200 vacuum tubes, 18,000 crystal diodes (semiconductors), and 300 relays. It weighed 29,000 pounds (13 metric tons), consumed 125 kilowatts of power, and could perform 1,905 operations per second (FLOPS) while running on a 2.25 million cycles per second (2.25 MHz) clock. The processor and memory unit occupied a space of 4.3 m × 2.4 m × 2.6 m. UNIVAC I's footprint was 35.5 m^2 of floor space (a large room). The appearance is shown in Fig. 16.

UNIVAC was designed with an eye for customers' convenience; it was outfitted with a wide range of peripheral devices, including printers, magnetic tape drives, card-to-tape and tape-to-card converters, high-speed tape copiers, and more. It was marketed between 1951 and 1954. CBS News used it in 1952 to predict the results of the presidential election. The UNIVAC I was sold to major insurance companies, market research firms, and the government, notably to the U.S. Census Bureau, which signed the first contract to purchase the machine. For the most part, universities did not have the money to buy it. In total, 46 machines were sold, ranging widely from the initial, naïve, and unrealistic price of $160,000, which contributed to EMCC's financial demise, to the eventual price between $1.2 and $1.5 million (approximately $14 to $17.6 million in 2025). Despite naysayers' predictions, it was a commercial success, rewarding Mauchly and Eckert for years of hard labor in an atmosphere of veiled hostility and financial uncertainty.

UNIVAC's commercial success prompted other companies to enter the computer business, notably IBM, which had, after all, the right name for it: *International Business Machines*. Indeed, that was what UNIVAC was — a *business* machine.

In the opening chapter of my previous book, I compared the progress in car-making and computers, using the UNIVAC I as a starting point for the comparison on the computer side and the 1954 Chevrolet Corvette on the car side. Let me provide an excerpt here, slightly rephrased:

> Today, all computers, including the smallest ones disguised as smartphones or smartwatches, contain a processor with billions of transistors. They operate at several billion cycles per second (several GHz) rather than MHz clock rates, i.e., they have 1000 times higher clock rates than the UNIVAC I. They can perform hundreds of billions of operations per second, over 100 million times "faster" than UNIVAC. With such numbers, the analogy with the speed of a car fails. Indeed, if the vehicle in 1954 could reach a maximum speed of 150 kilometers per hour (or 42 meters per second), it would now be moving over ten times faster than the speed of light, which is impossible. As a physicist, I cannot write down a speed greater than 300,000,000 meters per second (the speed of light), or I will lose my job[17] and be forced to sustain myself on the uncertain sales of this book. Apart from speed, modern computers can perform a wide range of tasks beyond simply calculating numbers; we are all familiar with their capabilities, and this book will explore some of their applications. As for the price, a desktop, laptop, or smartphone in 2021 costs several hundred dollars, which is approximately 10,000 times less than UNIVAC, when adjusted for inflation using *constant dollars*. So, in our admittedly silly analogy, if cars were progressing in the intervening time as much as electronics, a car costing a few cents would take you from New York to Los Angeles in a few thousandths of a second. In other words, it would not be a car but an incredibly cheap *teleportation device*!
>
> As I mentioned, I wouldn't mind driving a 1954 Chevrolet Corvette. Many of us feel this way about old cars. But can you imagine anyone — a collector of vintage technology, a history lover, or a tech enthusiast — wanting to use a UNIVAC-1 instead of a modern computer? And even if such a person existed, found a large room to station the UNIVAC, and was willing to replace burnt tubes daily, paying about $150 in daily electricity costs, what would they do with it? Do you agree then that

[17]Now that I am retired, I could.

> there is no comparison between what was achieved in the last 67 years in electronics and the same period in transportation technology? Or, for that matter, in all the other technologies? (Gurvitch 2023, Vol. 1, Chapter 1)

All these early electronic computers contained tens of thousands of vacuum tubes, crystalline diodes, relays, capacitors, and resistors. Vacuum tubes were the *active* elements, switching between *low* and *high* voltage levels, or 0s and 1s. The typical vacuum tube switching time was 1–2 microseconds, about 10,000 times faster than relay switching times but about a million times slower than today's transistors. Early computers weighed many tons, occupied large spaces, and required hundreds of kilowatts of power. Teams of engineers and technicians looked after them, ensuring they worked more or less continuously.

Vacuum tubes were the most expensive and power-hungry, yet the least reliable, components of those early electronic machines. One of the most tedious tasks was replacing burned vacuum tubes. The earliest machines had to be manually reprogrammed for each new problem. Many years later, when personal computers (PCs) appeared, an old-timer from that era, Harry Read, remarked that ENIAC was a kind of *personal computer*, too, in the sense that you *lived* inside it (Dyson 2012, p. ix (Preface)).

The tendency of electronic systems to become smaller in subsequent generations was already apparent at that time. We can compare early computers by weight: ENIAC weighed 30 tons, EDVAC 8 tons, and UNIVAC I 7.3 tons.

Manchester computer group

Back in Great Britain, not all computer work was secret. After the War, important pioneering work was performed at the Victoria University of Manchester (now the University of Manchester). We refer to the world's first *electronic stored-program* computer, built there between 1946 and 1948. Theoretically, the concept of a stored-program machine originated with Alan Turing; however, it was practically developed by von Neumann, Mauchly, and Eckert. The Brits beat the Americans by building a working machine that incorporated that idea. Max Newman, one of Turing's Cambridge professors and a leading British mathematician, originated the Manchester project in 1946. It was called the *Manchester Small-Scale Experimental Machine*, nicknamed "Baby." Frederic C. Williams (the project's chief engineer), Tom Kilburn, and Geoff Tootill were its creators. Rather than demonstrating the stored program, their Baby was intended as proof of concept for the first electronic memory invented by Williams and Kilburn. That memory was based on a *cathode ray tube* (CRT).

The Baby was born (demonstrated) in June 1948, and the group started working on a full version called *Manchester Mark I*, which was finished in 1949. The Manchester Mark I, in turn, inspired the development of the first British general-purpose commercial electronic computer, the Ferranti Mark I. Ferranti International was a Manchester-based firm that collaborated with the Victoria University team. It delivered the Ferranti Mark I in 1951, the same year the first commercial computer, the UNIVAC I, was produced in the U.S. The Manchester group continued doing first-rate pioneering computer work throughout the 1950s, 1960s, and beyond. In 1952, they began replacing vacuum tubes with transistors and solid-state diodes. The first version of this machine was finished in 1953, and the second in 1955. This academic work inspired a British firm, *Metropolitan-Vickers*, to produce in 1956 a commercial computer without vacuum tubes, the *Metropolitan-Vickers 950*. Later, the Manchester group pioneered some of the fastest computers of their time. All in all, it was an outstanding and brilliant group of early computer scientists and engineers.

John von Neumann and the Stored Program Concept

We have already mentioned John von Neumann on numerous occasions. Let us now talk about this great and prolific man, who, in his lifetime wore so many hats: mathematician, economist (father of game theory), physicist (involved in the mathematics of quantum mechanics), designer of nuclear weapons (not my favorite hat), foundational computer scientist, top-level U.S. government advisor and consultant, and member of the Atomic Energy Commission. His impact on computers was immense, and it was more practical than that of Alan Turing.

John von Neumann (Fig. 17, left) was born in 1903, as Neumann János (Yonah) Lajos, in Budapest, Hungary, to secular (non-observant)

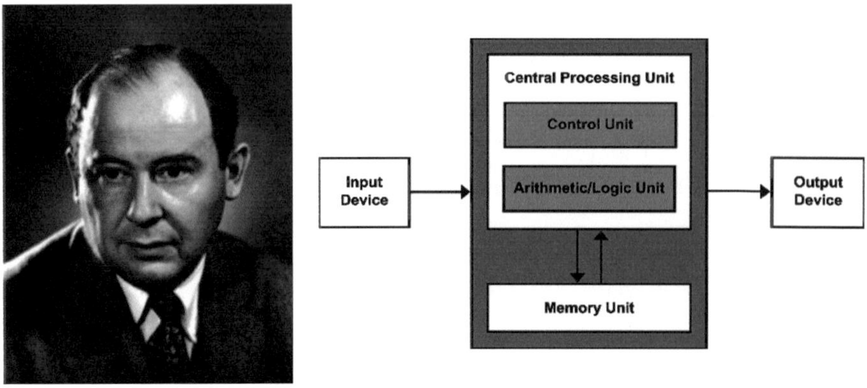

Fig. 17. John von Neumann (1903–1957) and schematics of the *von Neumann architecture.*

Sources: Left image reproduced by permission of Los Alamos National Laboratory (copyright holder) and US government; right image by Kopooht, Lic. CC BY-SA 3.0.

Jewish parents. His father was a wealthy banker with a doctorate in law. In 1913, Emperor Franz Joseph elevated János' father to the nobility for his service to the Austro-Hungarian Empire (and, one assumes, for his considerable wealth). That new status was reflected in the German-style "von" added to the family name, equivalent to the French "de." The family was strong on education and culture, possessing an excellent library (which occupied one of the 18 rooms in the apartment where the family was *cramped* in Budapest) and hiring governesses and private tutors for their three children. János (he changed his first name to John, or Johnny, when he emigrated to the U.S.) was a genuine child prodigy, often referred to as the Mozart of mathematics. At the age of six, he was able to multiply eight-digit numbers in his head. At eight, he knew calculus (the mathematics of differentiation and integration), which ordinary people usually study with difficulty at university. He had a "photographic memory," reciting pages from any book he read. From a young age, he was fluent in Classical Greek, Latin, German, French, and English.

At eight, János was admitted to Budapest's best school, Fasori Lutheran Gymnasium. This school, in the space of a quarter-century, saw within its walls a constellation of young geniuses: Theodore von Kármán (outstanding applied mathematician and one of the creators of modern aerodynamics), George de Hevesy (Nobel Prize in Chemistry), Leó Szilárd (a physicist who, in 1933, came up with the idea of a nuclear chain reaction), Dennis Gabor (Nobel Prize in Physics, creator of holography),

Eugene Wigner (Nobel Prize in Physics), Edward Teller (a theoretical physicist, one of the principal creators of the hydrogen bomb and later *Star Wars*),[18] and Paul Erdős (one of the greatest mathematicians of the 20th century). Some school graduates later emigrated from Hungary to the U.S. (notably von Karman, von Neumann, Szilard, Wigner, Teller, Gabor, and Erdős), where they became collectively known as "The Martians." They were called so by Leo Szilard, who humorously responded to a question posed by Enrico Fermi. Fermi wondered why the highly intelligent aliens from the other worlds (who, he argued, are likely to exist) weren't seen on Earth. Szilard answered Fermi by saying, "*They are among us, but they call themselves Hungarians*." Later, a whimsical story was composed about how the aliens (Martians) came to Hungary and sent their children to the Fasori Gymnasium, and the name has stuck.

Among this brilliant constellation, von Neumann was still the first. We will not list his many achievements; he worked simultaneously on a wide range of topics, publishing a significant paper every month. I want to conclude this tribute to von Neumann's talents with quotes from those who knew him (there are many more such quotes). One comes from his ETH Zurich professor, Hungarian mathematician George Polya, who said:

> Johnny was the only student I was ever afraid of. If in the course of a lecture I stated an unsolved problem, the chances were he'd come to me at the end of the lecture with the complete solution scribbled on a slip of paper.

Edward Teller, who was not known for the lavish praise of his colleagues, said of von Neumann:

> I never could keep up with him… Nobody knows all science, not even von Neumann did. But as for mathematics, he contributed to every part of it except number theory and topology. That is, I think, something unique…If a mentally superhuman race ever develops, its members will resemble Johnny von Neumann. (Dyson 2012, p. 158)

[18] I once spoke with Edward Teller at a scientific conference in Washington, D.C., in 1987 or 1988. After my talk, Teller started a conversation; we sat down and talked for about half an hour. He wanted to learn more about high-temperature superconductivity, which I was researching at Bell Labs. He asked questions and listened very attentively to my answers. He was 79 or 80 years old at the time.

Great David Hilbert noticed young von Neumann, and for two years, from 1926 to 1927, he was Hilbert's postdoctoral fellow at Göttingen. Then, he became *Privatdozent* ("private lecturer," a position similar to that of our assistant professor) at the universities of Berlin and Hamburg. His fame grew, and in 1930, Princeton University in the U.S. invited him to teach a course on quantum theory.

In 1933, von Neumann was a visiting professor at Princeton and a lecturer in Hamburg and Berlin. Then something happened that sparked his renewed interest in Princeton: in January 1933, Adolf Hitler was appointed Chancellor of Germany.[19] Understanding what was happening in Europe, Americans offered von Neumann a permanent professorship, not at Princeton University, but at the newly established *Institute for Advanced Study* (IAS), also located in Princeton. One did not have to be a genius (which he was) to figure it out: Von Neumann relinquished his German academic posts and emigrated to the U.S. without a second thought.

The IAS was a new scientific center closely related to Princeton University, both in terms of its location and faculty. It was established in 1930, benefiting greatly from the exodus of some of the best European scientists fleeing from Hitler's Germany and the rest of troubled Europe. Albert Einstein joined the Institute in the same year as von Neumann. Among the other Europeans were Kurt Gödel, Hermann Weyl, and J. Robert Oppenheimer, who headed the Manhattan Project during the Second World War. Oppenheimer was later appointed Director of the IAS and remained in that position from 1947 to 1966.

In the late 1930s, von Neumann developed mathematical methods for complex computations related to explosives, specifically modeling pressure profiles produced by *shaped charges*. In his usual fashion, upon entering that field, he quickly became the leading authority in it, which made him popular with the U.S. military. Von Neumann was in high demand as a consultant, which eventually led him to the Manhattan Project. Von Neumann was a valuable presence at Los Alamos, which he frequently visited, traveling back and forth between Princeton and New Mexico. He traveled willingly and easily all his life, crisscrossing the

[19]When Hitler came to power in 1933, von Neumann remarked, "If these boys continue for only two more years...they will ruin German science for a generation — at least." Little did he know that "these boys" would continue for 12 long years and will eventually ruin all of Germany and most of Europe "for a generation — at least."

country by train or airplane and sometimes driving; his daredevil driving habits were well known around Princeton. In Los Alamos, his main contribution was designing the explosive lenses needed to compress the plutonium core and, later, the deuterium-tritium core in the hydrogen bomb. While solving that complex computational problem, von Neumann traveled to various emerging computational facilities, encouraging and influencing the emerging field of computer science.

Von Neumann's computer architecture

One significant idea he promoted became known as *von Neumann's computer architecture* (Fig. 17, right), an overall computer organization that made sense. Von Neumann described it in the "*First Draft of a Report on the EDVAC*" in 1945 (although this report was not published, it was widely circulated) and in another paper written with Arthur Burks and Herman Goldstine a year later.

A great deal of controversy surrounds the question of priority concerning the *stored-program* concept. As we mentioned earlier, Konrad Zuse first expressed this idea in 1936. Still, it remained unknown to other computer scientists, and it did not influence how things were done a decade later in the U.S. and Britain. Then, of course, Turing's abstract machine stores data and instructions on the same memory tape; however, this mathematical abstraction did not influence practical computer designers. It was known to von Neumann, however. Stibitz, in his *Complex Number Calculator*, used punched paper tape, which contained numerical data and operational instructions. The tape in his calculator could be punched repeatedly, which allowed the machine to modify its instructions (Dyson 2012, pp. 77–78). Eckert and Mauchly planned to store data and program instructions in the same memory when they designed their ENIAC in 1943–1944, before they met von Neumann.[20] The Manchester Mark I team at the Victoria University of Manchester became the first to build a working machine implementing the stored program concept in England.

So, the idea was in the air and even on the ground, in hardware. What was von Neumann's role then? Why is it called *von Neumann's architecture*? Computer scientist Robert Rosin summed it all up in his entry in the

[20] Eckert felt rather strongly about the issue, remarking that "von Neumann architecture" should be rightfully called "Eckert architecture."

comprehensive *Encyclopedia of Computer Science* published in 1976. The first phrase in the passage quoted below hits the nail on the head:

> The most influential paper in the history of computer science, *whether or not anyone expressed a similar idea earlier* [emphasis ours], was written in 1946 by John von Neumann...with Arthur W. Burks and Herman H. Goldstine. Its title is "Preliminary Discussion of the Logical Design of an Electronic Computing Instrument," and the ideas it contains collectively known as the *von Neumann machine*, have provided the foundation for essentially all computer system development since that date. Central to the von Neumann machine is the concept of the stored program — the principle that instructions and data are to be stored together in a single, *uniform* [emphasis ours] storage medium, rather than separately, as was previously the case.... (Rosin, 1976)

So much has been said about the stored-program concept that it would be worthwhile to understand its significance. Initially, programs were imposed on computers from the outside. In Babbage's *engines*, program instructions were encoded on punch cards, a stack of cards fed to the machine. Programming in the first electronic machines, such as Colossus and ENIAC, required external cables and switches. Each program required a new configuration that took days or weeks to set up and debug. Compared to that, storing a program in the computer memory was a significant improvement. Both program instructions and data are stored in the same memory with the same *address structure*. That is what Rosin means by "uniform storage medium" in the passage above. This *uniformity* provides significant benefits. One is to shorten the time needed to access a program command. Another is using the memory most effectively: Some problems require short programs and large amounts of data, while others require extended programs and few data points. With uniform storage, both situations are accommodated while utilizing all available memory. Data and instructions exist in computer memory as strings of 1s and 0s. In the case of data, these strings are binary numbers; in the case of program instructions, they are strings of *true* and *false* logic values. The CPU "knows" which is which because each type has a specific marker that indicates "I am data" or "I am an instruction." Such a system allows for additional flexibility, including *program modification*: The program can act on itself, modifying its instructions based on the intermediate results obtained in a computation.

Computer Programs

It may be a good opportunity to briefly discuss computer programs. As we mentioned, an algorithm is a set of mathematical and logical operations designed to achieve a specific goal or solve a particular problem. A computer *program* is an algorithm written in a *language* a computer can understand.

Historically, numerous programming languages have been developed, including *Ada* (named in honor of Ada Lovelace), *Algol*, *BASIC*, *C*, *C++*, *COBOL*, *FORTRAN*, and *Pascal*. *Python* is the dominant language for AI development, thanks to its simplicity and a vast ecosystem of libraries (such as *TensorFlow* and *PyTorch*) that cater to machine learning, data science, and research. Beyond Python, several newer or increasingly popular languages are making their mark:

- *Julia* is designed for high-performance numerical and scientific computing, offering an excellent balance of speed and ease of use. Its syntax is friendly for mathematicians and engineers.
- *Rust* was created for system-level programming with a strong emphasis on safety and concurrency,[21] Rust is increasingly used in AI projects. Its growing popularity is driven by its ability to manage memory without sacrificing speed.
- *Kotlin* and *TypeScript* are not AI-specific; they are used for general-purpose and web development. They are popular among programmers for building applications that interface with AI services.
- There are also experimental languages, such as Google's *Carbon* (proposed as a successor to C++).

These are all *high-level* languages; they are studied by programmers, who proudly list the ones they mastered on their resumes. Sometimes, computer languages use understandable symbols and plain English words, such as IF…THEN…ELSE or GOTO, but for the most part, the vocabulary of a program is not English. The machine does not work directly with these commands. The hardware within the processor "understands" *machine language*, which consists of bits that are fed as inputs into gates, such as NAND and NOR. These gates then spit out the results, again in the form of

[21] In programming, "concurrency" means structuring a program so that multiple operations can progress independently, even if they're not all executing at exactly the same moment.

logical 0s and 1s. In this way, the CPU performs logic operations, resulting in actions expressed by IF...THEN...ELSE and other high-level instructions. A computer must have a way of translating high-level language into machine language, which is the job of *interpreters* and *compilers*. A general-purpose computer will work with various high-level languages.

Returning to sequences of 1s and 0s, let us review the relevant terminology. In computer literature, we often see the terms "bit," "byte," and "word." We know that a bit is a single digit, either 1 or 0. We also know that a *byte* is an 8-bit string of bits, for example, 01110010. We often see abbreviations such as kb, Mb, and Gb referring to 1,000 bits, 1,000,000 (million) bits, and 1,000,000,000 (billion) bits. We also see the abbreviations kB, MB, and GB, with the capital "B" referring to bytes. The term "word" indicates a standard-length piece of data or a set of logic symbols. A word may consist of 8, 16, 24, 32, or 64 bits. In modern general-purpose computers, word sizes are usually 32 or 64 bits. For example, 100110 10010011101010100011011100 is a 32-bit word. CPUs and computer *registers* are designed to work with a specific word length. They treat 32-bit or 64-bit strings as single units, "swallowing" data and logic in 32 or 64-bit chunks.

The IAS computer

In 1945, von Neumann initiated a computer project at his home institution, the IAS. It was not easy to convince the "ivory tower"-like IAS, home to Einstein and Gödel, to make a computer. Yet, von Neumann's powers of persuasion were such that he prevailed and went ahead with his plans. The strongest argument for allowing him to do what he wanted was that, otherwise, he might leave the IAS. The necessary funds — von Neumann wanted $100,000 per year for three years — were provided by the U.S. Army's Ordnance Department, the U.S. Atomic Energy Commission, and the Institute itself. And so, despite all odds, the *IAS machine* was built. The detailed story of its construction can be found in George Dyson's book *The Turing Cathedral* (2012). Von Neumann wanted to get Eckert to head the project, but Eckert was unavailable, as he planned to go into business with Mauchly at the time. Nonetheless, von Neumann assembled a strong team of six engineers (two of whom were from the ENIAC project) headed by the talented Julian Bigelow.

Von Neumann could freely implement his ideas in the machine built under his supervision. The IAS computer differed significantly from the EDVAC: It used parallel processing, while the EDVAC was strictly sequential, and it was faster. In addition to implementing his architecture, von Neumann made his computer more reliable than the EDVAC and the ENIAC. Once the IAS computer became fully operational in June 1952, von Neumann began distributing its detailed plans freely without concern for patent rights. As a result, a dozen clones or near-clones were built in the U.S., in government labs, universities, and by IBM, and at least a dozen more abroad, in various countries including Denmark, the Soviet Union, Israel, and Australia. The dominance of von Neumann's architecture in the computer world had been established.

Two new sciences

Together with mathematician and close friend Stanisław Ulam, John von Neumann helped to create two new fields at the intersection of mathematics, computer science, and biology. One was the study of *self-replicating systems*; the other, the theory of *cellular automata*. While these areas are not strictly part of computer history, they are deeply relevant to our discussion of AI. We will review them briefly in the following section and Chapter 4.

The theory of self-reproducing automata

After the War, von Neumann became interested in formulating a general theory of *reproduction*, seeking to define the conditions under which complex systems could replicate themselves. The axiomatic mathematical theory he created applies equally to living and artificial systems. In 1946, von Neumann wrote to the father of *cybernetics,* Norbert Wiener, "I can formulate the problem rigorously, in [*the same way in*] which Turing did it for his mechanisms."[22] Indeed, Turing's one-dimensional machine, which we discussed earlier, copies its "state" onto a tape. Von Neumann wanted to go further. While Turing's machine could only copy its "state"

[22] Despite possessing strong linguistic skills, Von Neumann never lost his Hungarian accent, and his written English often "asks" to be completed or corrected. Grammarly goes crazy when given a text from von Neumann.

in software, von Neumann considered machines that could make material copies. It would not necessarily be a copy of the machine itself. In one implementation, the machine builds two copies of *anything* described in sufficient detail. In von Neumann's own words:

> ...there exists an automation B which has this property: If you provide B with a description of anything, it consumes it and provides two copies of the description...His [Turing's] automata are purely computing machines. Their output is a piece of tape with zeros and ones on it. What is needed ...is an automation whose output is other automata...Small variations of the foregoing scheme also permit us to construct automata which can reproduce themselves and, in addition, construct others." (quoted in Dyson 2012, p. 284)

Von Neumann proved this ability to make material copies mathematically. He used the same mathematical methods that Turing and Gödel had used in the 1930s. All he did was substitute memory cell addresses with the locations of physical components. The machine then identifies these components and assembles what it needs from them. It produces physical objects constructed from these components, rather than ones and zeros.

We are writing about von Neumann's reproducing automata in the chapter about the history of computers because computers are a subclass of these automata. Whatever can be said about von Neumann's automata also applies to computers. Thus, computers, or computer-powered assemblers, can reproduce under certain conditions — a profound insight which may be of great importance to the future of AI and robotics. And which, I might add, may be bad news for our species.

This sounds like science fiction, something we expect to hear from Stanisław Lem (whom we will discuss in Chapter 5), but not from a mathematician and a computer scientist. Yet, von Neumann managed to dress this fantasy in exact mathematical robes. He did it in 1946–1948, before Watson and Crick discovered the DNA structure and explained how replication occurs in living organisms.

Von Neumann also considered the *minimum complexity* required to initiate replication. We will leave this aspect for Chapter 4, where we discuss AI in more detail.

The Theory of Self-Reproducing Automata was never completed by von Neumann, who had planned to collaborate with his friend and colleague Stanisław Ulam as a co-author and write a comprehensive book on this subject. After his death, logician Arthur Burks collected and published von Neumann's lecture notes and manuscripts under this title (von Neumann 1966).

Cellular automata

To formulate his theory of self-replicating machines, von Neumann needed a precise, abstract model in which to implement them. The physical world was too messy and uncooperative; he required a discrete, rule-based system that could support both construction and computation. Working with Stanisław Ulam, he devised a new mathematical framework known as *cellular automata.*

A cellular automaton is a regular grid of cells — each in one of a finite number of states — which evolve over time according to simple, local rules. Each cell updates its state based on the states of its immediate neighbors. Despite their simplicity, cellular automata can exhibit surprisingly rich and complex behavior. They provide a model for exploring the emergence of structure, order, and replication from purely local interactions.

Von Neumann's cellular automaton, developed between 1949 and 1951, was a two-dimensional grid with 29 possible states per cell — just enough to support his logical machinery for self-replication. Within this grid, he designed and proved the existence of a *universal constructor*, a self-replicating machine capable of reading a symbolic blueprint and building another machine described by it, including a copy of itself (see above). This construction was the first rigorous demonstration that self-replication was possible in an artificial medium, long before DNA was understood as nature's version of such a code.

Cellular automata were not merely a tool for implementing von Neumann's theory; they became a field of study in their own right. Later simplified versions — such as John Conway's *Game of Life* (1970) — captured the public imagination, demonstrating how complexity and even apparent intelligence could emerge from straightforward, deterministic rules. These models are now central to the field of *artificial life* and continue to inspire research in emergent computation, biological modeling, and distributed AI.

In von Neumann's hands, however, cellular automata were not merely a curiosity. They were the foundation for a deep and formal attempt to bridge biology and computation and to show how machines might grow, evolve, and multiply under the governance of logic alone.

The last years

The range of von Neumann's interests was terrific, from modeling explosions to mathematical economics and from the foundations of quantum mechanics to building computers. At the end of his life, von Neumann became interested in neuron modeling and the parallel between computers and the human mind. His last book, *The Computer and the Brain* — an unfinished collection of notes for a series of planned lectures that were never given due to his illness — was published posthumously in 1958.

John von Neumann died of cancer in 1957, when he was 53, possibly as a result of radiation exposure during the Manhattan Project, but of course, no one can say this with certainty. The disease hit him hard in 1955 and spoiled the last 18 months of his life, and many things he started he was unable to finish. His premature demise denied him the Nobel Prize in Economics, which he would surely have received for his contributions to game theory (the award cannot be given posthumously). Another Los Alamos alumnus, Enrico Fermi, also died at 53 from cancer.

Computers are Maturing

Computer compatibility

After 1950, new computers began to spring up like mushrooms after a radioactive rain. The UNIVAC I in the U.S. and the Ferranti Mark I in Great Britain, both introduced in 1951, were the first commercial computers, and their debut was a triumph. Their success prompted IBM — later to become the world's largest computer manufacturer — to develop its first electronic computer, the IBM 701, which was introduced to the market in 1953. The military used it first — Los Alamos was the inaugural buyer — and soon after, universities and large corporations began buying or renting IBM computers. The IBM 701 also played a role in the early history of AI, as described in Chapter 4.

Between 1953 and 1956, the first computers built entirely from transistors and diodes, without vacuum tubes, began to appear. As mentioned earlier, the first of these all-solid-state machines was constructed at Victoria University in England.

By the early 1960s, numerous computers were available on the market, and they were a diverse group. Each was designed as a one-of-a-kind creation; the more original, the better — more glory to the designer.

But by the beginning of the 1960s, that attitude was becoming obsolete. IBM was the first to recognize the importance of continuity and uniformity in computer design. By 1963, the company had no fewer than seven mutually incompatible computer lines, each serving a different part of the expanding market. IBM was preparing to launch an eighth incompatible line, the IBM 8000, when everything changed (Swedin and Ferro 2005, 71).

The idea of compatibility was ripe, but it always takes one visionary to make others see clearly. In this case, the visionary was Robert O. Evans, IBM's manager. Despite resistance from within the company, he canceled the IBM 8000 project. Instead, he championed the creation of what became known as the New Product Line: System/360 — a family of machines in which compatibility was deliberately built into both the hardware and the software.

From that point on, all System/360 and later IBM computers employed identical instruction sets and input/output architectures. Software no longer needed to be rewritten for each new machine. It was a turning point — not only for IBM but also for the entire industry.

Operating system

At the same time, computers were being equipped with an *operating system*. The one developed for IBM's System/360 was simply called "OS." Other computer companies quickly followed suit. For example, in the 1960s, *Control Data Corporation* (CDC) created a series of operating systems called SCOPE, standing for *Supervisory Control of Program Execution*. The name itself offers a succinct explanation of what an operating system is meant to do.

An operating system (OS) is the manager of all software on a computer. It oversees the execution of programs and applications, deciding when each runs and whether they operate sequentially or in parallel. The hardware and software on a given machine must comply with the OS installed on it.

That is why *Microsoft Windows*, the dominant desktop operating system, has maintained such a tight grip on the computer industry since its introduction in 1985. That grip later loosened somewhat when, under pressure, Windows was made more compatible with other standards.

As of 2025, the market share for desktop operating systems is approximately:

- Windows: ~70–75%
- macOS: ~15–18%
- Chrome OS: ~7–10%
- Linux: ~2–3%

Chrome OS is a distinct operating system developed by Google, built on the Linux kernel and designed primarily for cloud-based applications.

Today, all computers — and more broadly, all "smart" electronic devices — use an operating system of one type or another.

Clock cycle

The computer is timing itself with a *clock cycle*, which can be likened to the *rhythmic unit* in a dance sequence, ensuring that every movement (operation) is completed within a structured tempo. A group of interrelated operations (moves in a dance) must be completed by the end of a cycle (the beat). The shorter the cycle, the faster the "dance."

Computer engineers specify the frequency corresponding to the cycle's time interval. For example, if the clock cycle is one billionth of a second (0.000000001 seconds, or a nanosecond), the frequency is one billion cycles per second, or one gigahertz (GHz).

> The reader may recall that in Z3, the world's first Turing-complete electromechanical relay-based computer, built by Konrad Zuse in 1941, the clock frequency was 5 Hz, or a clock cycle of 0.2 seconds — one of the slowest computer cycles ever. In the world's first fully electronic, vacuum-tube-based computer, ENIAC, which was completed in 1946, the clock frequency was 5,000 Hz (5 kHz), and the clock cycle was 0.2 milliseconds, or 200 microseconds — a thousand times shorter than in Z3. The first electronic computer with a synchronous master clock cycle and stored-program

architecture was the British Manchester Mark I, which came online in 1949. The cycle was 1 millisecond, corresponding to a clock frequency of 1 kilohertz.[23] In the first "commercial" computer, UNIVAC I, made in 1951, the frequency was already measured in millions of hertz (MHz) and clock cycles in microseconds. In today's computers, the clock frequency is several billion hertz (GHz), with cycles measured in fractions of a nanosecond. So, today's computers "dance" a billion times faster than Zuse's machine did in the 1940s.

In modern computers, each cycle is divided into four steps. The first step, *Fetch*, gets an instruction from the random access memory (RAM) and puts it in a cache memory closer to the CPU. Step two, *Decode*, is performed by the part of the CPU called the Control Unit. It reads the instructions from the cache, transforms them into commands, and sends them to the ALU. Step three is *Execute*: The ALU acts according to the received command, performing calculations or executing logic operations and writing the results again into the cache. Finally, step four, *Store*, involves storing the result in main memory.

We will now discuss computer memory; a modern computer has several distinct memory components.

Thanks for the memories

Creating computer memories has always been the most challenging problem in computer engineering. The evolution of clever technologies that led to several types of computer memories could be the subject of a separate book, but we will be very brief here. Memories are also commercially significant; semiconductor memory accounts for approximately 30% of total semiconductor industry revenues, amounting to hundreds of billions of dollars annually.

In the Atanasoff–Berry calculator, ABC, only one type of slow electromechanical capacitive memory was used to hold the data. The program for solving linear algebraic equations was not in the same memory; it was

[23] Directly comparing "clock cycles" in ENIAC and Manchester Mark I suggests that the former, made in 1945, was faster than the latter, created in 1949. This comparison, however, does not capture the full picture of their performance and design trade-offs.

hardwired. In EDVAC, there were two types of memory: fast memory for storing data and the program, and slower memory for additional data. By *fast* and *slow*, we mean short or long memory access times, which refer to the time it takes to read the content of a memory cell. Today, there are four or five types of memory found on every computer. They all store programs and data, but they do it in different ways — they specialize. They are distinguished by their speed (access time) and whether they will retain or lose their contents without power (*permanent* versus *volatile* memory). The cost is also an important consideration. As ENIAC designer John Mauchly has put it, "…the fast memory was not cheap, and the cheap memory was not fast."

Memories and memory access times have undergone significant changes since the first electronic computers emerged in the late 1940s. In 1946, the first electronic memory, *Williams' tube memory*, had an access time of 25 microseconds (25×10^{-6} seconds). Today, the fastest memories have access times (we should say, *memory reading times*, to be more accurate) of about one nanosecond (10^{-9} seconds), which is shorter by a factor of 25,000.[24] One consequence of this short access time is that the fast memory must be located close to the CPU; otherwise, it will lose its speed advantage. Placing memory a foot away from a CPU is not an option; it would be like living in San Francisco and ordering pizza from Los Angeles.

The speed of propagation of electrical signals in typical computer interconnects is close to 60% of the speed of light in a vacuum. The rule of thumb followed by engineers states that a distance of 6 inches (15 centimeters) inside the computer delays the signal by one nanosecond. Suppose one nanosecond reading-time memory is placed 6 inches away from a processor. The processor sends a request, and it takes about a nanosecond to read the content of the memory cell. The information from the cell must then reach the processor. Therefore, the total access time for the fastest memory, with a reading time of 1 nanosecond, located 6 inches from the processor, will be approximately 3 nanoseconds. It would make sense to place such a fast-reading memory closer to the processor.

[24] Driving from my home on Long Island to Manhattan takes me about 1.5 hours. If it were 25,000 faster, it would take me 0.2 seconds. Twenty-five thousand is a large number!

The fastest single-CPU computers must be small (while multiple-CPU machines can be large), and engineers know that. However, a fast device dissipates high power, and a small computer tends to overheat. This is the central conundrum computer designers face today: They want to increase computational speed, but high speed causes excessive heating. We later discuss this critical problem.

In the following special section, we are listing the types of memories found on today's computers and providing some of their parameters.

Random Access Memory (RAM)

RAM, also known as *main memory*, stores data and instructions (programs). It is a *volatile* memory made of semiconductors. It is a "working memory" containing everything a computer needs to perform its immediate tasks. The main memory is divided into *dynamic random access memory* (DRAM) and *static random access memory* (SRAM), collectively referred to as RAM. The *data access time* for RAM is measured in tens of nanoseconds — it is a relatively fast memory. The operating system is loaded into RAM from a hard drive (see the following) when a computer is *booted* (started). The RAM capacity is one of the essential characteristics of a computer.

There is a legend that Bill Gates said in 1981, with RAM in mind, that "640K ought to be enough memory for anybody." He repeatedly explained that he never said this, but the legend refuses to die. In 1981, it was at least *possible* to say something like this. Today, the absolute minimum RAM is 4 GB (gigabytes, or 10^9 bytes), which exceeds 640 kB by four orders of magnitude, and even 4 GB isn't "enough for anybody." To run Windows 10 and modern applications, one needs at least 16 GB.

Cache Memory

Cache memory ("cache" means "hidden" in French) is a high-speed memory that sits between RAM and the CPU, assisting the CPU in retrieving data and instructions from RAM. Its data access times are 10–100 times shorter than those of RAM. Accordingly, a cache is the closest memory to the CPU, typically integrated into the CPU chip. The data cache and the instruction cache are separate; there are several subdivisions (layers) of cache labeled L1, L2, and L3, with L1 being the fastest. If RAM contains what is needed for a computer to perform its tasks, the cache contains what is *genuinely* required; it is like memorizing a list of essential groceries — the ones about which your wife said, "do not come back without them."

Processor Registers Memory

This memory is similar to *cache* memory. If the cache resides directly inside a CPU, *processor registers* are placed outside, on a separate chip, but very close to it. Access times are similar to cache times. Processor registers can be compared to a list of groceries that your wife wrote on a piece of paper when sending you to the store.

Read-Only Memory (ROM)

ROM is non-volatile, hard-wired, and slow. Its data access time is measured in hundreds of milliseconds — about ten million times slower than a typical RAM and a hundred million times slower than cache or registers. ROM does not change its content; it is intended for reading only. ROM can be likened to thick cookbook kept on a shelf in the kitchen. ROM holds special programs called *firmware*, which are necessary for computer operations, and it also contains permanent data types such as look-up tables. When stored on an optical (or compact) disk, it is called a CD-ROM.

Secondary Memory

It is a non-volatile memory containing programs and data used long-term, all of which are organized into *files*. The secondary memory may be located on a magnetic *hard disk*, an *optical disk*, or a *solid-state drive* (SSD) made of semiconductors. Secondary memory is only about a factor of 10 faster than ROM, and, like ROM, it is considerably slower than RAM. Secondary memory is significantly larger than main memory; its capacity is measured in tens of terabytes (TB = 10^{12} bytes). To continue our "buying groceries — cooking" analogy, if ROM is a cookbook, then secondary memory is the collection of all the cookbooks kept on that shelf in the kitchen and elsewhere in the house. There may be several secondary memory locations: inside the computer, on the computer disk, and outside, on peripheral devices such as hard drives. This memory spreading is dictated by the desire to protect data by creating *backups*. When a computer works, it copies files from a secondary memory location into its main memory (RAM) as it needs them, on a "need to know" basis, directed by an OS.

Computer Processors: CPUs, GPUs, and TPUs

Memory is essential, but a computer's heart (or, perhaps, a better metaphor, its brain) is its processor, central processing unit, or CPU. That is where all the calculations and logical decisions are made. Babbage found a good name for it: the *Mill*. Babbage's Mill, as shown in Fig. 18(a), was a bulky mechanical construction intended to be some 15 feet tall, not unlike an actual grain-grinding mill. In the first electronic computers of the 1950s, the CPU was a large unit packed with vacuum tubes, resistors, capacitors, and wires (a small section of a vacuum tube processor is shown in Fig. 18(b)). With the advent of integrated circuits (ICs), it

Fig. 18. Computer processors through the ages: **(a)** Reconstructed model of Babbage's Mill. A full-size Mill would be 15 feet (4.5 meters) tall. **(b)** Vacuum tube unit (not a complete processor), circa the 1950s. **(c)** The first microprocessor, the Intel 4004, was released in 1971. **(d)** Intel 233 MHz Pentium processor, released in the late 1990s, mounted on a board and compared to an inch scale. **(e)** AMD Ryzen 9 12-core 5900X 3.7 GHz processor, introduced in October 2020; it is built on 7 nm IC technology, and its size is about 3 × 3 inches. **(f)** A portion of 2024 NVIDIA H200 NVL PCIe graphics processing unit: the GPU package (chip plus high-bandwidth memory), several inches across; the silicon die (black rectangle) is over an inch on a side. Such GPUs are made for AI-training computers, for example, Elon Musk's Colossus (see later).

Sources: (a) Image from London Science Museum, Lic. CC BY-SA 2.0; (b) photo by Jitze Couperus, Lic. CC BY 2.0; (c) photo by Thomas Nguyen, Lic. CC BY-SA 4.0; (d) image in the public domain; (e) image courtesy of AMD; (f) image courtesy of NVIDIA.

became a *microprocessor* (Fig. 18(c)), and eventually, it turned into a square chip a couple of inches on the side (Figs. 18(d)–(f)) called a "central processing unit" or "geometric processing unit" (GPU).

As can be seen in Figs. 18(a)–(f), as electronic complexity increases, the visual appearance becomes increasingly muted, devoid of features and approaching the menacing pitch-black monolith from *2001: A Space Odyssey*.

The **CPU** handles general-purpose computations, from running an OS to executing applications. It consists of 2–64 cores and is optimized for sequential processing. It has a large cache memory for quick data access. CPUs are found in all PCs and servers. CPUs are best suited for handling numbers rather than one-dimensional arrays of numbers (vectors) or multi-dimensional arrays (matrices and tensors), which are more efficiently handled by GPUs and TPUs.

Graphics accelerators appeared in the 1980s, followed by specialized graphics processors in the early 1990s. The term graphic processing unit (**GPU**) was first introduced by NVIDIA in 1999. Initially designed for rendering graphics, GPUs excel at parallel processing, making them ideal for tasks such as deep learning, scientific simulations, and Bitcoin mining computations. They contain thousands of smaller cores working in parallel and are well suited for performing calculations with *matrices* (two-dimensional arrays of numbers, numerical tables). All of that makes them ideal for training deep neural networks, something we will discuss in Chapter 4.

Another type of CPU, the *tensor processing unit*, or **TPU**, has been designed for handling *tensors*. Tensors are used in math and physics. The complexity grows from numbers to vectors to matrices to tensors, which are multi-dimensional arrays. Mathematics can define rules for dealing with such complex objects; for example, it is possible to multiply them. The higher the tensor *rank* (dimensionality), the harder it is to deal with. And that is where GPUs and TPUs come in handy. The *large language models* used in AI require dealing with vectors, matrices, and tensors.

While most CPUs can be purchased for $500–$1,500, GPU prices start at hundreds of dollars and rise to $30,000–$40,000 for more advanced models, such as the renowned NVIDIA H200 used in AI training. TPUs are not available for direct purchase; instead, they are accessible via cloud services, where an hour's usage of a single TPU chip may cost a few dollars. The catch is that to do anything useful, one needs many of them working simultaneously. Google offers *TPU Pod configurations* for large-scale machine-learning tasks. A TPU v4-Pod with 4,096 TPU v4 chips is priced at $32,200 per hour. So, good luck with doing your own AI training and paying for cloud services.

Energy Cost and the Speed of Computations

When we compute on any machine, whether mechanical, electromechanical, electronic, or even in our minds, there is always some energy associated with computation, known as the *energy cost* of computation. The amount of energy depends on the type of operation and the computing machine. In a mechanical calculator, the energy is used to rotate the wheels, move the levers, and display the results. In an electromechanical device, the energy source is electricity. For example, in a relay, the energy is dissipated by passing the electrical current through an electromagnet's coil, and additional energy is drawn to create the core's magnetic field. In a vacuum tube, the lion's share of energy goes into heating the cathode for *thermionic emission* of electrons, and some energy is also consumed in driving the current between the anode and cathode and in the outside circuits. In a transistor, the energy is again used to drive the current, and some energy is stored in the electric and magnetic fields present within the device.

The amount of energy per operation (for example, for the multiplication of two numbers) is greatest for a mechanical machine — Babbage planned to power his Analytical Engine with a steam engine!

If an operation requires energy ε and is completed in time t, we say that the *power* is $P = \varepsilon/t$. In physics, the unit of energy is the *joule* (J), and the unit of power, measured in joules per second, is called a *watt* (W), named after James Watt, the inventor of the practical steam engine. Babbage's Difference Engine II could add two 31-digit numbers in about a second, performing one *operation per second* (OPS). I do not know how much power was required, but it was a lot. Early vacuum-tube computers also used considerable power to heat vacuum tube cathodes. The UNIVAC I used 125 kW of power and performed 1,905 OPS, which meant it required 125,000/1905 = 65.6 W per operation. It is the power of a medium incandescent light bulb, which is very high considering that it relates to merely adding or multiplying two numbers, even if they consist of many digits.[25] Computer engineers usually quote *operations per watt-second* or *operations per joule* (because a watt-second equals a joule). In these terms, UNIVAC I performance can be cited as 1905/125,000 = 0.015 operations per joule. If we compare this with the performance of today's computers, we will find numbers differing by a factor of about a trillion (10^{12})!

For example, a Fujitsu chip produced in Japan in 2005, called *FR-V VLIW/vector processor,* achieved 51 Giga-OPS with 3 W of power, resulting in 17 billion operations per watt-second (from the Wikipedia article entitled "Performance per watt"). Dividing 17×10^9 by UNIVAC's parameter of 0.015, we get a factor of 1.1×10^{12}.

Instead of OPS, computer scientists nowadays characterize computer performance in terms of FLOPS, an abbreviation for *floating-point operations per second.* A floating-point operation is an arithmetic operation performed on numbers represented in what we (non-computer scientists) call *scientific notation* — the notation we use in this book. When we divided 17×10^9 by 1.5×10^{-3} and obtained 1.1×10^{12} in the special section above, we performed one floating-point operation.

Simple hand-held calculators may perform arithmetic operations with only 10 FLOPS — enough speed for a user who requires one operation at a time and does not mind waiting for 0.1 seconds for an answer. Computers, of course, work much faster. The number of FLOPS for one core inside a processor is calculated by taking the number of FLOPS that can be made in that core in one cycle and multiplying it by the number of cycles in one second: $(\text{FLOPS})_{\text{core}} = (\text{FLOP/cycle}) \times (\text{cycles/second})$. In the case of multiple parallel cores, processor performance can be expressed as $(\text{FLOPS})_{\text{CPU}} = (\text{FLOP/cycle}) \times (\text{cycles/second}) \times (\text{cores/processor})$. For example, if a CPU contains 24 cores and operates at a 3.9 GHz clock speed, and each core can perform 32 operations per cycle, then we obtain $(\text{FLOPS})_{\text{CPU}} = 32 \times (3.9 \times$

[25] 125 kW was the total power required to run UNIVAC-I, with all its tape drives, printers, memory, and everything else; it was not the power to run its processor, which must have been considerably lower. Still, a user needed about 66 watts to perform one operation on the UNIVAC-I.

10^9) × 24 = 3 × 10^{12}, i.e., 3 Teraflops. I simplified this calculation somewhat; computer engineers also talk about *nodes*, *racks*, and *sockets*, adding terms such as (sockets/nodes) to the equation, but we will not go there.

Some modern computers can perform record-breaking numbers of operations per second by engaging hundreds of thousands of parallel processing cores. These machines are called *supercomputers*, and we will return to them later for a more detailed discussion. For now, it is worth noting that some of them can execute over 10^{18} FLOPS — a truly staggering figure. This number is called a *quintillion*, and when used to describe computational performance, it is referred to as one *exaFLOPS*.

To appreciate this number, let me assume that I am making calculations by hand on paper and need 10 seconds to complete each operation. This would exaggerate my calculational abilities because I must handle two 32- or 64-digit numbers in binary notation in each operation. Nevertheless, let me assume I can add or multiply two such numbers in 10 seconds. Then, it will take me 10^{19} seconds to do what a supercomputer does in 1 second, meaning it will take me twenty times the age of the Universe, approximately 300 billion years.

There are several obstacles on the road to fast computers. As we mentioned above, a fast CPU should be small and placed close to its memory, or it will be limited in its performance by the speed of light — more accurately, the speed of signals in wires and interconnects. In a fast computer, this requirement clashes with energy dissipation. Densely packed fast-switching transistors dissipate considerable energy, and such circuits may overheat. People have discussed this problem since the 1970s, but it has only become truly acute in the last decade.

Due to Moore's Law, the amount of energy per transistor switching has steadily decreased; for the current 3 nm *technology node*,[26] it is about 25 × 10^{-18} J, or 25 aJ. (One attojoule is one quintillionth (1/1,000,000,000,

[26]As we discussed in Chapter 1, since the 2010s (the FinFET era, 22 nm and below), technology nodes no longer represent a single physical feature. Modern nodes (e.g., 7 nm, 5 nm, and 3 nm) do not correspond to any single dimension but represent overall scaling improvements. For example, in the 3 nm technology, the transistor gate length may be 10–14 nm.

000,000,000) of a joule.) The switching time has also dropped, currently being around 0.3 picoseconds (0.3×10^{-12} seconds) for the same 3 nm technology. However, the transistor is not switching at a frequency equal to the inverse of that time, which would be about 3 THz. Instead, at the maximum processor workload, it switches on average once in every three cycles, which at a 5 GHz clock frequency (5 billion cycles per second) corresponds to 1.7×10^9 switching events in one second. Thus, the energy dissipated by one transistor in one second equals approximately $25 \times 10^{-18} \times 1.7 \times 10^9 \approx 40 \times 10^{-9}$ J = 40 nanojoules. Then, the power (energy per second) dissipated by one transistor is 40×10^{-9} W = 40 nanowatts.

This seems like a small power, but let us not forget that modern chips contain tens to hundreds of billions of transistors, on the lower end of this range for CPUs and the higher end for GPUs and TPUs. The power dissipation of a fully engaged (high-workload) chip with 10 billion transistors, manufactured according to the 3 nm technology, would be $40 \times 10^{-9} \times 10^{10} = 400$ W! Removing half a kilowatt of power from a chip with a surface area of a few square inches is a very tall order. We usually associate such power with a red-hot cooktop on an electric kitchen range. Removing it from a chip can only be done with forced water or freon cooling, and even then, it is a serious engineering challenge.

Silicon transistors can operate at temperatures up to 200°C, but even 85°C is enough to shift their parameters significantly, potentially throwing an entire system out of whack. However, in chips that dissipate hundreds of watts, the primary concern isn't maintaining stability below 85°C — it's the danger of the system melting down or catching fire! Indeed, if left unchecked, the heat generated in a GPU can raise its temperature to the point where it would literally melt.

To fight this, engineers use forced water cooling and even freon cooling. Still, these methods are not always sufficient. As a result, they lower transistor densities, spacing transistors farther apart on the chip. But this, in turn, slows the system down, interfering with performance and undermining the very benefits Moore's Law is supposed to deliver.

So, engineers are becoming desperate. When they earn their computer science degrees, they seem to pledge allegiance to Moore's Law, vowing to fulfill it at any cost — but that promise is becoming harder and harder to keep.

To help their production brethren, scientists are working in two directions. First, they try to reduce the energy released during each switching event — but this is a difficult proposition, one that borders on reinventing the transistor itself. Second, some of them — those with better job security — are exploring entirely new methods of computation.

For example, in the 1970s and 1980s, several major exploratory programs aimed to replace transistors with superconducting Josephson junctions. I was involved in one such project while working at AT&T Bell Labs, as described in my first book (Gurvitch, 2023, Chapter 4 of Volume 2).

Fundamental energy associated with information transfer

Although the energy dissipated in an individual transistor is small, the cumulative heat generated by billions of them on a chip creates serious problems. This heating issue began to surface in the 1970s and has only intensified with the advance of transistor technology and the relentless scaling driven by Moore's Law.

This is, above all, a practical problem: Its details are not dictated by fundamental science but by engineering constraints. However, in this section, I want to focus on something more profound: nature's absolute limits on energy dissipation in computing. I do this partly because it is interesting and partly because some forward-looking computational architectures are approaching these natural limits — limits that no clever engineering can bypass.

These constraints arise from two major areas of physics: thermodynamics (the science of energy exchange) and quantum mechanics. One of the key insights comes from information theory.

Ever since Claude Shannon's seminal work in 1948, which formally linked entropy and information, physicists have understood that the transfer of information must be associated with a minimum amount of energy per bit. Whenever entropy changes, a specific quantity of heat must be either released or absorbed.

> Entropy entered physics in the 19th century out of the new science of heat engines. Sadi Carnot's 1824 work showed that heat engines have a fundamental upper limit to their efficiency, determined only by the temperatures involved. Carnot himself did not use the term "entropy." In the

1850s–1860s, Rudolf Clausius reformulated thermodynamics and introduced the quantity that is conserved in reversible processes and increases in real ones, defining it in 1865 as "entropy." Later in the 19th century, Lord Kelvin, James Clerk Maxwell, Ludwig Boltzmann, and Josiah Willard Gibbs clarified its physical and statistical meanings.

When Claude Shannon observed that something resembling entropy arises in the exchange of information, he effectively drew a conceptual connection between information and energy. That insight implied a profound conclusion: the transfer or manipulation of information must incur an energy cost. But how exactly are these costs realized?

For many years, it was assumed that each computational step must consume a minimum amount of energy equal to $(k_B T) \times \ln(2)$, where $k_B = 1.38 \times 10^{-23}$ J/K is the Boltzmann constant, T is the absolute temperature, and ln(2) is the natural logarithm (logarithm with base $e = 2.718\ldots$) of 2, equal to 0.693. This conjecture originated directly from Boltzmann's statistical interpretation of entropy. The "2" in the logarithm refers to the binary decision associated with processing a single bit of information: yes or no, 0 or 1.

At normal "room" temperature, taken as 300 K, this limit is approximately equal to 3×10^{-21} J, which is about 10,000 times smaller than the actual switching energy of 25×10^{-18} J in the smallest transistors we quoted above.

The scientist who finally clarified these issues was Rolf Landauer, who in 1961 published a seminal paper titled "Irreversibility and Heat Generation in the Computing Process." Landauer arrived at an unexpected conclusion: A computation done very slowly and carefully — what is called *reversible computation* — requires negligible energy. In fact, there is no lower bound: the slower the computation, the less energy it consumes. In the limit of infinitely slow computation, the required energy approaches zero. In such a case, entropy remains constant; it does not increase.

However, entropy does increase when we erase one bit of information. Landauer showed that erasing a bit requires a minimum energy of $(k_B T) \times \ln(2)$.[27] This fundamental energy threshold is now known as the *Landauer limit.*

[27] It makes intuitive sense: While computation may be done slowly and reversibly, erasing information is abrupt; you cannot erase something slowly and carefully.

So, for an infinitely slow process, the energy cost is zero, but who would want to compute infinitely slowly?

It can be intuitively guessed that the faster we attempt to write a bit, the more energy it must cost; in other words, the minimum required energy is inversely proportional to the time taken to write the bit. In the context of switching transistors, this corresponds to the switching time.

The answer to this question came in 1998 in a paper by Norman Margolus and Lev Levitin, titled "The Maximum Rate of Information Processing in Physical Systems." Their elegant result, based on quantum mechanics, established that the minimum energy required to transfer (write) one bit is given by $h/(4\tau)$, where h is Planck's constant and τ is the bit write time. This fundamental quantum constraint is called the *Margolus–Levitin bound.*

For modern transistor switching times — around 0.3 picoseconds — this bound implies a minimum energy requirement that is about five times smaller than the Landauer energy required to erase a bit.

Currently, the Landauer limit is approximately 10^4 times (four orders of magnitude) lower than the actual energy used by even the most miniature (3 nm technology) transistors. The Margolus–Levitin limit is lower still. So, for now, these limits do not pose practical challenges for computer engineers. However, they are of significant interest in forward-looking research in low-energy computing, including ongoing attempts to build quantum computers, a topic we will explore in the following chapter.

I'd like to close this section by connecting it to our main theme — AI.

Not long ago, a friend and I were discussing the future and potential dangers of AI. "Surely," he said (I'm paraphrasing), "AI can't develop as much as you suggest — it will be limited by energy. It already consumes a growing share of the world's electricity. To go much further, it would need vastly more energy, which doesn't exist — unless we build many more nuclear power plants or master controlled fusion."

I replied: "But do you realize how far today's transistors are from the theoretical limits? Future AI might reinvent computation itself, reducing energy consumption by factors of thousands, effectively removing energy as a constraint. And yes — AI is already helping to stabilize plasma in experimental fusion reactors."

Supercomputers

A supercomputer is a computer with a much higher than average performance and a much higher than average price. The secret is not in using technology exclusive to supercomputers but rather in the number of normal, commercially available processors that, while working simultaneously, do not create computational "traffic jams." Richard Feynman, back in the 1980s, said that people are just beginning to figure out ways to do parallel computing and that it is not easy. Apparently, they have figured it out; we see *millions* of parallel processors working in modern-day supercomputers.

Just as the 400 wealthiest people in the world are listed annually in *Forbes* magazine, the 500 most powerful computers are listed in November and June in a list called TOP500. The list was established in 1993 when a meaningful system of comparisons (benchmarks) was established. The highest performance in the TOP500 list has increased in line with Moore's Law, but with a shorter doubling time of 14 months. (For a long time, Moore's Law's doubling time was 18 months, and then it grew to 24 months.) All supercomputers are tested using a standard, challenging computational routine called the *High-Performance Linpack* (HPL) benchmark.

Supercomputers are scattered worldwide, but the overwhelming majority are still concentrated in the U.S. and China, with the former currently leading in the total number of systems. As of the latest TOP500 list, the U.S. has more than twice as many listed supercomputers as China. Lenovo, a Chinese company, remains one of the top global supercomputer vendors, but the market share of Chinese manufacturers has declined in recent years due to international sanctions and export restrictions.

According to the latest TOP500 list compiled in November 2025, the three most powerful supercomputers are located in the U.S.:

1. *El Capitan*: Located at Lawrence Livermore National Laboratory in California, El Capitan (Fig. 19) was built by Hewlett-Packard Enterprise (HPE) in collaboration with Advanced Micro Devices (AMD). It achieved the top position with an HPL benchmark score of 1.809 exaflops (exa- denotes 10^{18}). To express this number fully: 1,809,000,000,000,000,000 FLOPS. El Capitan contains 11,039,616 combined CPU and GPU cores. Its power consumption is also notable: at peak performance, it can reach 29.6 MW.

Fig. 19. El Capitan, currently the fastest supercomputer in the world, as certified by the HPL benchmarks. El Capitan has a footprint of approximately 6,800 square feet (630 square meters) and weighs as much as 35 school buses. (I apologize for using these unusual weight units; that is the only figure I could find on the internet.)

Source: Image from the DOE.

2. *Frontier*: Situated at Oak Ridge National Laboratory in Tennessee, Frontier was developed by HPE in collaboration with AMD. In November 2025, Frontier secured the second spot, with an HPL score of 1.353 exaflops. It has 9,066,176 cores.
3. *Aurora*: Housed at Argonne National Laboratory in Illinois, Aurora was developed by Intel in partnership with HPE. It holds the third position, with an HPL score of 1.012 exaflops. It features 9,264,128 cores.

It is notable — though not unprecedented — that the same three computers remained at the top of the TOP500 list for three consecutive listings since November 2024. While the rankings often change with each new list, periods of stability at the top have occurred before, especially during transitions between technological generations.

As was noted above, a common misconception is that supercomputers are built using entirely novel technologies. In reality, they utilize the same types of transistors and IC designs as regular computers, adopting new technologies as they become available. Supercomputers are often constructed by the same companies that manufacture standard computers; essentially, they are well-organized assemblies of conventional computers working seamlessly in parallel.

Beyond the TOP500 machines, a new class of supercomputers has emerged, specifically designed for deep learning and training large language models — key components in the development of AI. These dedicated AI supercomputers are not evaluated using the HPL benchmarks and therefore do not appear in the TOP500 list. Nonetheless, they are formidable machines.

One notable example is Colossus,[28] built in Memphis, Tennessee, by Elon Musk's company xAI. The Tennessee Valley Authority initially approved an arrangement to supply the facility with on the order of 150 MW of electrical power; subsequent expansion has pushed the required capacity into the several-hundred-megawatt range, with long-term plans pointing toward gigawatt-class infrastructure. For perspective, a large metropolitan area such as San Francisco has an average electrical load on the order of a gigawatt, placing the planned scale of Colossus within the range of a major city's power demand.

Despite not appearing on the TOP500 list, xAI's Colossus is among the most powerful AI supercomputing installations in the world. It began operation in 2024 with about 100,000 NVIDIA H100 GPUs and was expanded to roughly 200,000 accelerators in 2025. Rapid subsequent build-out has brought the total into the several-hundred-thousand range by late 2025–early 2026, with a mix of H100, H200, and newer Blackwell-generation (GB200) systems, and with publicly stated ambitions to scale to on the order of one million GPUs. At contemporary prices — typically tens of thousands of dollars per accelerator — this corresponds to a multibillion-dollar investment and represents an unprecedented scale for an AI computing cluster.

Comparing El Capitan and Colossus is not easy, as they are optimized for different tasks. Colossus, explicitly designed for AI workloads, may theoretically perform certain operations faster than El Capitan; however, direct comparisons are challenging due to the differing architectures and performance metrics.

[28] The reader may recall that the first programmable electronic computers, built in Great Britain during the war, operated in secrecy under the name "Colossus." Naming the latest AI supercomputer after them serves as a fitting tribute to Britain's wartime codebreakers and their pioneering contributions to computing.

The history of supercomputers

Now that we have described the latest supercomputers, a few words about their origins. In 1955, the Los Alamos Scientific Laboratory requested that IBM develop a computer that would be 100 times more powerful than any existing machine. There were delays, but after about six years, IBM delivered the result: the IBM 7030, better known by its nickname, "Stretch."

The first Stretch was delivered to Los Alamos in 1961, and seven others were shipped to significant technology centers across the U.S. and Europe. Stretch was among the earliest all-transistor computers, featuring no vacuum tubes, and it showcased several cutting-edge innovations. One of the most important was a novel "look-ahead" instruction execution scheme, which allowed the machine to anticipate the next instruction based on the results of the previous one (Williams, 1997, pp. 392–393). This mechanism resembled a primitive form of AI, enabling Stretch to "think ahead" during computation.

Its arithmetic unit was optimized for speed through parallel processing and included 450 separate arithmetic circuits (Williams, 1997, p. 393). Stretch also featured a then-fast 2-microsecond RAM — a remarkable achievement for its time.

In the end, Stretch did not quite reach its original target of being 100 times faster than existing systems. But it did achieve a tenfold performance improvement, making it one of the fastest computers of its era.[29] More importantly, its innovations had a lasting impact. The computing industry quickly adopted many of Stretch's architectural advances; in that sense, this first supercomputer helped shape the design of future generations.

To conclude this section, I'd like to share the story of one other historically significant supercomputer — one that I saw with my own eyes — and to say a few words about its creator, Seymour Cray (Fig. 20, left), often called "*the father of supercomputing.*"

I saw Cray-1 (Fig. 20, right) in 1979 or 1980 at the National Center for Atmospheric Research (NCAR) in Boulder, Colorado. A few months earlier, I had started working at Bell Labs on a project that *could have* resulted in the creation of an entirely new kind of supercomputer based on

---

[29] The pace of progress in electronics cannot be decreed. One might want to increase a performance parameter by a factor of 100, but what actually happens depends on what can be achieved, not on what is desired. In the case of Stretch, a tenfold improvement was possible; a hundredfold was not. Evolution proceeds as fast as it can — and not any faster.

Fig. 20. Seymour Cray (1925–1996) and Cray-1 which was installed at NCAR.

Sources: Left: image from the Minnesota Science&Technology Hall of Fame website. Right image by Bubba73, taken at the Computer Museum of America, Lic. CC BY-SA 3.0.

superconductors (it didn't). So, when our group — composed mainly of IBM and Bell Labs scientists — attended a summer workshop in Boulder, we were taken to see our "competition": the best supercomputer of the time, the Cray-1.

Seymour Cray began his career at a company called Control Data Corporation (CDC), where he and his colleague Jim Thornton created two of the fastest machines of their time: the CDC 6600 (released in 1964) and the CDC 7600 (1969). These machines redefined what people meant by a *supercomputer*. The CDC 6600, for example, was three times faster than the IBM Stretch. That fact did not sit well with IBM's president, Thomas Watson Jr., who famously wrote an internal memo to his engineers — now widely known as "*The Janitor Memo*":

> I understand that in the laboratory developing the 6600 system, there are only 34 people, including the janitor. I fail to understand why we have lost our industry leadership position by letting someone else offer the world's most powerful computer.

The CDC 6600 was also the first computer to use silicon transistors, marking a critical milestone: it was the moment silicon technology first flexed its muscles. Its successor, the CDC 7600, was about 10 times faster than the 6600 — and thus roughly 30 times faster than IBM's Stretch — achieving a peak performance of 36 megaFLOPS (36×10^6 floating-point operations per second). That was very impressive in 1969, but it is about 50 billion times slower than El Capitan, the fastest supercomputer today.

In 1972, Seymour Cray left the CDC and founded his own company, Cray Inc. (also known as Cray Research), in his hometown of Chippewa Falls, Wisconsin, where he was born and had lived his entire life. NCAR in Boulder became the first official customer of Cray Research. In 1977, they purchased Cray-1 serial number 003, paying almost nine million dollars for it, and that was the machine I saw in 1979.

The Cray-1 operated at a clock speed of 80 MHz, and it was one of the first computers to use integrated circuits — a step up from the CDC machines, which relied on individual silicon transistors. The ICs were manufactured by Fairchild Semiconductor and Motorola, and they contained about 200,000 logic gates.

The Cray-1 was also among the first machines where heat dissipation became a significant problem. To manage it, the system used forced circulation of Freon, the same cooling fluid used in refrigerators. That was the very first thing we were told during our visit to NCAR, and it made us proud: We were working on a project that aimed to reduce energy dissipation by several orders of magnitude.

I also vividly remember how a meteorologist from NCAR told us that, in principle, they already knew how to predict the weather accurately 10 days in advance. They could feed observational data into a set of differential equations and solve them to produce a forecast. The catch? *It took 10 days to do the computation.* So, no one was doing it — the results would arrive too late to be of any help. I often wondered afterward: Was that really true? That ten-day computational delay would shrink to nothing as computers grew more powerful. And indeed, today, ten-day forecasts are standard, but are they accurate?

Cray, Inc. remains a vital computer company. As mentioned earlier, in collaboration with AMD, Cray co-developed Frontier, which, as of November 2025, is the world's second-fastest supercomputer.

Data Centers and Cloud Computing

Predecessors of modern-day data centers were *computer rooms* with *time-shared* mainframe computers. Timesharing was proposed in 1959 in a memo at MIT by Professor John McCarthy, one of the pioneers of AI, who is best known for organizing the 1956 Dartmouth Workshop on Artificial Intelligence, which we will describe in Chapter 4. McCarthy, tired of the difficulty of obtaining computer time, proposed that a computer should be able to serve multiple users simultaneously by rapidly switching between them.

As a result of McCarthy's proposal, two years later, in 1961, *MIT's Compatible Time-Sharing System* (CTSS) was developed by Fernando Corbató and his team at MIT. The IBM 7090 mainframe could be accessed remotely via terminals. Soon afterward, timesharing was deemed helpful by the Advanced Research Projects Agency (ARPA), which funded the next stage of timesharing at MIT called *Project MAC*. Timesharing took root at MIT, Dartmouth, and other places; the era of on-demand remote computing has arrived.

Centralized computing continued to scale up, and over the past two decades, we have seen the rise of *hyperscale data centers*. A typical large hyperscale campus today spans several million square feet, but some announced AI campuses are planned to reach nearly 10–11 million square feet — effectively small industrial cities. One of China Telecom's complexes in Inner Mongolia is already in that range, at 10.7 million square feet.

Power use has also climbed: Where 50–100 MW used to define the high end, recent projects now discuss 200–300 MW campuses, and the very largest AI data centers are being designed to reach well above 1 GW (1,000 MW) once all phases are online.

Costs have skyrocketed accordingly: Whereas a hyperscale facility used to be a multi-billion-dollar project, the Meta AI data center in Louisiana — Hyperion — is now being discussed at up to $50 billion, making it the among the most expensive data center projects ever announced.

Inside such a center are rows (streets, really) lined with powerful computers and servers containing hundreds of thousands to millions of CPUs, GPUs, and TPUs. The memory capacity in a single center would be measured in petabytes (PB, 10^{15} bytes) for RAM and exabytes (EB, 10^{18} bytes) for permanent storage memory. For example, Google Cloud Storage provides over 10 exabytes of memory.[30] Cooling of all this power-hungry equipment requires millions of gallons of water daily.

By 2025, there were just under 1,200 hyperscale data centers worldwide, with hundreds more under construction. In 2024, data centers consumed approximately 415 TWh of electricity — roughly 1.5% of global demand — and this figure reached around 540 TWh in 2025,

---

[30]Ten exabytes (10 EB) is a truly massive amount of data. All the books ever written in human history could fit into 10 EB more than 20,000 times.

accounting for nearly 2% of the world's electricity use. In the U.S., where AI development is most intense, it accounts for approximately 4% of the country's electricity, and U.S. agencies warn that this load could double by 2030 if current trends continue.

Data centers' primary uses are *cloud computing* and *AI training*, which we will discuss in Chapter 4. Cloud computing is a hybrid of the internet and data centers: It is the on-demand delivery of computing resources over the internet, such as storage, data processing, and applications.

In discussing the latest supercomputers, data centers, and cloud computing, we jumped by a half-century from where we were in our story. Let us now return to the 1970s to discuss the rise of the *personal computers*.

Computers Get Personal

From the 1940s to the 1970s, computers were bulky machines — mainframes housed in large cabinets and packed into even larger rooms. They served the needs of government, the military, big business, and academia. These machines were so complex that they required teams of specialists to operate, program, and maintain them. As described in the previous section, since 1959, their use has been governed by carefully orchestrated time-sharing, a strategy developed to increase productivity.

Then, something changed — not because someone woke up one morning with a brilliant new idea but because technology had finally made it possible. PCs emerged when the semiconductor industry developed commercially available microprocessors and compact memory technologies: DRAM and ROM. Intel and IBM were pioneers in this transformation. In 1970, Intel introduced the *1103 DRAM*; in 1971, it released the *4004 microprocessor* (Fig. 18(c)). That same year, IBM launched the 8-inch floppy disk, which became an essential tool for loading and transferring computer programs. These breakthroughs opened the gates: Suddenly, it became possible to build your own computer.

The first microcomputers had to be assembled by the user, but soon factory-assembled PCs began to appear, transforming how people lived and worked. Anyone could become a computer hobbyist — even complete unknowns such as Steve Jobs and Steve Wozniak, who in 1975 were members of the Homebrew Computer Club. Before long, they had developed their own product, but more on that later.

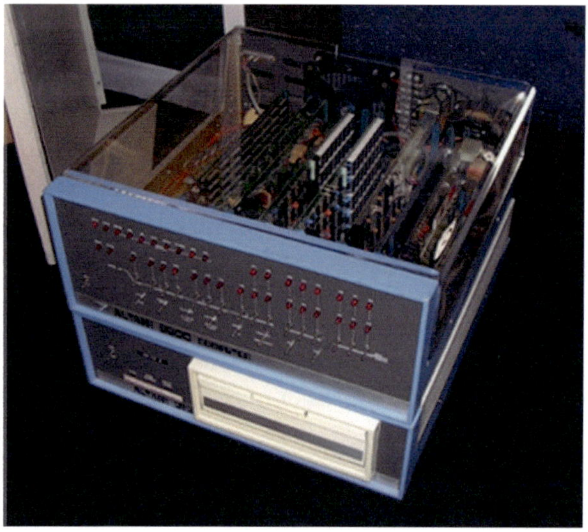

Fig. 21. Altair 8800 home computer with 8-inch floppy disk drive.
Source: Image by Michael Holley, in the public domain.

The descent from Olympus to the living room was made possible by the introduction of the *Altair 8800* kit (Fig. 21), the first widely available home computer, sold in 1974 for $397 (about $2,500 in today's money). It was offered by a small Albuquerque-based company, *Micro Instrumentation and Telemetry Systems, Inc.* (MITS), founded in 1969 by Ed Roberts and Forrest Mims. Roberts was the designer of the first home computer, and his name is rightly deserving of remembrance. Hobbyists flooded MITS with orders, and the Altair 8800 earned the cover story in *Popular Electronics* in January 1975.

Then, a series of small miracles followed in rapid succession. Paul Allen and Bill Gates read the magazine article and began writing software for Altair, eventually developing *Altair BASIC*. They even moved to Albuquerque to work for MITS, and in July 1975 they founded a small company called *Microsoft*. BASIC became Microsoft's founding product.

One year later, in 1976, a computer enthusiast from the Homebrew Computer Club, Steve Wozniak, single-handedly designed the Apple I. He created the hardware, circuit board layouts, and the OS. Unlike the Altair 8800, which required manual switches, the Apple I included a built-in keyboard and video display. Wozniak initially offered the design to his employer, Hewlett-Packard (HP), but was turned down — not once, but five times![31] Upon hearing this, Steve Jobs, his friend, suggested they start their own company to sell the machine. Wozniak sold his HP scientific calculator, and Jobs sold his Volkswagen van to fund the venture. In April 1976, along with Ronald Wayne, they formed *Apple Computer Company*.

The Apple I was unique for its time. It featured a single circuit board, a $25 MOS 6502 CPU,[32] 256 bytes of ROM, 4K or 8K of RAM, and a video controller capable of displaying 40 characters × 24 rows. Unlike other hobbyist computers, it included video display circuitry, allowing it to connect directly to a standard monitor or television. Approximately 200 units were produced and sold for $666.66 (roughly $3,700 in 2025 — not an inexpensive price).

Building on the success of the Apple I, Wozniak designed the Apple II (Fig. 22, in the middle), which was introduced in April 1977. This model featured several innovations, such as color graphics, a sleek plastic

Fig. 22. The first factory-made American personal computers released in 1977; left to right: Commodore PET 2001, Apple II, and TRS-80 Model I.

Source: Photo by Springsgrace, Lic. CC BY-SA 4.0.

[31] We have been informed that Bill Hewlett, then president and CEO of Hewlett-Packard, who passed away in 2001 at the age of 87, was sentenced to burn in Hell for that refusal. He tried to appeal, but his appeals were blocked by the deceased HP shareholders.

[32] This uniquely inexpensive yet effective and capable CPU was produced by MOS Technology, Inc., a semiconductor company founded in 1969. This processor, which was used in "iconic" computers Apple I and II, Commodore 64, Atari 2600, NES, and some others, was designed by Chuck Peddle and Bill Mensch.

casing that housed electronics, a keyboard, and a power supply, and, very significantly (Job's idea), eight expansion slots, allowing users to add peripherals and expand the system's capabilities. The computer came with the BASIC programming language in permanent memory, enabling users to write and run their own programs.

The first four PCs, entirely built by the manufacturer, appeared on the market in 1977; three were offered by American companies (Fig. 22) and one by a Japanese company. They were the Apple II (by Apple Computer), the TRS-80 (by Tandy–RadioShack Corporation), the PET 2001 (by Commodore International), and the M200 Smart Home Computer (by Sord Computer Corporation of Japan). They all had a slick design, integrated screens, diskette drives, and, in some models, an integral keyboard.

The mouse and the graphic user interface

Making PCs user-friendly was the key to winning consumers' hearts and opening their wallets. In that, the pointing device controlling a cursor — the *mouse* — and the *graphical user interface* (GUI) played a central role. Although they were invented earlier, they found their true calling only with the rise of PCs.

The initial concepts were proposed in 1963, long before the PC era: the mouse by Douglas Engelbart at the Stanford Research Institute in Menlo Park, California, and the prototype of a GUI, which was called *Sketchpad*, by Ivan Sutherland at MIT. Sketchpad allowed users to interact with graphical objects directly on a display using a *light pen*. The GUI, as we know it — where a mouse controls a cursor to select clickable icons on a desktop — was developed in the 1970s at Xerox's Palo Alto Research Center (PARC) by Alan Kay and David Canfield Smith.

In 1973, Xerox Palo Alto researchers developed the *Xerox Alto* computer. The Alto featured a GUI with windows, icons, menus, and a pointing device (mouse). However, it had a strange fate: Thousands of machines were manufactured but never sold commercially; instead, they were used inside various Xerox offices and at some universities. When Steve Jobs saw it at Xerox PARC in 1979, he immediately grasped the importance of the GUI for PCs and told his Xerox PARC hosts, "You're sitting on a goldmine; I can't believe Xerox is not taking advantage of this." Jobs certainly took advantage of the GUI when introducing the Macintosh line of Apple computers in 1984 (Fig. 23). When Jobs was later

Fig. 23. Steve Jobs in 2008, holding a MacBook Air laptop; Apple logo in the background.

Source: Photo by Matthew Yohe, Lic. CC BY 3.0.

accused of stealing Xerox's ideas, he quoted Pablo Picasso: "Good artists copy, great artists steal," adding about Xerox management, "They just grabbed defeat from the greatest victory in the computer industry. Xerox could have owned the entire computer industry" (Isaacson 2014, 363–364).

The success story of personal computers

In 1977, approximately 48,000 PCs were sold; by 1979, that number increased to around 580,000. In 2024, approximately 245 million traditional PCs (desktops and laptops) were shipped globally. Including tablets, the total number of units would be higher, though exact figures are not specified.

The tremendous success of the PC — and of the people who introduced it — is the stuff of legends. But note this: Steve Jobs is far more famous than his partner, Steve Wozniak, who actually designed the first

Apple computer. And both are far more famous than the almost-forgotten Ed Roberts, the man who created the first PC.

Steve Jobs was one of the great heroes of that era. He was neither a programmer like Bill Gates nor an electronics expert like his partner Steve Wozniak. Instead, his genius lay in dreaming up product concepts that were elegant, intuitive, and user-friendly. In 1976, Apple Computers began in Jobs's garage in Los Altos, California, and it went on to become one of the most successful companies in history — the world's largest information technology company by both revenue and market capitalization.

Jobs's partner, Steve Wozniak (known affectionately as *the Woz*), designed the Apple I, Apple II, and much of the Macintosh. Together, Jobs and Wozniak formed one of the most productive and celebrated collaborations in the history of technology.

In 2001, Apple introduced the iPod, a portable music player so irresistible that even I — someone who had previously stayed away from Apple products — bought one and enjoyed using it. Then, in 2007, Apple released the iPhone, and once again, I got hooked — as did millions of users around the world.

When PCs first appeared, no one had a clear idea of what they would be used for. Their early promoters gave it some thought and proposed three main applications:

1. Managing personal finances: balancing checkbooks, preparing taxes, and keeping household budgets.
2. Home automation: controlling lights, heating, blinds, and other systems at pre-set times.
3. Playing computer games.

They couldn't think of much else. Time would reveal which ideas had staying power. Home automation proved to be of limited interest, likely because it was harder to implement than promised. Doing taxes on a home computer also turned out to be relatively unimportant; H&R Block did not go out of business. But the third prediction was spot on: Playing computer games became a hugely popular pastime — even an obsession — for children and adults alike.

The early designers of home computer games amused themselves and, in the process, pushed the limits of what these machines could do. Some of the famous AI scientists we'll encounter in Chapters 4 and 5 began their journey by playing and programming games. For many, gaming was not just fun — it was a boot camp in programming, design, and systems thinking.

Computer games

PCs are merely one of several platforms for playing computer games. Games can also be played on mainframe computers, dedicated gaming consoles, and machines installed in public arcades. The first computer games date back to the early 1950s, in the era of big mainframes. The very first computer game was *tic-tac-toe*.

The first computer game to gain popularity was *Spacewar!*, which debuted in the early 1960s on the PDP-1, a computer developed by Digital Equipment Corporation (DEC). It was a two-player shootout between spaceships caught in a star's gravitational pull. The game required both a monitor and significant computational resources — particularly memory — and appeared as soon as the PDP-1 made those possible. It spread rapidly among students at MIT, becoming an early cultural touchstone for the nascent hacker and gaming communities.

By 1972, the first arcade games and home consoles were introduced. After 1977, PCs joined the ranks of gaming platforms. Soon, clones of mainframe and arcade classics appeared on PCs. But then something new happened: Users began recreating the games they loved, writing their own code to bring them to life on personal machines. As noted earlier, with a PC, anyone who wanted to become a programmer could — and many did. Writing games became a hands-on school for an entire generation of future software developers and AI scientists.

I had never thought I'd be interested in computer games until 1994, when my graduate students introduced me to *Doom*. It was a first-person shooter in which you fought your way through increasingly difficult "levels" filled with zombies, monsters, and devils. It was challenging, but you were well-armed and had plenty of ammunition. So, wandering through half-lit dungeons wasn't as scary as it could have been, though still intense.

I remember playing for hours. It was addictive. After a long session, I would walk down the corridor of our physics department and, upon seeing an "Exit" sign, instinctively press an imaginary "Enter" button in my palm — a deeply embedded reaction from the game, where "Exit" marked the much-anticipated end of a *level*.

Many otherwise rational people were similarly consumed by this obsession. I recall a Russian physicist who aspired to relocate to the U.S. When people told him it might be difficult, he responded confidently: "No problem for me — I passed 27 levels of Doom!"

Today's games are played on high-speed machines with high-resolution monitors, 3D graphics, and surround sound. I resolved long ago not to try modern games, fearing instant addiction. But I broke that resolution once, in the summer of 2017.

While visiting Moscow, I played a virtual reality game courtesy of Ivan Branitsky, the son of my good friend Svetlana Branitsky. Ivan, who was at the time working in marketing for a VR game company with teams in England and Russia, invited me to try one of their new games.

Two players play the game, but not against each other; instead, they support each other (and boy, did I need that support; thank you, Ivan!). Each player wears a helmet, electronic goggles, and several transmitters on their arms and legs. A player holds a weapon — a futuristic machine gun that can shoot various types of ammunition. The game is played in a room equipped with electronic gadgets capable of sensing the position of the player's body and the direction in which shots have been fired.

The headset delivered 360-degree visuals and spatial audio, dynamically adjusting as you moved. The world it created was disturbingly familiar: abandoned warehouses, crate-filled storerooms, eerie catwalks, and narrow, rail-less bridges — a darker, more immersive cousin of Doom. The inhabitants of this world were also familiar: zombies, monsters, and devils — now fully three-dimensional and, because of that, more terrifying. They jumped out from behind corners and charged at you while you desperately tried to reload and fire before they reached you (do not forget about reloading — very important!).

People susceptible to vertigo should not play the game; at times, it was positively uncomfortable. No amount of reasoning convinces your vestibular apparatus that you are still on the flat concrete floor of a large, empty room and not on a swaying plank over a bottomless abyss.

All in all, it was an exciting, terrifying, and unforgettable experience. Virtual reality has a big future, for better or worse.

Other uses of personal computers

Since their introduction in 1977, personal computers have steadily expanded their repertoire of applications. In addition to the uses discussed earlier, PCs are now widely employed for the following:

- *Word processing*: creating, editing, and formatting documents.
- *Spreadsheets*: managing data and performing calculations.
- *Presentation software*: designing and displaying visual aids for meetings and lectures.
- *Web browsing*: accessing online information and services.
- *Email*: communicating quickly and globally.
- *Video conferencing*: holding remote meetings and calls.
- *Social media*: sharing content and staying connected.
- *Multimedia playback*: listening to music, watching videos, and streaming content.
- *Digital art and design*: creating graphics, animations, and visual effects.
- *E-learning*: participating in online courses and accessing educational content, including the priceless Wikipedia.
- *Programming and software development.*
- *Scheduling and planning*: managing calendars and organizing tasks.
- *Network management*: setting up and overseeing home or small business networks.
- (And the latest) *accessing chatbots* powered by large language models.

And I've probably missed a few. Among these, web browsing, email, word processing, and access to chatbots are among the most universally used.

Word processing

Word processing predates the PC. In 1964, IBM combined its Selectric typewriters with magnetic tape storage, producing a system known as the Magnetic Tape/Selectric Typewriter (MT/ST). This allowed users to automatically retype and edit stored text — a novel idea at the time. In marketing this product, IBM coined the term "word processing."

The next leap came in 1972 when companies introduced display monitors, allowing users to see and edit text on-screen, thereby eliminating the need to retype an entire page to correct a typo. By the mid-1970s, magnetic tape had been replaced by floppy disks, which could store approximately 100 pages of text — a significant improvement over tape's one or two pages.

Dedicated word-processing computers began to appear around 1975. These were large, expensive machines with limited editing capabilities — good enough for offices but still far from today's standards. All the essential components were in place when PCs hit the market in 1977. However, it took nearly another decade for software to catch up: Programs with flexible editing, text formatting, and spell-checking capabilities became available only after computers had sufficient memory to support them.

The success of word processing stemmed from the convergence of hardware that was powerful enough and software that was practical and user-friendly. Two of the most popular programs in the field were WordPerfect and Microsoft Word. WordPerfect (originally developed by Satellite Software International in 1979) dominated the market through the 1980s before being acquired by Novell in 1994 and then by Corel in 1996. Microsoft Word was first released in 1983 for Xenix and MS-DOS, with its first Windows edition in 1989, gaining real traction after Windows 3.0 in 1990; since 1992, it has been bundled in Microsoft Office. As of early 2026, consumer Microsoft 365 subscriptions are approaching 95 million worldwide.

Bill Gates

This is a good moment to pay tribute to another prominent figure of the personal computer era: Bill Gates (Fig. 24), co-founder and long-time chief technology officer of Microsoft Corporation.

Gates is a programming prodigy. He wrote his first computer program at the age of 13 in 1968. After graduating from high school in 1973 with an SAT score of 1590 out of 1600, Gates entered Harvard University. While at Harvard, he demonstrated his mathematical talent by solving an unsolved combinatorial problem, known as the *pancake sorting problem*. It is exceedingly rare for a student to solve a problem that had previously defeated professional mathematicians[33] even more so to produce the fastest known algorithm for it when implemented on a computer.

Despite this, Gates dropped out of Harvard (technically, he was on leave but never returned to complete his studies). In 1975, he and his

[33]The reader may recall that in the 1997 film *Good Will Hunting*, the protagonist, played by Matt Damon, solves an unsolved combinatorial problem while working as a janitor at MIT (Harvard's close neighbor). I've always suspected that this part of the plot was at least loosely inspired by Bill Gates's story.

Fig. 24. Bill Gates in 2013.

Source: Image in the public domain.

childhood friend Paul Allen founded Microsoft, driven by the opportunity to create a programming language — *Altair BASIC* — for the first PC, the Altair 8800, about which we wrote earlier.

In Microsoft's first five years, Gates either wrote, reviewed, or rewrote every line of code produced by the company. His technical and managerial involvement was total.

In 1980, as IBM prepared to launch its IBM PC, it approached Microsoft to supply an OS. Microsoft responded with what became MS-DOS, securing the IBM PC's success and its own future dominance.

In 1985, Microsoft released the first version of its Windows OS. After dozens of iterations, Windows is used by billions of people worldwide.

In the 1990s, Microsoft and Gates were accused of creating a business environment that ensured Windows OS would dominate the PC market. The U.S. Department of Justice brought antitrust litigation against Microsoft in 1998–2000, threatening to break up the company. Although Microsoft's appeal softened the final ruling, it was clear that the company had not been entirely innocent: It had actively suppressed competition in the 1980s and 1990s. Then again, one might ask: Who among those with the power to do so didn't?

In 2006, Gates formally stepped away from his role at Microsoft to focus on philanthropy. He and Melinda French Gates ran what is now called the Gates Foundation, the world's largest private philanthropic foundation, with assets of roughly $85–90 billion as of 2026.

Gates was the world's richest person for a record 18 years — from 1995 to 2007, again in 2009, and from 2014 to 2017.

He is widely respected for his leadership in the personal computing revolution, his professional and administrative work at Microsoft, his global philanthropic efforts, and his public engagement. He was among the few to publicly warn, years in advance, about the high likelihood of a future global pandemic. In 2021, he published a book about fighting climate change. We will return to Gates's views on artificial intelligence in Chapter 5.

Since I know Russian, I often encounter Russian-language posts and news links on social media. Russian journalists and internet trolls seem obsessed with attacking Gates — he is one of their favorite targets. They accuse him of an array of outlandish crimes, ranging from plotting to implant chips in people's brains to profiting from vaccines. These ridiculous accusations are a badge of honor for Bill Gates. He is in good company — the same people are also attacking George Soros, who spent a significant amount of his own money helping post-Soviet science when it was in deep trouble in the 1990s. Soros, like Gates, was also accused of profiting from his philanthropy. As the saying goes, "No good deed goes unpunished."

Computers "Holding Hands" Across the World: The Great and Powerful Internet

Earlier, we discussed the unforeseen rise of personal computers. Before the 1970s, nearly all predictions about the future of computing assumed that large, powerful mainframes would remain center stage, growing ever larger and more powerful. Then, small PCs appeared and completely overturned that vision.

A similar surprise occurred in the 1990s, though its origins stretch back thirty years earlier. This next unexpected and transformative development was *the internet*.

As early as the 1960s and increasingly through the 1970s, 1980s, and 1990s, computers began to be connected into networks, first in the U.S. and later worldwide.

No one predicted that the sheer *quantity* of interconnected computers would create a wholly new *quality* in the system, but that is what happened. This sudden leap from quantity to quality was famously described as one of the laws of dialectics by Georg Wilhelm Friedrich Hegel in the early 19th century. In the case of the internet, that pivotal quantity was the number of users.

By 1973, ARPANET — the U.S. government-funded precursor to the internet — had roughly 2,000 users, mostly researchers and academics. Twenty years later, by 1990, the global user base had grown to an estimated 2.6 million. The 1990s then saw an explosion in internet adoption, spurred by the World Wide Web and the appearance of user-friendly web browsers.

By the year 2000, the number of internet users had surged to approximately 413 million. By early 2026, about 5.7–5.8 billion people — roughly 70–72% of the world's population — were using the internet (Fig. 25).

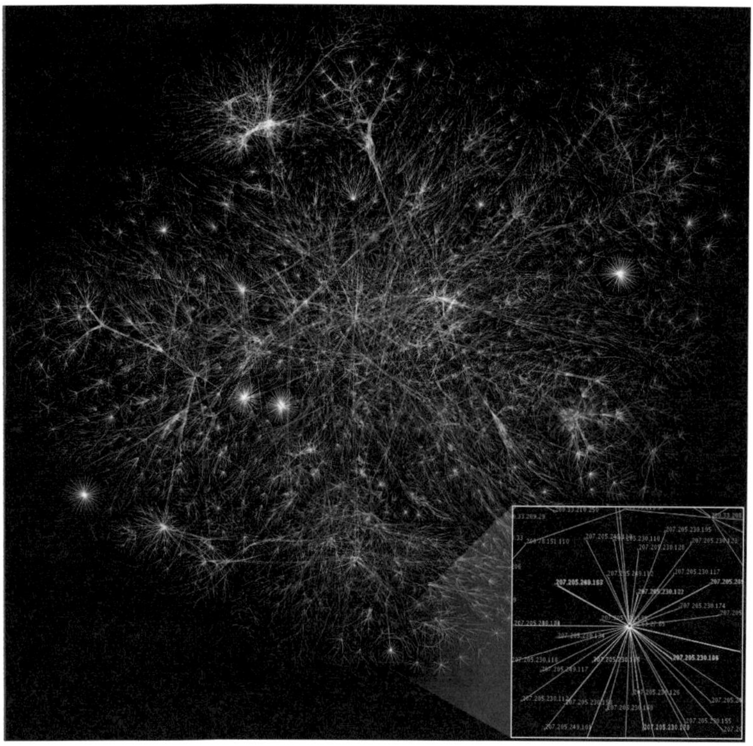

Fig. 25. Visualization from the Opte Project of the various routes through a portion of the internet in 2005. This "map" of the internet is supposed to depict its complexity. I cannot say I learned something from this lovely picture, but what do you do to illustrate the internet?

Source: Image from British Wikipedia, Opte Project, Lic. CC BY 2.5.

Hegel was right: Connecting two, ten, or even ten thousand computers didn't change the world. But billions of interconnections, forming a living, evolving global nervous system, have. The internet became an Earth-shaking phenomenon.

Here's what Wikipedia tells us about its early development:

In the 1960s, the *Advanced Research Projects Agency* (ARPA) of the U.S. Department of Defense funded research into computer time-sharing. Around the same time, the concept of *packet switching* — a foundational internet technology — was independently developed by Paul Baran in the U.S. and Donald Davies in the U.K.

After the 1967 *Symposium on Operating Systems Principles*, packet switching from the British *National Physical Laboratory* network was incorporated into several resource-sharing networks.

Computer networking expanded during the 1970s. In 1982, the introduction of the *Transmission Control Protocol* (TCP) and *Internet Protocol* (IP) — collectively known as TCP/IP — provided a standardized method for transmitting data across networks.

Still, the internet as we know it today did not truly emerge until 1995, when restrictions on its commercial use were lifted. And perhaps that's the moment when the Hegelian threshold was crossed — when the accumulated *quantity* tipped into a new *quality*.

The Internet of Things

While people were connecting their computers and forming networks, inanimate objects looked on with envy. By the early 1980s, they decided to take action. Thus began the system we now call the *Internet of Things* (IoT).

The IoT refers to a network of physical objects, such as devices, vehicles, and appliances, embedded with sensors, software, and connectivity features that allow them to collect and exchange data. These "smart" objects can communicate not only with centralized systems but also with each other, enabling automation, monitoring, and control — often without the need for direct human intervention.

It all started with Coca-Cola. In 1982, researchers at Carnegie Mellon University connected a campus Coca-Cola vending machine to the internet. The machine could report inventory status and beverage temperature, making it the world's first internet-connected appliance.

The conceptual foundations for the IoT were laid in 1991 when American computer scientist Mark Weiser introduced the idea of ubiquitous computing — a vision in which computing devices would be embedded invisibly into everyday objects and environments. In other words, he foresaw the coming of the IoT.

The term "Internet of Things" was formally coined in 1999 by Kevin Ashton, who then worked at Procter & Gamble. Ashton proposed using radio-frequency identification (RFID) tags to track products through the supply chain, an idea that eventually became a cornerstone of modern IoT applications.

Radio-frequency identification is a technology that uses radio waves to automatically identify and track objects equipped with small tags containing electronically stored information. It is especially valuable in supply chain management, where real-time tracking and accurate inventory management are crucial.

RFID systems consist of tags attached to products or packaging. These tags store unique identifiers and may also include additional data about the items. One of RFID's key advantages is that it does not require a direct line of sight between the reader and the tag — unlike *barcodes* — because it uses long-wavelength radio waves that can penetrate packaging and other materials.

The information collected from RFID tags is processed in real time, enabling businesses to monitor inventory levels, product locations, and movement histories with unprecedented precision. As a result, the use of RFID has led to significant improvements in inventory accuracy and overall logistics efficiency.

Internet organization

Let us define a few basic terms that we often hear and use, yet may not fully understand — myself included.

A *network* connects two or more computers, enabling them to communicate and share resources. These connections can be established through various means, including cables, telephone lines, electromagnetic waves (popularly known as Wi-Fi), satellites (which transmit electromagnetic waves), or infrared light beams (which are also electromagnetic waves of specific frequencies). Networks can vary in scale and complexity, ranging from simple *local area networks* (LANs) that connect devices within a single location to *wide area networks* (WANs) that span large geographic areas, linking multiple LANs.

The internet is the sum of all interconnected networks worldwide, a vast "network of networks" consisting of linked private, public, academic, business, and government networks. These networks communicate using the standard internet protocol suite (TCP/IP), which we discussed above.

The infrastructure supporting the internet comprises various hardware components for data transmission, storage, and security:

- *Servers* are computers that store and process data and applications over a network. There are various types of servers, including web servers, database servers, mail servers, and others, each dedicated to specific tasks.
- *Routers* direct data packets between networks, determining optimal paths for data transmission.
- *Switches*[34] connect devices within a single network, facilitating internal data communication.
- *Modems* modulate and demodulate signals for data transmission over various media, including telephone lines and cable systems.
- *Firewalls* monitor and control incoming and outgoing network traffic based on predetermined security rules, establishing a barrier between trusted and untrusted networks.
- *Network interface cards (NICs)* connect computers to a network, allowing them to communicate with each other.
- *Data centers* (of which we have already spoken) house groups of servers and storage (memory) systems, providing centralized data processing, storage, and dissemination.
- *Transmission* is achieved through wired (physically connected) or wireless (electromagnetically connected) media.

[34]Not to be confused with the simple electric or electromechanical switches we discussed in Chapter 1. In the context of the internet and networking, a *switch* is a hardware device that connects multiple devices within a local area network (LAN) and efficiently manages data traffic by forwarding data only to the specific device that needs it.

The internet's many roles

Priceless gifts poured from the internet as from the *Horn of Plenty*: email, search engines, file sharing, cloud computing, computer telephony, Wikipedia, online education, online shopping, social media, and more keep flowing, as if from a bottomless source. Guessing the next big internet trend is difficult, but history shows that those who do often become billionaires.

The internet has brought new opportunities and new challenges to society. It is hard to imagine life without email and Wikipedia. On the other hand, no one cheers when encountering malware, viruses, cyberattacks, or some of the less desirable aspects of social systems (such as when they do not put "likes" under my Facebook posts). I will not discuss any of that here; it is a different subject, as these issues involve a combination of technical, social, and psychological factors.

In his 1997 book, George Dyson suggested that the internet is evolving into a form of *global intelligence* (Dyson, 1997). He says that the internet is not merely a tool for human use but is evolving into a self-organizing, intelligent entity. This view has some appeal. We are propagating a similar view, except that we are considering all of the electronics (what I call *meta-electronics*) rather than just its part (the internet) as the object of our analysis, and that we are trying to find specific parallels between the developing electronics and Darwinian evolution (Gurvitch 2023, Vol. 2, Chapters 5–7) rather than do poetic justice to their similarity. Now that we see real AI emerging, we wouldn't attribute it to the massive networking of the internet but instead to advances in neural networks, deep learning, and large language models.

In all its forms, the computer stands as the most significant electronic device. The interconnection of millions of computers through the internet has given rise to a dynamic electronic ecosystem. As noted earlier, computers are the foundation of AI development and are integral to the themes explored in this book. Therefore, dedicating the initial chapters to the history and technology of computers is both appropriate and essential. I hope this story did not seem boring to anyone.

Chapter 3
Quantum Computers

So far, we have discussed *classical computing* (CC). The foundational ideas and concepts of CC can be traced back centuries and even millennia. The history of CC presented on these pages can be seen as a long ladder toward deep learning and AI training. Leaving all topics related to AI for Chapters 4 and 5, I would like to tell you about an entirely different approach to computing called *quantum computing* (QC). The machines that are supposed to carry out quantum computations are called *quantum computers* (QCs).

I am talking about them in this book for several reasons:

1. The attempts to develop these techniques are in the public eye, and we often hear that QCs will be of utmost importance for the future of computing in general and AI in particular. We will see in the following that if fully developed QCs become a reality, they may solve *some* problems inaccessible to our most potent supercomputers. However, the class of problems in which QCs will have a significant advantage is relatively narrow. In particular, the training and implementation of AI using QCs seems unlikely. Due to the frequent claims to the contrary, I felt obliged to make that point. We will discuss the kinds of problems where QCs will be very helpful, and we will see that they are interesting enough without the need to make unrealistic broader claims.
2. Like all physicists, I greatly respect and admire Richard Feynman, who formulated the ideology of quantum computing in 1982. When

I become disillusioned with the slow progress in quantum computing, I remind myself that Feynman thought it should be possible.
3. In my professional life, I was involved with superconductivity, quantum tunneling, and Josephson junctions, which form the basis for one of the best QC hardware implementations. It is hard for me to resist discussing these fascinating subjects.
4. Whether or not QCs will take their place in the future history of computers (and some physicists consider that unlikely), they still have that chance, and I think avoiding them altogether in a book about computers would be wrong. Besides, writing about quantum mechanics and quantum computing is darn interesting.

In quantum computing's 40-year history, the barriers to building practical machines have often seemed insurmountable. Yet, the researchers in the field have not lost their enthusiasm — or at least, if they have, they haven't shown it. If I read the signs correctly, the prospects for quantum computing have brightened over the past couple of years, largely due to advances in quantum error correction (see the following). With improved fault tolerance, we may finally begin to see tangible results from this decades-old technology.

Since quantum mechanics and quantum computing are, at least for now, outside of mainstream AI, this section will be brief. If you want to learn more about these fascinating subjects, you may consult my previous book, where you will find a considerably longer dedicated chapter (Gurvitch 2023, Vol. 1, Chapter 7). And, of course, there are many other books about quantum mechanics and quantum computing at all levels. There are also excellent presentations on YouTube and elsewhere in which the best experts explain these subjects. I would especially recommend public lectures by Brian Greene and his World Science Festival guests.

The processor in a quantum computer is supposed to act according to the rules of quantum mechanics, which are different from the classical rules. The logical *truth values* with which a quantum computer operates when making a calculation or a logical decision are also different from the binary classical bits. These quantum relatives of classical bits are called *qubits*. And they are *rich* relatives. Mathematically speaking, they are *infinitely* wealthier than bits.

Bits and Qubits

Let us reiterate: In 1948, Claude Shannon introduced an elementary unit of information, a *bit*, which stands for "binary digit." A bit is the amount of information gained when a question has been asked about a situation that logically allows only two outcomes, and a yes-or-no answer has been received.

For example, suppose a quarter-dollar coin is lying flat on a tabletop. It may be lying with "head" (George Washington) up or with "tail" (the other side, US Eagle) up. We may ask, "Is it a head?"[1] An answer can be only "yes" or "no." Once we heard the answer, for example, "no," we gained one bit of information. That answer — one bit — may be represented as a binary "1" or "0"; for example, "yes" may be defined as "1" and "no" as "0." Simple enough. All classical computations are based on these "ones" and "zeros" or "true" and "false" values.

Consider now a coin toss that has preceded that definite outcome. Before ending on the table, the coin had been tossed in the air, where it flew up and then down under gravity, rotating and wobbling wildly. Asking "Is it a head?" while the coin is in flight would be meaningless: The outcome has not materialized yet, and no one can tell what it will be. All we can say is that there are equal probabilities, ½ - ½, that the coin will have heads up or tails up after landing on a table.

Let us imagine that we have a high-speed camera that can take instantaneous photos of our coin. If we take such a photo while the coin flies and find that its flat parts are parallel to the tabletop with George Washington looking up and the eagle looking down, we will call such a state "1" (i.e., we will call it the same way as when the coin lies flat on the table with Washington up). The opposite position, with the eagle looking up, is "0". But of course, while it is in flight, there will be many more photos (in principle, infinitely more) in which we will find the coin neither in a state "1" nor "0" but in some *superposition state*,[2] with the coin's

[1]An equally meaningful question would be, "Is it a tail?" Since we need a question to which an answer is either "yes" or "no," the question "What is it, head or tail?" would not do here.

[2]*Superposition of states* is one of the central concepts of quantum mechanics. Here, we are trying to give it a plausible classical analogy. We realize that it will be only a partial, imperfect analogy.

plane at some angle to the tabletop. We might say that the flying coin is "pregnant" with "1" or "0" outcomes.

This classical analogy illuminates the difference between a bit and a qubit. The flying, wildly rotating, and wobbling coin is our analogy of a quantum system; we cannot say what state it is in, and in some loose sense, we may say that it is in all possible states simultaneously. Once the coin lands on a table — once a measurement has been made, or once our coin interacts with a large (macroscopic) object, the table — the outcome can be described with a classical bit. But not when the coin is in flight. Thus, a qubit is much more complicated than a bit. It is a *superposition state* with two yet unrealized outcomes. It can be manipulated as long as we do not disturb it. Once disturbed for whatever reason, the superposition state collapses into one of the two possibilities, which can be described with a classical bit. In our flying coin analogy, placing a *fair* bet assumes that nobody interferes with a tossed coin while it is in flight; if someone does in any way, all bets are off.

Let me try to broaden the analogy for a *measurement.* A measurement does something dramatic to the *wavefunction*[3] — collapsing it, forcing certainty on the previously uncertain situation. So far, our analogy of a measurement consisted of landing on a table. But what if we capture the coin in mid-flight and call it a measurement, too? I can place the palm of my hand under the coin and capture it, or I might take a flat plank and insert it horizontally under the flying coin. Then, the coin will land on my palm or the plank with either Washington up or eagle up. Thus, our measurement forces a definite outcome whenever it is made, like in quantum mechanics.

Is the flying coin a good analogy for a qubit? Yes, as far as classical analogies go. But why is it only an analogy? Why is the flying coin not identical to the qubit? In quantum mechanics, the outcome of a superposition state is impossible to predict, whereas in the classical scenario, it is merely hard to predict. If we knew the precise conditions of tossing a coin,

[3]The *wavefunction* is a fundamental concept of quantum mechanics. It is a mathematical construction existing not in real space but in the space of complex numbers. It is connected to the real world by a mathematical procedure that predicts probabilities of finding the physical parameters of a quantum object when they are measured. Erwin Schrödinger introduced wavefunctions in 1926.

and its precise shape, and the properties of the surrounding air, and the exact shape and texture of the tabletop, and plugged all that information into a supercomputer that "knew" the laws of mechanics and aerodynamics, then it could predict the outcome, head or tail.

Another significant difference is that the quantum superposition state is much more delicate than the classical coin-in-flight state. In a quantum system, we could not take a photo of a system and determine what is going on mid-flight; that would immediately destroy the system's *coherence*,[4] while the flying coin can be photographed with impunity; its flight would not change because of these photographs.

In quantum computations, things get interesting when there are at least several *entangled* qubits.[5] They may not be entangled from their "birth;" it may be necessary to make them interact with each other to create the entanglement. Entanglement guarantees that N qubits will encode 2^N superposition states. 2^N is an *exponential* function; it grows faster and faster with increasing *argument* N. For $N = 2$, the number of states is 4; for $N = 10$, the number of states is approximately 1,000; for $N = 30$, it is close to 1 billion. At 50 qubits, $2^{50} \approx 10^{15}$ (1 with 15 zeros, *quadrillion*, or thousand trillion) bits. At that level, a quantum computer becomes capable of performing calculations that no classical computer can match. Anticipating such spectacular results (which so far have not materialized), QC practitioners named them *quantum supremacy*. It is widely believed that the key to quantum supremacy lies in the phenomenon of entanglement.

Pros and Cons of QCs From a Physicist's Perspective

Adding one extra bit to a large number of bits for a classical computer means nothing; you will need to double the "size" of a classical computer (processor, memory) to double its computational power. But if you have a QC with N qubits, adding one more entangled qubit will double its

[4] *Coherence* is another quantum concept that essentially indicates that the system under consideration is undisturbed while it is doing its quantum "magic."

[5] *Entanglement* is yet another quantum term indicating that objects have been interacting with each other, putting them under the spell of quantum mechanics. Entangled objects are described by a common wavefunction.

computational power, $2^{N+1} = 2 \times 2^N$. As we noted earlier, the exponential nature of the function 2^N makes it enormously attractive because even for modest values of N, such as $N = 50$, 2^{50} is a large number, approximately equal to 10^{15}, or a thousand trillion. Ah, if we could only make a computer capable of accessing so many states.... This prospect was probably the single most persuasive argument in favor of developing QCs.

However, one must be cautious. Here is what Professor Michel I. Dyakonov, a world-renowned expert in spin physics and one of the harshest critics of quantum computing, says about it, taking the example of an *electron spin*, which forms one of the simplest qubits. I am keeping Dyakonov's italics and bold fonts in the following quote. I also keep his depiction of spin states as arrows pointing up and down instead of our 1s and 0s. I apologize for including this text in the book, where we have so far avoided more specialized content. Whether or not you know the physics language Professor Dyakonov is using, I think you will capture the seriousness of his fundamental objections:

> The *arbitrary* spin state is described by the *wave function:*
>
> $$\psi = a\uparrow + b\downarrow.$$
>
> The quantum amplitudes a and b depend on ***our*** choice of the basic states $\uparrow$ and $\downarrow$, and
>
> $$|a|^2 + |b|^2 = 1.$$[6]
>
> In contrast to the **classical bit**, that can be only in ***one*** of the two states, $\uparrow$ **or** $\downarrow$, the qubit can be in a ***continuum*** of states defined by quantum amplitudes a, b. **The qubit is a continuous object!**[7]
>
> Note that being *continuous variables*, a and b can practically never have exact prescribed values.... *Quantum computer* has a **continuum** of states described by 2^N continuous quantum amplitudes....
>
> **Errors in the state of a quantum system do not only consist in the fact that some qubits, which should be in their "up" state, are in their "down" state (or vice versa). Rather, the general errors consist in the fact that the 2^N quantum amplitudes $a_1, a_2, \ldots, a_2{}^N$, describing**

[6]This is the *normalization* condition — it guarantees that the probability of finding an electron in *some* spin state is unity.

[7]Recall our flying coin analogy: When in flight, it passes through a continuum of angles.

> **the general state of N qubits, do not have the prescribed values (e.g., $a_1 = 1$, $a_2 = a_3 = \ldots a_8 = 0$).**
>
> Error correction cannot consist simply in rectifying the state of N independent qubits. The prescribed values should be imposed on 2^N quantum amplitudes describing the whole system. Obviously, for large enough values of N this will never be practically possible (Dyakonov, 2024).

If I were a braver person, I would stop right here; I would say, "Okay, if Professor Dyakonov is correct, then attempts to develop practical QCs are doomed from the start; let's move on to the next chapter." But I am not *that* brave, and so I will continue.

Let's continue exploring the perceived advantages of quantum computing. In classical computers, as discussed in Chapter 2, computation speed is limited by the signal propagation time between components. However, in a quantum computer with entangled qubits, spatial separation does not impose such a constraint — entangled correlations are not bound by the speed of light.[8]

Finally, quantum computers consume very little energy. If they became widespread and applicable to a broad range of computational tasks, they could help mitigate the growing energy demands of information technology.

However, the near-total omission of energy considerations in quantum computing theory is a concern for physicists. Here is what Professor Dyakonov says about it:

> The notion of **ENERGY**, of primordial importance in physics, both classical and quantum, plays absolutely no role in QC theory. Because of this, the QC theory misses the fundamental fact of Quantum Mechanics: if the energies of some quantum states entering in a superposition differ by ΔE, there will be a **free evolution** of the system (of the wavefunction describing our system) with frequency $\Delta E/\hbar$ (where $\hbar$ is the Planck constant).

---

[8]This fact was extensively discussed in the 1930s; Einstein thought it was impossible and criticized quantum mechanics in which such seemingly instantaneous correlations appeared, calling them "spooky action at a distance." Eventually, Einstein lost that argument.

To repeat: The existing QC theory ignores the fundamental fact of Quantum Mechanics: ***If the energies of some quantum states (entering in a superposition) differ by ΔE, there will be a free oscillation of the system with frequency $\Delta E/\hbar$, where $\hbar$ is the Planck constant*** (Dyakonov, 2024).

How and Why the Interest in QCs Arose in the 1990s

As we mentioned, the field of quantum computing was initiated in the 1980s, particularly after the 1982 Feynman paper outlined the advantages of modeling quantum systems with quantum processors. Initially, in the 1980s–early 1990s, the field developed slowly, partly because it presented severe technical challenges and partly because the promise of superior modeling of atomic and molecular systems did not sufficiently excite government funding agencies and private companies. Additionally, it did not help that this goal was stated in the total absence of devices capable of quantum calculations; even their parts, such as qubits or quantum gates, did not exist anywhere except on paper. So, for some time, that theoretical goal provided a weak magnet to all but a handful of curious scientists.[9]

However, interest increased gradually. The boost to the QC field came when scientists identified significant mathematical problems that resisted classical computations but could be solved on a quantum computer.

Feynman, in his 1982 paper, in addition to rooting for quantum modeling of molecular systems, mentioned that there might be mathematical problems that require exceedingly lengthy classical computations, with their solution being *exponentially* faster on a QC. It took a decade to identify the first such problem: In 1992, David Deutsch and Richard Jozsa found a mathematical problem that was impossible to solve classically but could be solved using QC (Dowling, 2013, pp. 102–106). That was proof of principle and confirmation of Feynman's intuition, but the problem identified by Deutsch and Jozsa did not excite anybody. The first exciting problem of this kind was discovered two years later, in 1994, by Peter Shor, a mathematician at Bell Labs.

---

[9]Richard Feynman was one of them. In the 1980s, he consulted for the Boston-based *Thinking Machines Corporation* and delivered several talks about quantum computations. I was in the audience when he spoke at Bell Labs in 1986.

Shor's problem is well known in mathematics; it is a problem of finding prime-number multipliers for an arbitrary integer, known as a *factorization* problem:

Any integer (positive whole number) can be uniquely represented as a product of several prime numbers. For example, 15 = 3 × 5. Here, 3 and 5 are prime numbers, i.e., whole numbers that can not be further divided into more whole numbers. Prime numbers are central to the mathematical discipline called *Number Theory*. They are indivisible entities from which all the other numbers can be constructed by multiplication, similar to the ideal indivisible atoms of Democritus from which all material things are built (we say "ideal" because the real atoms turned out to be divisible).

Before Shor's work, no general solution existed for the factorization problem. Mathematicians lacked a practical algorithm for efficiently finding prime factors. While some algorithms improved upon brute-force *trial and error*, they did not accelerate the process sufficiently.

Factorization is an example of the so-called "one-way" mathematical problem, in which a solution is exceedingly difficult, but checking the validity of that solution is simple. It is challenging, almost impossible, even with the help of the most powerful computers, to find prime multipliers for huge integers (by huge, I mean integers consisting of several hundred digits). However, if we know the prime multipliers, we can quickly check that their multiplication will reproduce the original colossal number.

When checking prime multipliers for numbers consisting of hundreds of digits, the verification is more involved than performing a straightforward multiplication on a computer. Modern computers typically operate with 64-bit words, so inputting a hundred-digit multiplier directly into them is impossible. Instead, people use *arbitrary-precision arithmetic*. Libraries like GMP (GNU[10] Multiple Precision Arithmetic Library) allow computers

[10] *GNU* stands for *"GNU's Not Unix."* It is a recursive acronym, a playful nod to its roots and its difference from Unix systems. The GNU Project, initiated by Richard Stallman in 1983, was created to develop a completely free and open-source operating system that would be compatible with Unix but composed entirely of free software.

to work with numbers consisting of thousands of digits by representing them as arrays of smaller chunks (e.g., 64-bit words).

Shor proved that factorization can be performed exponentially faster on a quantum computer than on a classical one, and he proposed a specific algorithm to achieve this goal.[11]

So, you might ask, why would the theoretical, distant-future ability to factorize large numbers on non-existent quantum computers excite people in 1994? Were they so interested in number theory? Don't be silly; of course not. The reason for the excitement was that the factorization of huge integers is the key to the universally accepted *public-key RSA encryption*, where the initials stand for the creators of that scheme, Ronald Rivest, Adi Shamir, and Leonard Adleman, who published their algorithm in 1977. All the valuable secret communications — from financial transactions to military secrets — are protected by the scheme in which anyone might know a giant over 600 digits integer — this integer called the *public key* — but only those who are supposed to read the encrypted messages know the two prime numbers, each approximately 300 digits long, which, when multiplied, will reproduce that integer. These two prime numbers are referred to as the *private key* and are a closely guarded secret.

We are discussing two prime multipliers because, for encryption, such cases are preferred over those with multiple (more than two) multipliers. Integers that have two prime factors are called *semiprimes*. There are online programs that factorize integers, such as *Calculator.net/prime-factorization*. In that particular program, the maximum number of digits is 14. I tried several 14-digit numbers and, to my surprise and delight, on a third or fourth try, found a semiprime; let me present it here:

$$98979897969893 = 72493 \times 1365371801.$$

Checking the multipliers written above with the same program, I convinced myself they were indeed primes. As can be seen, the number of

[11]That proof consisted of a combination of a new theorem in number theory and a new algorithm intended for quantum computers. In the words of Brian Greene, it was a "tour de force." Shor received several prestigious awards for his work, including the Gödel Prize in 1999 and the Dirac Medal in 2017.

digits in the two multipliers is 5 and 10. Perhaps that would disqualify my semiprime from being used in the RSA encryption scheme, which works best when the multipliers consist of approximately equal numbers of digits.

Encouraged by finding a semiprime quickly, I tried at least a dozen more 14-digit integers, but none were semiprimes; they all had more than two prime factors. It is fun to play with numbers, but the truth is that a 14-digit public key would be laughable. The largest public key ever cracked consisted of 250 decimal digits or, if written in the binary system, 829 bits. This was done in 2020 using massively parallel processing on a supercomputer.

The difficulty of factorization increases exponentially with the number of digits. For a 14-digit number, using readily available online programs, one can find the result in a fraction of a second. Factorization of a 250-digit decimal number would take thousands of years if performed on a single processor. For a number consisting of 617 digits or 2048 bits, using all the computers in the world, the age of the Universe would not be enough to perform such an operation. The estimate I have seen on the internet says that it would take 300 trillion years, which is approximately 20,000 ages of the Universe. Cryptographers use semiprimes for public keys with 617 digits or 2048 bits, hoping the bank they are consulting will complete all its financial transactions before 300 trillion years are up.

So, now that we have some understanding of the problem, what would it take for a quantum computer using Shor's algorithm to break the state-of-the-art 2048-bit (or 617-digit) encryption? It has been estimated that it will take a QC with 20 thousand *logical* (noise-compensated) qubits[12] about 8 hours — eight hours versus 300 trillion years; not bad for a speed-up!

If the modern RSA encryption scheme based on semiprimes and their prime multipliers one day becomes susceptible to being cracked, this would devastate the world's financial system and, by extension, everything else. Without reliable encryption, finances, commerce, and the military cannot function. That is the primary reason why, in the mid-1990s, after Shor's publication, QC research was noticed by private investors, government funding agencies, and, last but not least, the National Security Agency of the USA.

---

[12]We will explain the difference between *physical* and *logical* qubits in the following.

Encryption experts started looking for a genuinely unbreakable encryption scheme as soon as QCs threatened the well-being of the RSA. It did not matter that the threat was purely theoretical, awaiting the arrival of practical quantum computers with many thousands of qubits, which were not likely to appear any time soon. When it comes to guarding trillions of dollars and atomic weapons, one can't be too careful.

For 30 years, cryptographers have sought a one-way mathematical problem that remains secure against both quantum and classical attacks. In response to this challenge, the National Institute of Standards and Technology (NIST) launched a competition to standardize post-quantum cryptographic algorithms. Among the leading candidates are *lattice-based* cryptosystems, which rely on the hardness of problems like *Learning With Errors* (LWE) and *Shortest Vector Problem* (SVP), leveraging random noise to ensure security.

Another significant mathematical problem that would be solved much faster on a QC than on a classical computer was identified in 1996 by Bell Labs scientist Lov Grover. It belonged to the *NP-class* problems, also called *decision* problems or *traveling salesman* problems.

NP-class problems are all mathematically equivalent to the original Traveling Salesman Problem, or TSP. The problem is formulated as follows: A salesman must visit several cities connected by straight roads and return to the town where he started. He travels by car, using gasoline in proportion to the length of his route. What is the optimum route he should take to visit each city only once, in any order, while saving as much as possible on gasoline? It is an optimization problem or a scheduling problem, and many similar issues come up in various fields. For example, the optimal routing of a telephone call or an email on the internet is akin to the TSP. The optimization quickly gets out of hand for a large number of cities, requiring a very long time for its solution on a classical computer. It gets exponentially harder because the number of possible routes grows fast with the number of cities. A salesperson would be well advised to forget trying to save on gas — consider it a business expense — and choose the route according to your intuition. An alternative is to wait 100 years for an optimal solution.

Some algorithms allow the solution of TSP in less time than what would be required in the brute force approach, which consists of testing all possible routes one by one, but these algorithms are limited in terms of the number of cities they can handle. Like prime factorization, TSP is a one-way type of problem — checking whether the proposed route is truly optimum is much easier than finding it.

The reader can probably see that this problem must be amenable to a solution on a QC. A quantum salesman, spreading his wavefunction, can be simultaneously in all cities. Glover showed that the optimal route can be determined *quadratically* (rather than *exponentially*) quicker on a QC than on a CC. If the number of cities is considerable, substituting CC with a QC and the resulting quadratic speedup will not help the salesman; calculating the optimum route will still take too long.

I was curious to find out about the approximate thresholds, so I asked ChatGPT-4o about them, and this is what I received:

For 100–200 cities, classical exact solvers might take minutes to hours.

For 300+ cities, even with a quadratic quantum speedup, exact solutions may still take more than 24 hours.

For over 500 cities, even an ideal quantum computer would not be able to provide an exact solution within a reasonable time.

The range of practical applications for NP-class problems is much broader than applications for Shor's factorization problem. However, the computational speedup in the Glover algorithm is much less impressive than in Shor's.

I recently listened to a very informative discussion between a Professor of Physics at Colombia University, a string theorist, a celebrated author of popular books about theoretical physics, Brian Greene, and a renowned computer scientist, Scott Aaronson, who is the Chair of Computer Science Department and the Founding Director of the Quantum Information Center at the University of Texas in Austen. Among other subjects, they discussed the set of problems that can be solved more quickly on a quantum computer than on a classical one, and, according to Aaronson, such issues are relatively scarce. Let me quote here parts of that conversation in which Aaronson said:

> …the key point for a layperson to understand is that it [QC] is not just a general-purpose accelerator of everything; *quantum* is not just a synonym for *awesome*…it's really for problems with a very particular structure that we know how to get a dramatic speed up over a classical

> computer...The biggest examples here are the simulation of quantum physics and chemistry...that was the original application that Richard Feynman had in mind when he talked about this 40 years ago...and that's probably the most important application economically...even to this day...
>
> ...now a huge discovery in the 1990s that made a lot more people interested in Quantum Computing was the famous discovery of Peter Shore that a quantum computer can also give huge speedups for certain problems that have nothing to do with quantum mechanics and that among these are breaking the public key encryption systems that currently protect the internet...and a quantum computer could do that... quickly finding the prime factors of huge composite numbers [and] solving other problems in Number Theory and Group Theory that underly modern cryptography. But it's crucial for people to understand that...to develop those algorithms, Peter Shore had to take advantage of very, very special properties of these problems...and we don't know how to get the same kind of speed up for most of the problems we care about in computer science, like *training neural networks* [italics ours] or financial portfolio optimization, or DNA sequence alignment... (Aaronson, 2024).

I am providing this rather lengthy quote to dispute the often-made claim that QCs, if and when they become practical, will be used in training the future AI, adding, so to speak, gasoline to the already raging fire. According to Scott Aaronson, that is not the case. However, quantum modeling of atomic and molecular systems, plus the promise of factorization of large numbers and speedup in solving NP-class optimization problems, provide lucrative enough targets for QC development.

Universal Quantum Computers and Quantum Logic Gates

Excited by Shor's algorithm and its promise of exponential speedup, many scientists after the mid-1990s began working toward a general-purpose quantum computer capable of simulating quantum systems and solving complex mathematical problems. This effort required fundamental decisions at every level, from the physical realization of qubits and quantum gates to the development of quantum programming languages and algorithms.

Classical computers rely on logic gates such as AND, OR, NOT, and NAND, which are combined into logic circuits, as discussed in Chapter 2. But does a quantum computer, which operates on qubits instead of bits, require entirely new types of gates? Deutsch and others demonstrated that a universal quantum computer can be built using three fundamental quantum gates, each operating on two- or three-qubit inputs and producing a single-qubit output. We will not describe them here, but we will provide their names: CAT gate, ENT gate (also known as Controlled-NOT (CNOT) gate), and RAT gate. It is funny that the CAT gate is named after Schrödinger's cat, while the RAT gate is named after a mechanical *ratchet* rather than the cat's favorite prey.

It turned out that one needs about N^3 ($N \times N \times N$) gates to handle N qubits. For example, for 50 qubits, one would need approximately $50^3 = 125{,}000$ gates. All these gates must act much faster than the average lifetime of an entangled state, which tends to be short due to the inevitable interactions of a quantum processor with the environment. This lifetime would be *extremely* short if computer elements were macroscopic,[13] like a transistor. Even when qubits are microscopic objects such as charged atoms (ions) or electronic spins, the lifetime of an entangled state is still relatively short — seconds or, at best, minutes.

Errors and Error Correction Schemes

The central difficulty facing quantum computing scientists is that errors plague their systems because qubits are delicate and inherently noisy. High error rates prevent quantum computers from executing large-scale computations, which require millions of sequential operations to be performed reliably and without interruption. For a long time, the outlook was grim. However, over the last two years, the situation has begun to improve, offering hope that quantum computers will eventually live up to their promise.

Classical computers make mistakes, too — a "1" may be randomly replaced by a "0," or vice versa. Engineers correct these occasional errors

---

[13]There is one remarkable exception to this rule: superconductivity, which is the manifestation of quantum mechanics on a macroscopic scale. We shall discuss superconducting qubits in the following. They enjoy infinite lifetimes despite being large.

by introducing redundancy: A bit may be copied into multiple locations, and the copies are compared at intervals. If one of them changes — for example, if a "1" flips to "0" — the discrepancy raises a red flag, and the remaining consistent copies are used to correct the error. In addition, classical error-correcting codes are designed so that minor errors do not affect the final result.

In quantum computing, mistakes arise primarily from *decoherence* — the tendency of qubits to lose their delicate quantum superposition states over time due to interactions with the environment. Since decoherence is a fundamental and persistent feature of quantum systems, such errors are not rare; they are pervasive. To make things worse, one cannot simply copy qubits to introduce redundancy. This limitation is captured in a result known as the *no-cloning theorem.* As explained in Ball (2018, pp. 267–268), the issue is technically subtle, but the bottom line is that an arbitrary quantum state cannot be duplicated without destroying it.

It's understandable — no one knows what's going on inside a quantum superposition until the state is measured, at which point the superposition collapses. A quantum state cannot be copied, nor can it be examined for errors without altering or destroying it. Even identifying a quantum error involves making a measurement, which disrupts the system. You can't make backup copies, and you can't directly observe mistakes. For a while, quantum error correction seemed impossible.

Then, in 1995–1996, Peter Shor and Andrew Steane independently developed the first successful methods for detecting and correcting quantum errors. They introduced the idea of quantum error-correcting codes; see Ball (2018, pp. 268–269) for a brief explanation of what this involves.

The cost of implementing this scheme is paid in qubits. One can create a perfect, noise-free "logical qubit" by combining noisy "physical qubits" in such a way that they collectively behave as a single fault-tolerant unit. The key question is as follows: *How many physical qubits are required to construct one logical qubit?*

There are two interrelated challenges:

1. The *fidelity* of physical qubits — that is, how reliably they perform quantum operations.
2. The *overhead* — the number of physical qubits needed to build one logical qubit.

Until recently, *two-qubit gate fidelities*[14] hovered around 99.8–99.9%, which limited error-free sequential operations to just a few hundred. By 2025–2026, a significant advance was made by MIT researchers, who demonstrated a two-qubit gate fidelity of 99.94% using *superconducting fluxonium qubits* (see the following). More importantly, they showed that this high fidelity could be sustained continuously over several weeks, providing a realistic foundation for fault-tolerant quantum computing.

When engineers were constrained to 99.8% fidelities, they needed a staggering 1,000–10,000 physical qubits to construct a single reliable logical qubit — an overhead that made quantum computing effectively impractical. More recently, with higher-fidelity qubits, companies such as Quantinuum and Microsoft have demonstrated a reduced overhead of about 100 physical qubits per logical one. This is still high, but it is orders of magnitude more feasible than what was required just a few years ago.

How Close Are We to Cracking the RSA Encryption Scheme?

Recent advances in quantum error correction are promising. However, quantum computers remain very far from being able to crack the problem of large-number factorization. To understand the gap, let us consider what is required to use Shor's algorithm to factor the widely used 617-digit public key known as RSA-2048. The number *2048* refers to the number of binary digits (bits) used to represent the key.

According to a well-known estimate by Gidney and Ekera (2021), factoring an RSA-2048 number within 8 hours using Shor's algorithm would require roughly 20,000 logical qubits. If we assume a physical-to-logical qubit overhead of about 100 — which is optimistic and not guaranteed at the scale of billions of operations — this would translate into approximately 2 million high-quality physical qubits. As of the beginning

[14]In quantum computing, a "two-qubit gate fidelity" measures how accurately a quantum gate performs compared to its ideal, error-free version. It is expressed as a percentage: A fidelity of 99.9% means that the gate works correctly 999 times out of 1,000. Two-qubit gates are generally more prone to errors than single-qubit gates and often serve as the benchmark for overall system reliability.

of 2026, the number of physical qubits in functioning quantum computers remains in the range of tens to several hundreds (and in some platforms approaching 1,000), depending on the platform and fidelity. This leaves us with a gap of nearly 10,000× between today's reality and the required hardware.

The number of logic gates needed to execute Shor's algorithm at this scale is equally daunting. For RSA-2048, the computation is projected to require roughly 3 billion Toffoli gates — a class of three-qubit gates essential for quantum arithmetic. In comparison, current quantum systems can support only thousands of gates per run at best. Finding detailed numbers on this is notoriously difficult — perhaps intentionally so — but even a generous assumption of 3,000 usable gates implies a gap of *1,000,000*× in this category.

A similar ratio holds when we consider the number of sequential gate operations. Executing Shor's algorithm on RSA-2048 is estimated to require around 10 billion operations. The highest number of error-free, sequential two-qubit gate operations demonstrated so far is approximately 5,000, reported by IBM's Heron R2 (156-qubit) superconducting processor in November 2024. As of early 2026, no system has surpassed this milestone. Thus, in terms of operational depth, we face a shortfall of about *2 million*×.

In sum, the road to factoring RSA-2048 — and thus to breaking modern cryptographic schemes — remains long and steep. Quantum computers would need to advance from four to six orders of magnitude (to put it simply, QCs must be 10,000 to 1,000,000 times better then they are now) in hardware scale, gate fidelity, and architectural sophistication before such a feat becomes plausible.

Worldwide QC Effort and Funding

The question of how to build a quantum computer comes down to choosing the physical implementation of qubits and their controls (gates). Here is an incomplete list of qubit technologies and organizations pursuing them:

1. *Trapped ion qubits* [IonQ, Quantinuum, Alpine Quantum Technologies (AQT), HQS Quantum Simulations, University of Innsbruck].
2. *Neutral atom qubits* [ColdQuanta (now Infleqtion), Atom Computing, Pasqal, QuEra Computing] Amazon (AWS Center for Quantum Computing).

3. *Superconducting qubits* [Google Quantum AI, IBM Quantum, Raytheon BBN, Intel, Rigetti Computing, D-Wave Systems, Quantum Circuits, Inc. (QCI)].
4. *Photonic qubits* [Xanadu, PsiQuantum, University of Bristol].
5. *Topological qubits* [Microsoft Quantum (StationQ), Research at Yale University and Weizmann Institute of Science, University of Copenhagen].
6. *Quantum dots (semiconductor qubits)* [Intel, HRL Laboratories, Toshiba, CEA-Leti].
7. *Qubits based on diamond nitrogen-vacancy centers* [Quantum Brilliance, Q-CTRL, University of Stuttgart].

The "List of companies involved in quantum computing, communication, or sensing," accessed at the end of 2025 on Wikipedia, enumerates 300+ entities spanning startups, academic spin-offs, tech giants, and national laboratories.

As noted earlier, funding for quantum computing (QC) development began in the mid-1990s, primarily driven by excitement surrounding Shor's factorization algorithm. Over the ensuing three decades, more than 30 governments worldwide, notably the United States, China, and the European Union, have collectively committed over $40 billion in public funding for quantum technologies.

A major boost to U.S. federal support for QC came with the *National Quantum Initiative Act* (NQI) of 2018, which formally launched coordinated QC efforts across a wide range of agencies, including NIST, NSF, DOE, DOD, DHS, and NASA.

Until recently, private investment in quantum computing grew in parallel with government support. While it is difficult to estimate precisely, global private investment in QC over roughly the past decade is believed to total on the order of tens of billions of dollars (approximately $15–25 billion). A year-by-year private investment in QC is somewhat uneven: In 2024 it reached a "local minimum" of 1.1 billion, to raise again in 2025 to the level of several billion worldwide.

Following several high-profile exits in 2023–2024, including those of Alibaba, Baidu, NQCG, and Zapata AI, no further major closures have occurred as of early 2026. On the contrary, the winding-down phase seems to have halted, with renewed momentum evident. In Europe, governments and institutions have launched initiatives to increase private-sector investment in quantum significantly, and notable acquisitions, such as Ion Q's purchase of Oxford Ionics in June 2025, suggest consolidation rather than contraction in the space.

Two Examples of Quantum Hardware

Trapped ions

An atom is a natural choice for physically implementing a qubit. Richard Feynman, when he gave a talk about quantum computing at Bell Labs in 1986, which I attended, spoke about atoms as the medium for quantum computing. Quite a few of the 300-some QC organizations currently active in the field have chosen to work with atoms, some with neutral atoms, and others with ionized (charged) atoms. The latter distinction lies in how these atoms are handled: with optical methods for neutral atoms or electric fields for ions.

Atoms are quantum objects, yet they are large enough compared to elusive free electrons to be manipulated with confidence. Their quantum nature is most evident in the behavior of their orbiting electrons, which occupy discrete energy levels unique to each atomic species. These electronic states are well defined, stable, and relatively unaffected by external influences.

Imagine an ocean liner with multiple decks. Though the ship (an atom) may navigate rough seas, with wind and waves (the environment) rocking it, its internal deck structure (electronic energy levels) remains unaffected so long as it avoids a collision with an iceberg or another vessel. The liner can be repositioned and moved while maintaining its deck arrangement intact.

Imagine a sailor named Bob (representing an electron) aboard the ship. He can stand on deck zero (the lowest deck) or deck one (a higher deck). However, this is where our analogy breaks down — Bob, weighing 180 pounds (approximately 80 kg), is a macroscopic object, whereas an electron, with a mass of roughly 10^{-30} kg, is a fundamental particle. Unlike Bob, who must be on one deck or the other, an electron can exist in both states simultaneously. When we are uncertain whether the electron is on deck zero or deck one, we have a qubit in superposition. However, the moment we make a measurement, the qubit collapses into one of the two definite logical states: zero or one.

An atomic qubit consists of two chosen atomic states such that the energy difference between them is relatively small, and they are shared by one electron. Let us denote a lower energy state $|0\rangle$ and a slightly higher energy state $|1\rangle$. An atom can be in one of the definite states, $|0\rangle$ or $|1\rangle$, or it may be in a *superposition* state, a $|0\rangle$ + b $|1\rangle$, where a and b are complex numbers (the same as in Dyakonov's text we quoted above).

The states $|0\rangle$ and $|1\rangle$ should be long-lived because the qubit must survive for the time required to complete a quantum computation. The state $|0\rangle$ is usually chosen to be the lowest energy state in the electronic energy spectrum of an atom (the *ground state*), and thus, it is naturally stable; an electron will stay in it forever unless disturbed. The higher energy state, $|1\rangle$, must be metastable; it must last for several seconds or even minutes, which is incredibly long by atomic standards and long enough by computer science standards to complete a quantum calculation. This is referred to as a *long decoherence time*.

A few words about electronic energy spectra: A free electron has a continuous energy spectrum, meaning it can possess any energy from zero to infinity. However, everything changes when an electron becomes bound to an atomic nucleus. As part of an atom, an electron must obey the constraints imposed by Schrödinger's equation and other principles of quantum mechanics. The most significant change in the electron's 'life-style' is the transformation of its energy spectrum: gone are the days when it could take on any energy at will — now, its energy levels are discrete, forming a spectrum of sharp lines. Moreover, these energy levels are not just electrons'; they belong to the entire atom.

This holds for all elements, from hydrogen to the heaviest atoms at the end of the periodic table. Quantum mechanics predicts the detailed energy spectra of all elements and governs the distribution of electrons across these levels. By the late 1920s, physicists realized they had uncovered the key to interpreting chemistry and the periodic table. For instance, what chemists had long referred to as *valency* turned out to be simply the number of electrons occupying the highest energy level of a given element.

The lowest possible energy state of an atom is called the *ground state*, while higher energy levels may be occupied or vacant, depending on the number of electrons in the atom. In hydrogen, the simplest element, a single electron typically resides in the ground state, with all higher levels unoccupied. In more complex atoms, electrons fill several lower levels, leaving higher ones vacant. Shining light (photons) onto an atom allows an electron to absorb energy and transition to an unoccupied higher level.

However, quantum computers do not rely on these higher, unoccupied energy states to form qubits. Instead, they exploit a subtler quantum mechanical phenomenon known as *hyperfine splitting* within the ground state. This effect, which lies beyond the scope of our simplified picture, causes the ground state in certain atoms to split into two or more sublevels separated by slight energy differences. An electron in the ground state has an almost equal probability of occupying either of these finely split levels, a property that makes them ideal for encoding qubits.

Several elements can be used to create qubits in quantum computers, but among the most commonly used are *ytterbium* (Yb) and, more recently, *barium* (Ba). These atoms are particularly well suited for quantum computing because their ground states are split into two closely spaced *hyperfine* levels, which can serve as a qubit's logical '0' and '1' states. An electron (or, more precisely, the atomic system) in the ground state can be *coherently manipulated* to occupy either of these metastable levels, meaning it can maintain its quantum state for extended periods without decoherence. Yb and Ba also have other practical advantages that make them easier to work with in experiments.

In ion-trap quantum devices, atoms such as ytterbium and barium are ionized by removing one electron, creating Yb^+ or Ba^+ ions that can be confined and manipulated using electric fields. The transition between the quantum states $|0\rangle$ and $|1\rangle$ in Yb^+ corresponds to a photon frequency of 12.6 GHz, which lies in the microwave range; Ba^+ qubits operate on a similar principle, though the exact transition frequencies differ depending on the isotope and configuration.

For qubits to function effectively, the atoms they reside on must be *entangled*. This is typically achieved by using lasers or microwave fields to induce collective oscillations, much like a ship's captain might build crew unity by having everyone dance in unison before real work begins.

In atomic quantum computing, quantum gates are created through carefully controlled laser and microwave interactions that excite and manipulate atoms to produce superposition, entanglement, and the logical operations required for quantum computation.

The ion trap and its contents are placed in an ultra-high vacuum and cooled to less than one-thousandth of a degree Kelvin — so close to absolute zero that virtually all thermal motion ceases. The vacuum prevents air molecules from colliding with the qubit-forming ions, which would otherwise destroy their delicate quantum states.

Control is achieved through precisely tuned, narrow-beam lasers that target selected ions at specific frequencies and for carefully timed intervals.

The concept of trapped-ion quantum computing was proposed in 1996, and while it appeared promising in theory, its practical implementation proved highly challenging. It took a decade before the first experimental system, featuring just two qubits, was demonstrated. Another 10 years passed before such systems could reliably perform even the simplest computations.

It is essential to clarify that when we say a quantum computer performs the 'simplest computations,' we do not mean basic arithmetic operations. In fact, QCs are highly inefficient at arithmetic and were never designed for such tasks.

While future quantum computers *may* prove valuable for specific applications — some of which we have discussed in these pages — as of today, they have not demonstrated practical advantages in any domain outside of inherently quantum mechanical tasks, such as generating genuinely random numbers.

Notably, the fact that QCs struggle with even the most basic arithmetic is not widely publicized by quantum computing researchers. However, this limitation has been repeatedly pointed out by Michel Dyakonov, whom we quoted earlier.

In Fig. 1, we see an ion trap chip manufactured by the US company IonQ. The blue dots above the chip are the artist's conception of trapped ions, which are, in reality, very close to the chip's surface and invisible.

Fig. 1. Ion trap on a chip made at IonQ, Inc. Eighty ytterbium ions are suspended above a radiofrequency ion trap and would not be visible in a photograph; the blue dots above the chip are the artist's conception. The ion-ion separation is 3 μm.

Sources: Image courtesy of IonQ co-founder Professor Christopher Monroe; photograph by Kai Hudek (IonQ) and Emily Edwards (University of Maryland/JQI).

Qubits made from superconductors

Another popular QC technology is based on the phenomenon of *superconductivity*, a manifestation of quantum mechanics on a *macroscopic*[15] scale.

Superconductivity is manifested by the absence of electrical resistance, which becomes identically zero below a temperature specific to a given element or a compound. The absence of electrical resistance is absolute. In one experiment, physicists induced a current in a superconducting ring and observed it circulating around the ring without decreasing its value for over three years. In a normal metal ring (the same ring warmed above its superconducting transition), the current would decay to zero in a tiny fraction of a second. That persistent current in a superconducting ring can be likened to the perpetual motion of the Moon around the Earth. The Moon moves without friction, which allows it to stay in its orbit with a never-changing orbital velocity for billions of years, or, as far as we are concerned, forever.[16]

---

[15]The term *macroscopic* refers to the realm of "big things" in which we live.

[16]The Moon's orbit shifts slightly because of its gravitational pull on Earth's oceans (tides), but we'll set that aside for this analogy. In this way, the supercurrent in a ring is more stable than the Moon!

Zero resistance to electrical current suggests that electrons move inside the superconducting material without "friction."[17] Although superconductivity was discovered in 1911, the mechanism of this remarkable phenomenon was not fully understood until 1957, when American physicists John Bardeen, Leon Cooper, and Robert Schrieffer explained it. They were awarded the Nobel Prize in Physics in 1972 for this groundbreaking work. It was John Bardeen's second Nobel Prize in physics, as discussed in the section about transistors in Chapter 1.

The explanation of superconductivity was based on quantum mechanics and solid-state physics. In essence, in a superconductor, wavefunctions of individual free electrons join together and become a collective wavefunction of all free electrons. Due to this, individual electrons cannot deviate from the collective motion that comprises an electrical current, which eliminates electrical resistance. An interested reader will find a more detailed, yet accessible, explanation in Gurvitch (2023, Vol. 2, Chapter 4).

Superconductivity is observed at low temperatures.[18] In the metal *niobium* used in superconducting electronics and QCs, superconductivity sets in below 9.2 K.[19] Such low temperatures are obtained using liquid helium, which boils at 4.2 K. A device made of niobium will become superconducting if immersed in liquid helium, but for quantum computing, the temperature of 4.2 K is prohibitively high. In QCs, the temperatures are made as low as technically possible, usually a few thousandths of a degree above absolute zero,[20] to avoid quick decoherence of a quantum state.

In addition to zero resistance, superconductors exhibit a peculiar magnetic behavior. One of the aspects of that behavior is that inside any

---

[17]The origin of electrical resistance in normal (non-superconducting) materials is electronic scattering, where moving electrons encounter obstacles that change their motion. In a superconductor, electron scattering does not occur. In our "big things" world, scattering can be compared to friction.

[18]Except for the high-temperature superconductors, which are not used in quantum computing.

[19]Kelvin temperature scale (abbreviated K) starts at *absolute zero*, the lowest possible temperature at which all atomic and molecular motion stops. Zero Celsius (°C) corresponds to 273 K; zero Kelvin is negative 273 °C. The temperature of boiling helium, 4.2 K, corresponds to negative 268.8 °C.

[20]Obtaining such ultra-low temperatures is a science and an art that we do not have the luxury of discussing on these pages.

superconducting ring, the *magnetic flux*, which is the product of magnetic field strength and the area of the ring, must be equal to an integer number of *flux quanta*, the smallest magnetic flux allowed by quantum mechanics. Observing flux quantization on a macroscopic scale suggests that superconductivity is fundamentally quantum mechanical.

When superconductivity is used in QCs, it is often (but not always) implemented through the ingenious device called SQUID[21] (*Superconducting QUantum Interference Device*). A SQUID consists of a superconducting ring (loop) interrupted by one or two *Josephson Junctions*.

In the early 1980s, this author and his colleagues at Bell Labs developed the technology for reliable and reproducible Josephson junctions. It was called the "trilayer" technology because Josephson junctions were built from three metal layers: niobium, aluminum, and again niobium. The tunnel junction was created via a controllable oxidation of the aluminum layer, forming the Nb/Al-oxide-Nb structure. Niobium was used because it is a strong (high melting point, *refractory*) metal with the highest transition temperature into a superconducting state among all the elements in the periodic table. It enabled Josephson junctions to operate in liquid helium at 4.2 K. Aluminum was incorporated into the package due to its excellent tunneling oxide properties. This fortunate combination of properties formed a perfect technology that replaced all earlier versions of Josephson junctions. It remains in use in superconducting electronics worldwide, more than 40 years after its introduction.

The QCs inherited our trilayer junction technology for making superconducting qubits. This technology was in all early QC processors using superconductivity.

However, in QCs, niobium — the "supporter" of superconductivity in the Nb/Al-oxide-Nb system — is unnecessary. Aluminum would not be superconducting in liquid helium at 4.2 K, but it superconducts at a few thousandths of a degree, which is the operating temperature of a superconducting QC. Eventually (after 2007), QC technologists disposed of niobium and made simpler Al-oxide-Al Josephson junctions and SQUIDs based on them. So, I can no longer brag that all superconducting qubits are based on

[21] The SQUID was invented in 1964 by a group of researchers at Ford Scientific Laboratory in Detroit, Michigan.

"my" technology of trilayer niobium-aluminum junctions. All I can say now is that they *originated* from our technology, which does not sound so glorious.

SQUIDs are incredibly sensitive to the value of magnetic flux threading their ring, which makes them very attractive for ultra-sensitive measurements of currents and magnetic fields and, more recently, for qubit fabrication.

There are several methods for creating qubits using SQUIDs and stand-alone Josephson junctions. In one implementation, states $|0\rangle$ and $|1\rangle$ correspond to supercurrents flowing around the loop in opposite directions: clockwise and counterclockwise. Which, come to think about it, is mind-boggling. We have already grown accustomed to the idea that electron spins or atoms can exist in two states simultaneously; after all, they are so small that our everyday notions do not apply to them. We acknowledge that atomically small things are unusual, and we believe in the principles of quantum mechanics. However, here we have a large object — a loop made of a superconducting material containing trillions of atoms and electrons — and we must assume that when no one is looking, the current in it flows simultaneously in both clockwise and counterclockwise directions!

Superconducting technology is the oldest among all QC technologies; it has been developed for over forty years, the first proposals dating to the early 1980s. The first superconducting qubit was built in 1999. So, a lot has been tried and explored. There are three types of superconducting qubits: *flux* qubits (the kind we discussed above, where currents flow simultaneously in two opposite directions), *charge* qubits, and, more recently, *transmon* qubits. The latter were introduced in 2007 by Jens Koch and collaborators at Yale University and have since become the dominant qubit type used in IBM, Google, and Rigetti quantum processors.

A transmon qubit is a nonlinear superconducting resonator built using a Josephson junction, a large capacitor, and a superconducting circuit, forming an LC (inductor-capacitor) oscillator. A transmon operates in a quantum regime, where its energy levels are quantized. The two lowest energy states form the qubit.

In addition to qubits, one needs quantum gates. Several types of quantum gates have been developed. They exercise qubit control through precisely tuned microwave radiation pulses; some also use magnetic fields, currents, and voltages.

The latest on superconducting quantum computing

On December 9, 2024, Google introduced its latest quantum chip, Willow, with 105 transmon superconducting qubits. In the announcement accompanying this release, Google QC scientists stated that they had broken the *error correction threshold* with the new chip.

This chip created a stir in the excitable QC community; this is how it was described in *HP Wire*:

> Google today introduced its latest quantum chip — Willow (~100 qubits) — coinciding with two key achievements run on the new chip: breaking of the so-called Quantum Error Correction Threshold (error rate decreases as number of qubits rises) and posting a new benchmark for quantum performance against classical computing; Willow ran a benchmark task in five minutes that Google said would take ten septillion years (10^{24}) on Frontier, which until a few weeks ago was the fastest supercomputer in the world (hpwire 2024/12/09).

It all sounds great, but before you uncork champagne bottles to celebrate QC victory over classical supercomputers, perhaps you should know that the *benchmark task* to which the text above refers has been carefully chosen to highlight the strengths of QCs; it has almost no independent value in the mainstream of computer science, applied math, physics, or anything else. The benchmark is based on *Random Circuit Sampling* (RCS). This task involves generating random quantum circuits and sampling their outputs, a computationally infeasible process for classical computers. It is as if an English-speaking person were invited to compete with a native Chinese speaker at solving a crossword puzzle in Chinese, and the inevitable win of a Chinese player would be hailed as a milestone.

Let me put it this way: QCs excelling in the RCS benchmark task would not get much attention from a traveling salesman trying to save on gas when visiting many cities or a hacker attempting to break the RSA encryption scheme.

Since Google's *Willow* announcement, IBM has articulated its plans:

- In 2025, IBM announced plans for a chip called "Nighthawk," which will contain 120 superconducting qubits. This chip is designed to handle up to 5,000 error-free two-qubit gate operations.
- In 2026, IBM will follow with "Kookaburra," a modular processor designed to connect multiple chips. It will be equipped with an advanced error-correction scheme.
- In 2027, IBM plans to unveil "Cockatoo," a system that will demonstrate entanglement between separate quantum chips using universal adapters — a step toward building truly scalable quantum systems.
- All these efforts are intended to culminate in IBM's "Starling" platform, which is expected to be released by 2028–2029. Starling will be running on around 200 logical qubits, constructed from thousands of physical qubits.

QCs in Quantum Chemistry Calculations

When he envisioned quantum computing, Richard Feynman argued that the modeling of molecular structures performed on a quantum machine should be superior to that performed on a classical computer simply because the quantum machine would be based on the very principles according to which that molecular structure has been "constructed" by nature. Let us examine the achievements in molecular quantum modeling, also known as quantum chemistry, where, according to Feynman, QCs are expected to find their most natural application.

So far, QC scientists have utilized their creations to aid classical computers in calculating molecular energy spectra, selecting some of the simplest molecules for this purpose. They couldn't do these calculations entirely on QCs because, as we noted earlier, QCs cannot perform some of the simplest mathematical operations. They may have an edge over their classical brethren in certain specific tasks, and that is where they are utilized in the so-called *hybrid classical-quantum techniques*.

An inquisitive reader may ask the following: "Why molecular spectra? Why not calculate the atomic energy spectra first?" Good question. That has not been done because atomic energy spectra of all atoms in the periodic table of elements have already been calculated with utmost accuracy,[22] and, unlike molecules, atoms do not allow QCs to exhibit their advantages. So, QC scientists focused on the more challenging molecular calculations, where classical calculations may be less precise.

The first molecule whose energy spectrum — specifically, the *ground-state energy* and *equilibrium bond length*[23] — was calculated on a quantum computer was the *hydrogen molecule* (H_2). Being the simplest of all molecules, H_2 was a natural choice for this proof-of-principle calculation. In 2017, researchers at IBM utilized a superconducting quantum processor and a hybrid quantum-classical algorithm known as the *Variational Quantum Eigensolver* (VQE) to compute these properties. Their results came slightly closer to the accepted values of ground-state energy and bond length than those produced by comparable classical methods. To the best of my understanding, the differences involved were on the order of 1% or 2%, not a dramatic improvement, but significant enough at the time to be taken as an encouraging sign for the emerging field of quantum computing.

In the same year, IBM applied the VQE algorithm to a more complex molecule, beryllium hydride (BeH_2), extending the method beyond diatomics, but in this case, no greater precision than classical methods was claimed — only that the quantum-computed values were consistent with existing results.

In subsequent years, other groups used VQE and related hybrid algorithms to calculate the ground-state properties of molecules such as *diazene* (N_2H_2) (done at Google in 2020), *linear hydrogen chains* like H_{12} (also at Google in the same year), *hydrogen fluoride* (HF), *borane* (BH_3), and others. These studies were carried out using superconducting or trapped-ion platforms from Google, Rigetti, or IonQ. In all cases, researchers emphasized agreement with literature values as a validation of the method but refrained from claiming improved precision. In fact, after the first H_2 calculation, such claims were conspicuously absent. This pattern suggests that while quantum algorithms can reproduce known results, they have yet to demonstrate a clear advantage over classical methods.

If Richard Feynman were following this work from Heaven, I doubt he would be rejoicing and telling other angels that he was the one who initiated this field. Let us hope that quantum chemistry computations will give him more satisfaction in the future.

[22] By *accuracy*, we mean good agreement between measured and calculated values.

[23] The average distance between atoms in a molecule.

State of the Art in Quantum Computing in Early 2026

Let us summarize the current state of quantum computing as of March 2026 and compare it to where we were four years ago, when I discussed it in my previous book.

The number of physical (noisy) qubits has increased approximately tenfold from 2021 to 2026. However, it has become increasingly clear that performing meaningful computations with noisy physical qubits remains impractical. The complexity of problems solved by QCs has not significantly advanced, and the much-anticipated *quantum supremacy* — where a quantum computer solves a problem beyond the capabilities of classical computers — has only been demonstrated in a synthetic benchmark involving randomly structured quantum circuits, specifically designed to generate output patterns that are extremely difficult for classical computers to simulate.

Regarding Shor's factorization algorithm, its application in the presence of noise remains highly problematic (Cai, 2023). Even in an ideal, noise-free setting, Shor's algorithm requires far more qubits and quantum gates than are currently available. The largest integer factored using a compiled version of Shor's algorithm remains 21. This reduced version, proposed by Beckman and colleagues in 1996, relies on prior knowledge of the prime factors 3 and 7, allowing it to run with drastically fewer resources. It may look like cheating — and perhaps it is — but it remains the best anyone has achieved so far. As for factoring "interesting" large numbers, such as the 2048-bit RSA encryption key, the required number of logical qubits and quantum gates is so astronomically large that comparing it with present capabilities would be an exercise in futility.

Several quantum molecular models have been computed for simple molecules, all using hybrid algorithms that perform some steps on a classical processor and others on a quantum chip. The best one can say is that these calculations have reproduced accepted values. The breakthrough in quantum chemistry modeling envisioned by Richard Feynman more than 40 years ago has yet to occur.

To date, no quantum computer has demonstrated clear superiority over classical computers in any practical domain.

So, here we are. In 44 years of its conceptual existence — or 32 from the publication of Shor's algorithm — quantum computing has yet to produce tangible results demonstrating a practical advantage. The main challenge remains dealing with quantum errors. Achieving practical quantum computing will require many orders of magnitude improvements across multiple parameters. Not surprisingly, some physicists have begun to question whether a quantum computer of practical value is even possible.

And yet, our beloved Richard Feynman stood at the cradle of this idea, which is no small thing. Quantum computing may well be the most daring technological challenge ever attempted. If it succeeds, the speed-up for certain types of computation could be mind-boggling. Many talented and enthusiastic people are devoting their lives to this effort. If they do succeed one day, they will be forgiven for having raised everyone's expectations sky-high.

Chapter 4

The History of Artificial Intelligence

Early Attempts to Mechanize Human Reasoning: Llull and Leibniz

Suppose we skip our subject's mythological and fairy-tale prehistory. Indeed, in these concise and restrained pages, I could hardly do justice to the colorful legends — tempting as they are — such as the tale of the first giant bronze robot created by the god Hephaestus, Talos of Crete, or the story of the Golem brought to life by the great Rabbi Judah Loew ben Bezalel in 16th-century Prague. However, later, we shall still speak about the robots from Karel Čapek's play *R.U.R.*, about the lip-reading, murderous computer HAL from Arthur Clarke and Stanley Kubrick's *2001: A Space Odyssey*, and about many other fascinating things.

A long time ago, some people were trying to make, or at least envision, mechanisms that, when fed with some basic concepts, would "think" logically and provide thoughtful answers by mechanically combining the inputs. The first such person in history was a philosopher, theologian, poet, missionary, and former knight (the last being the most wonderful designation — who wouldn't want to be a former knight?) from the Kingdom of Majorca (who wouldn't want to live in the Kingdom of Majorca?), Ramon Llull (Fig. 1), who lived on that beautiful island from 1232 to 1315. Llull developed a method for arriving at logical conclusions using movable diagrams. He called his system *Ars Magna* (Latin for "Great Art"), or simply "Art," a concise name not unlike "AI."

Fig. 1. Ramon Llull (1232–1315) was the first to construct mechanisms intended for mechanical thought. This engraving, likely from a 16th- or 17th-century portrait, probably does not reflect any likeness to his actual face. The ribbon in his mouth says, "*Lux mea est ipse dominus*" ("My light is the Lord himself").

Source: Image courtesy Smithsonian Library, in the public domain.

Llull based his system on the notion that there is a limited number of undeniable truths in all fields of knowledge and that we could understand everything by studying various mechanical combinations of these elemental truths. It has been noted that Art's combinatorial mechanics resemble Zairja, a device used by medieval Arab astrologers and developed around the same time.

The Art was based on movable diagrams. One method, the *Lullian Circle*, consisted of two or more paper discs inscribed with letters and symbols around their circumferences. The discs could be rotated to generate combinations of these symbols. The results were supposed to show all possible truths about the chosen subject.

It is doubtful that such a system could produce meaningful answers, except accidentally. Yet, Llull was satisfied with his system, which he developed over many years, constantly modifying it and describing these modifications in numerous books. This satisfaction likely stemmed from the material he was working with — 13th-century Christian theology. If you are a Christian philosopher and ask your system, "Is God Trinity?" and receive an answer of "Yes, God is Father, Son, and the Holy Spirit," presumably all is well, and your system is working.

Llull believed that his mechanical system could "banish all erroneous opinions" and that it helped the practitioner to arrive at "true intellectual certitude removed from any doubt." One of the modern commentators called such thoughts and Llull's movable diagrams "semi-occult nonsense," and it is hard to disagree with that harsh judgment. Llull could not get any mechanical thinking going with his system. Yet, the method Llull used was so original that it made an impression on his contemporaries and many subsequent generations. Llull attempted to formalize human thought, and his intent (rather than his questionable material or his methods) was noticed and remembered in history. This former knight cast a long shadow.

Specifically, Llull's Art influenced Gottfried Leibniz — a testament to its significance, as there is no higher praise for a historical figure than having shaped the thinking of a certified genius. In 1666, at just 19 years old, Leibniz wrote *On the Combinatorial Art*, in which he expanded on Llull's idea of a finite set of basic concepts and applied it theoretically to all of philosophy and human thought. His central premise was that all truths could be expressed as appropriate combinations of simple ideas. Drawing an analogy to language, he likened complex thoughts to words, which arise from different arrangements of the same 26 letters.

Leibniz envisioned a universal symbolic language designed to combine ideas with perfect logical precision. Furthermore, he proposed that relationships between fundamental concepts could be manipulated mechanically, governed by logical rules, much like the relationships between numbers in arithmetic. Ultimately, he believed such a system could automate reasoning and even resolve disputes through calculation — an idea that foreshadowed modern computational logic and artificial intelligence (AI).

In 1714, two years before his death, Leibniz described his unfinished program in a letter to his friend Nicolas Rémond:

> I should venture to add that if I had been less distracted, or if I were younger or had talented young men to help me, I should still hope to create a kind of *spécieuse générale* [in Leibniz's writings, an alternative name for the *characteristica universalis*, universal formal language], in which all truths of reason would be reduced to a kind of calculus. At the same time, this could be a kind of universal language or writing, though infinitely different from all such languages that have thus far been proposed, for the characters and the words themselves would give directions to reason, and the errors (except those of fact) would be only mistakes in the calculation. (Leibnitz, 1714)

Leibniz never realized his vision of a universal language, yet as we noted in Chapter 2, during the 1770s and 1780s, he laid the foundations of *mathematical logic* — a discipline in which results emerge from the systematic combination and manipulation of logical symbols, akin to mechanically assembling elemental truths.

Leibniz was also the first to design and build a mechanical calculator capable of performing all four arithmetic operations: addition, subtraction, multiplication, and division. While it was not a universal language, it was a universal arithmetic machine — an embodiment of his belief in mechanizing thought.

When geniuses fall short of their ultimate vision, the remnants of their failure often prove magnificent. Leibniz's influence on computer science is universally acknowledged; Norbert Wiener even hailed him as the "patron saint" of cybernetics.

The Human Brain was a Gold Standard for Early AI Scientists

At the opening of Alan Turing's 1948 *Intelligent Machinery* report (Turing 1948), he wrote:

> The possible ways in which machinery might be made to show intelligent behavior are discussed. *The analogy with the human brain is used as a guiding principle* [italics ours].

This belief that to create "smart" machines, we must try to imitate the human brain was expressed by founders such as Turing and von Neumann. Still, soon the main thrust of AI development was redirected toward the hardware implementation of logical operations, in what would later be called "symbolic AI." And yet, the original impulse toward neural networks never disappeared, and over the decades, it has proven its value. A significant part of the entire history of AI is about this confrontation.

To build an electronic brain, the logical starting point was to imitate an organic one. However, the immense complexity of the brain, coupled with our still-limited understanding of its inner workings, makes this task extraordinarily difficult.

For centuries, the brain has been a subject of scientific inquiry, and modern research continues using advanced techniques such as functional magnetic resonance imaging (fMRI), computer tomography (CT) scans, and magnetoencephalography (MEG). Scientists seek to uncover every possible detail about its structure and function.

Since Turing wrote the words cited above three-quarters of a century ago, our understanding of the brain has expanded significantly. Some of these discoveries have shaped the development of AI, a subject we will explore shortly. But first, let us briefly summarize the essential facts about the human brain.

Neurons in the human brain

When I decided to precede the section on AI neural networks with an exploration of biological neurons and the workings of the human brain, I had no idea what I was getting myself into. The sheer volume and complexity of biological information on the subject is overwhelming. I'm beginning to suspect that our fascination with AI stems, at least subconsciously, from a desire to one day eliminate the molecular and electrochemical mess known as neuroscience altogether.

It doesn't help that most illustrations of neurons are profoundly unappealing, resembling skinned octopuses. To describe the brain, I will have to simplify things drastically. The biological terminology alone is enough to fell a horse, let alone an unsuspecting physicist. It's no wonder that some believe the human brain may be the most complex structure in the universe.

Neurons (Fig. 2) are the fundamental cells that enable communication — both between the organism and the outside world and among its internal systems. In the brain, vast networks of these neurons underlie all higher cognitive functions and memory, which are our primary focus here. Together with their supporting *glial cells*, neurons form the nervous system.

There are many types of neurons, including several distinct types within the brain. The human brain contains approximately 100 billion (10^{11}) neurons, each forming around 1,000 connections with other neurons. These connections, called *synapses*, total an astonishing 100 trillion (10^{14}) across the brain.

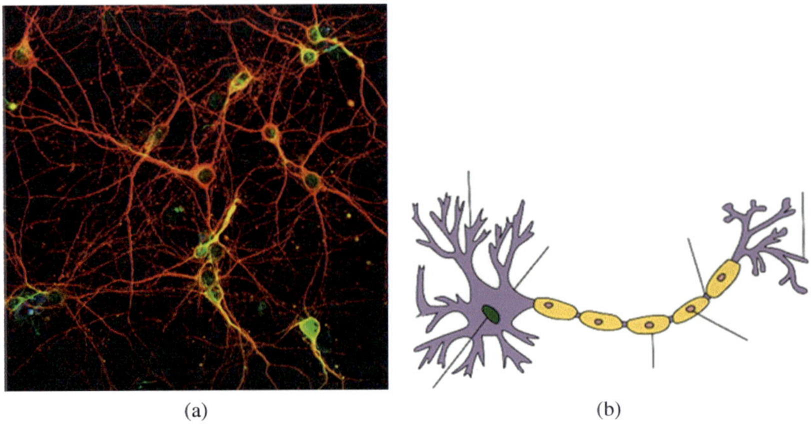

Fig. 2. (a) Neurons under the microscope, and (b) neuron schematic showing a nucleus, dendrites, and an axon.

Sources: (a) Image by AL188, Wikimedia Commons, Lic. CC BY 4.0; (b) image by Quasar Jarsz, Wikimedia Commons, Lic. CC BY-SA 3.0.

Connections between neurons are provided by *axons* and *dendrites*, which terminate in *synapses*. Each neuron has only one axon, which resembles the main electrical cable coming out of the neuron's cell body (*soma*). That single axon splits into thousands of branches (wires or interconnects, using an electrical analogy) connected to other neurons' *dendrites*. Synapses are electrochemical contacts at the ends of dendrites.

It should be noted that an electrical analogy, while useful, especially in the context of the upcoming discussion of artificial neural networks, overlooks an important biochemical component that is of primary importance in neuroscience. If you open any book about neurons written by a biologist or biophysicist, you will find that it is filled with chemistry and electrochemistry.

Neurons communicate electrically and chemically. Signals propagate between neurons via *nerve impulses* or *action potentials*, small voltages across cell walls that move along the connecting filament (axon, axon's branches, and dendrites; Fig. 2(b)) at speeds ranging from 1 to 150 meters per second. The synapse is "smart" in that it is selective: Depending on various factors, it may transmit either an *excitatory* or *inhibitory* signal, influencing whether the receiving neuron will fire its own action potential.

This selectivity plays a crucial role in regulating neural activity and shaping brain function.

At synapses, information is transmitted via chemical messengers called *neurotransmitters*. When an action potential reaches the axon terminal — the synaptic end of the signaling neuron — it triggers the release of neurotransmitter molecules. These molecules diffuse across the synaptic gap and bind to specific receptors on the membrane of the receiving neuron. This binding conveys either an excitatory or inhibitory signal, influencing whether the next neuron will generate its own action potential.

Each neuron receives excitatory and inhibitory signals simultaneously or nearly simultaneously across its roughly 1,000 synapses. Whether the neuron fires an action potential down its axon depends on the balance of these signals: if excitatory inputs outweigh inhibitory ones, the neuron fires; if inhibitory signals dominate, it remains inactive.

It is a bit more nuanced than that. A neuron fires an action potential when the combined effect of excitatory and inhibitory signals brings the axon's membrane potential to the threshold level. This process involves a complex integration of inputs through spatial and temporal summation rather than a simple count of excitatory versus inhibitory signals. The effectiveness of each input depends on its location. with signals from synapses closer to the axon hillock having a stronger influence, as well as its timing, with closely spaced signals reinforcing each other more effectively. Together, these factors determine whether the neuron reaches the threshold and fires.

Knowing this colossal complexity, can one hope to first fully understand the structure and functioning of the human brain, and only then build a "thinking machine" in its image and likeness? Presumably, everyone who has reflected on this has realized that such an approach would be impossible. Rather than trying to recreate all these thousands of types of neurons, with their subtle and largely not yet understood differences and functions, it would be easier to get married, have children, and thereby obtain new human brains.

So simplifications are necessary. One must attempt to create a model of a neuron that is capable of representing brain activity, yet is not unnecessarily complex. And no chemistry — we can manage with electricity alone. Thus, the promising direction is that of simplified modeling, and it was in this direction that two American scientists succeeded in the early 1940s, to whom we now turn.

Walter Pitts and Warren McCulloch, Pioneers of Neuron Modeling

An odd couple

The history of AI quietly began in 1942, five years after Turing's 1937 paper on the one-dimensional universal machine (discussed in Chapter 2) and four years before the completion of ENIAC, the first programmable electronic computer (also covered in Chapter 2). Yet, those who started this history were unaware of their role in it. Tell them then that they were laying the foundations of AI, and they would have asked: "What is AI?" — for the term itself did not yet exist in the 1940s. And then, when a visitor from the future explained the term to them, they would still be puzzled: what did it have to do with them? They had only wanted to model the workings of the human brain, to help neurophysiologists. What did computers have to do with it?

It all began in 1942 at the University of Illinois at Chicago, where two Americans met by chance. The two could not have been more different.

Professor Warren McCulloch, a 44-year-old experimental neurophysiologist and psychiatrist, held a B.A. in philosophy and psychology from Yale, as well as an M.S. and M.D. in psychology from Columbia. Throughout his career, McCulloch worked in laboratories studying the physical structures of human and animal brains. As an MD, he spent seven years practicing clinical psychiatry before becoming a professor of psychiatry at the University of Illinois College of Medicine in 1941, where he also served as director of the Illinois Neuropsychiatric Institute. Beyond science, McCulloch was a poet, a lover of philosophy, and a man who enjoyed his whiskey. In 1942, he lived comfortably with his wife and two daughters in Hinsdale, Illinois, an affluent suburb of Chicago.

In stark contrast, Walter Pitts was just 18 when he met McCulloch. He came from a working-class, uneducated family that saw him as an oddity, even a freak. In 1938, at age 15, he ran away from his Detroit home, never to return. As a child, Pitts was a prodigy, teaching himself multiple languages, including Greek and Latin. But that was only the beginning. Over the relatively short span of his life, he cultivated a staggering range of intellectual interests, including geology, music, art, poetry, history, philosophy, physiology, neurobiology, and, most importantly, mathematics. He also dabbled in camping, mountain climbing, sailing, and even, at one time, screenwriting for Hollywood. Thanks to his extraordinary memory, Pitts amassed encyclopedic knowledge in most of these fields (though perhaps less so in outdoor adventures and screenwriting). He later summarized his view on problem-solving with characteristic bluntness: "*A problem is either trivial or it is insoluble.*" For anyone else, such a statement might seem arrogant. For Walter Pitts, it was simply a fact.

Much of what we know about Walter Pitts in those early years comes from the recollections of his close friend Jerome Y. Lettvin, who, in 1938, was a pre-med student at the University of Chicago. Later in life, Lettvin became a professor of electrical engineering and biomedical engineering at MIT. In 1994, he was interviewed for "Talking Nets: An Oral History of Neural Networks," a book of interviews compiled and edited by James Anderson and Edward Rosenfeld, published in 1998. Here is what Lettvin said about young Pitts:

> The year before I had gone to medical school, I had spent at the University of Chicago. There I met Walter Pitts, who became and remained my best friend. Walter was an autodidact. He taught himself logic and mathematics and was able to read a fair number of languages, including Greek and Latin.
>
> He already had the peculiar history. At the age of twelve he was chased into a library by a gang of ruffians, and took refuge there in the back stacks. When the library closed he didn't leave. He had found Russell and Whitehead's *Principia Mathematica.*[1] He spent the next three days in that library reading the *Principia*, at the end of which time,

---

[1] *Principia Mathematica* is a monumental three-volume work on the foundations of mathematics, authored by Alfred North Whitehead and Bertrand Russell. First published between 1910 and 1913, it aimed to derive all of mathematics from formal logical principles. Renowned for its rigor, abstraction, and extreme formalism, it remains one of the most intellectually ambitious works of the 20th century. Its notation and structure are notoriously forbidding — it famously takes several hundred pages of symbolic logic to prove that $1 + 1 = 2$.

> he sent a letter to Bertrand Russell, pointing out some problems with the first half of the first volume; he felt they were serious.
>
> A letter returned from Russell inviting him to come as student to England — a very appreciative letter. That decides him; he's going to be a logician, a mathematician.
>
> He went to the University of Chicago, which is where I met him; but he never registered as a student…we became fast friends…The two of us became inseparable…
>
> During that year at the University of Chicago, Walter had gotten hold of Carnap's new book on logic.[2] This was in 1938. He walks into Carnap's office with his own annotated version of the book, pointing out some flaws. And he gives it to Carnap, talks to him a while, then goes out but doesn't introduce himself. Carnap spends the next couple of months hunting high and low for that "newsboy who knew logic." In the end, he did find Walter and persuaded the University of Chicago to give him some menial job. Walter had no funds, had separated himself from his family, so that was good. (Anderson and Rosenfeld, 1998, pp. 2–3)

It is remarkable that young Walter Pitts crossed paths with none other than Rudolf Carnap, the leading architect of logical positivism, who attempted to describe an algorithm — an explicit computational procedure — by which humans extract knowledge from elementary experiences. Carnap's theory conceptualized the mind as a *computational process*. As we will see, Pitts would soon provide a precise mathematical formulation of this idea.

Walter Pitts was, in many ways, a quirky individual, or at least an eccentric — a price nature sometimes exacts for granting exceptional abilities at an early age. He never signed documents, fearing to "make his name known," and, according to his biographer, he had "a lifelong fear of librarians" — so much so that he would rather steal books than face them at the checkout desk (Smalheiser, 2000). Despite attending university courses, engaging with professors, and publishing scientific papers, Pitts often had no income and, at times, was homeless.

---

[2]It must have been Rudolf Carnap's "*The Logical Syntax of Language*," originally published in Germany in 1934 and later translated and published in the U.S. in 1937. Carnap's *Logical Syntax* was the key work that attempted to formalize all of science through logic, just the kind of thing that would have fascinated Pitts.

At one point, theoretical biologist Wilfrid Rall took Pitts on as a Ph.D. student, fully aware that he had never been an undergraduate — an extraordinary act of academic generosity. Rall later recalled that Pitts was "an oddball who felt compelled to criticize exam questions rather than answer them" (Smalheiser, 2000). Despite his brilliance and involvement in Ph.D. programs at the University of Illinois at Chicago and later at MIT in Cambridge, Massachusetts, these eccentricities ultimately prevented him from earning a degree.

Yet, Pitts was not a recluse; he had a sharp sense of humor, a talent for practical jokes, and could be both a charming companion and a thoughtful, engaging colleague.

Before meeting McCulloch, Pitts had already developed an interest in neuron modeling. He regularly attended Nicolas Rashevsky's seminars[3] on theoretical biology, where mathematical modeling of biological systems was a central focus. Later, McCulloch, an authority on neuron physiology, helped steer Pitts more decisively in that direction.

In 1942, Walter Pitts was introduced to Professor Warren McCulloch, possibly by Nicolas Rashevsky.

Lettvin recalls:

> Warren was enchanted with him. Walter was homeless, and I had to escape from my family because the supervision was incredibly intense… And so early in 1942, Warren invites Walter and me to live with him and his family. Warren and his wife, Rook, were always enormously generous. We settled in, and it was in the evenings then that Walter and Warren got together on "A Logical Calculus of Ideas Immanent in Nervous Activity" [*the famous 1943 McCulloch-Pitts paper of which we speak below*]. Walter at that time was, if I remember correctly, about 18, something like that. Walter had read Leibniz, who had shown that any task which can be described completely and unambiguously in a finite number of words can be done by a logical machine. Leibniz had developed the concept of computers almost three centuries back, and had even developed a concept of how to program them.

[3]Nicolas Rashevsky was a theoretical physicist who emigrated from Soviet Russia during the Bolshevik Revolution and settled in the U.S. in 1927. He is considered the father of *mathematical biophysics*. In 1939, he founded and became chief editor of the *Bulletin of Mathematical Biophysics* (BMB), the journal in which Walter Pitts and Warren McCulloch published their foundational papers in 1942–1943.

> I didn't realize it at the time. All I knew was that Walter had dredged this idea out of Leibniz, and then he and Warren sat down and asked whether or not you could consider the nervous system such a device. So they hammered out the essay at the end of '42. (Anderson and Rosenfeld 1998, p. 3)

The two papers that started neuron network modeling

The two foundational papers appeared in 1942 and 1943. When discussing neuron physiology, we noted the two types of inputs: excitatory and inhibitory. The discovery of inhibitory synapses in 1941 by American physiologist David Lloyd was the final piece of the puzzle — the icing on the neuron cake, so to speak. The understanding of the behavior of physiological neurons reached a level where they could be modeled mathematically, a feat achieved by Pitts and McCulloch (Fig. 3).

This seminal work — the seed from which all artificial neural networks (ANNs) grew — was done in two stages. First, in 1942, Walter

Fig. 3. Walter Pitts (1923–1969) and Warren McCulloch (1898–1969), pioneers in neuron modeling and neural networks, photographed in 1949.

Source: Image in the public domain.

Pitts, working as a sole author, wrote and published a paper titled "*Some Observations on the Simple Neuron Circuit*" (Pitts, 1942). In this work, he demonstrated that neurons could execute basic Boolean logic operations, such as AND, OR, and NOT.

The natural next question was: If a single neuron can function as a logic gate, what can a network of such neurons achieve? The answer to that question was worked out in discussions between McCulloch and Pitts, witnessed by Lettvin. The result was a paper, "*A Logical Calculus of Ideas Immanent in Nervous Activity*" (McCulloch and Pitts, 1943). In this work, the unlikely collaborators explored the computational capabilities of neural networks. They rigorously demonstrated that an interconnected network of neurons, if sufficiently large and appropriately structured, could, in principle, perform any conceivable computation. In other words, they showed that neural networks had the theoretical potential to be *Universal Turing Machines* (UTMs; Chapter 2). This conclusion suggests that the brain functions like a calculator — a notion some scientists had suspected but could not prove. The fact that the McCulloch–Pitts result was an invitation to create calculators based on ANNs was realized only later.

In the model, a neuron receives a set of excitatory and inhibitory synaptic inputs (practically, positive and negative signals) and fires only when the net sum of these inputs exceeds a certain threshold. The McCulloch–Pitts neurons operate in a binary fashion: They either fire (representing the "on" state or the truth value of "1") or remain inactive (the "off" state or the truth value of "0").[4]

But there is more to the model than binary neurons. Notably, McCulloch and Pitts developed a novel method for representing logical truth values in a neural network. In their model, time is discrete, divided into fixed time units. When a neuron fires, it remains in the "on" state for the duration of a single time unit, referred to as the "time of latent addition," typically around one millisecond. The exact duration is not crucial for the mathematical model. Once this period ends, the neuron resets to the "off" state. In the early 1940s, the idea of assigning truth values of "1"

[4]At times, credit for this important idea has been mistakenly given to others. For instance, in his introduction to John von Neumann's last book, Ray Kurzweil attributes to von Neumann the realization that a neuron's output is binary (von Neumann, 1958, p. xxv). However, by the mid-1950s, it had already been known for many years, having been formalized in Pitts' 1942 paper and then in McCulloch and Pitts' 1943 paper.

and "0" based on the *presence* or *absence* of a pulse within a specific time interval was a novel concept. There is evidence that this idea originated with McCulloch, who explained it to Pitts (Geffer, 2015). If accurate, this significantly elevates McCulloch's contribution beyond what I initially thought.

> Interestingly, this concept resurfaced some 40 years later, in the early 1980s, with the development of Rapid Single Flux Quantum (RSFQ) logic for superconducting Josephson circuits. The creators of that now-famous logic scheme, K. K. Likharev and V. K. Semenov, then at Moscow State University, arrived at it independently. Had they been aware of the 40-year-old idea from a completely different field, they would undoubtedly have referenced the 1943 paper by McCulloch and Pitts.

These two papers, written with the aim of shedding light on the workings of the human brain, represented the first significant steps toward ANNs and AI in general. That could not have been seen at the time: The first programmable electronic computer, ENIAC, would appear four years later; Alan Turing's *Intelligent Machinery* report would be written seven years later, and the term "AI" would not emerge for another 14 years.

The significance of McCulloch and Pitts' work was evident to John von Neumann. In 1945, he wrote his "First Draft of a Report on the EDVAC "— a foundational document outlining the design of a stored-program digital computer and introducing the *von Neumann architecture*. In his report, Von Neumann had only one reference: McCulloch and Pitts' 1943 paper "A Logical Calculus of the Ideas Immanent in Nervous Activity."

The AI history acknowledges the role of McCulloch and Pitts' papers, but I do not think people see them as the *fathers of AI*. If only they had written in that 1943 paper, "*We feel that this computational technique could one day be used in machines designed to imitate the human brain.*" Alas, they didn't.

The central role of Walter Pitts, the mathematical genius behind these foundational papers, deserves greater recognition. The division of labor in the 1943 paper is clear: McCulloch wrote the *Introduction* and *Consequences* (the first three and last one and a half pages), while Pitts wrote the remaining eleven and a half pages of pure mathematics and formal logic. I do not wish to diminish Professor McCulloch's contribution, but it seems that for the future science of neural circuits, the application of mathematical logic methods was of greater importance than knowledge of neuron physiology. And let us not forget that an earlier paper on the single-neuron model was written entirely by Pitts.

While their neuron paper was arguably the pinnacle of their careers, McCulloch and Pitts continued to push the boundaries of neural modeling. In 1947, they proposed a model in which neurons could recognize universal forms, such as perceiving a circle regardless of its size or geometric perfection (Pitts and McCulloch, 1947). Although this highly mathematical model was later found not to align with how the human brain achieves such recognition, it was still a step forward from the perspective of AI.

Norbert Wiener meets Walter Pitts and Warren McCulloch

Norbert Wiener (Fig. 4) also stood on the same seashore where the young Aphrodite — AI — first stepped onto land, choosing to live among humans. A mathematician of the highest caliber, Wiener is best remembered as the father of *cybernetics* — the mathematical study of control

Fig. 4. Norbert Wiener (1894–1964), photographed in 1946.

Source: Image from Tufts Archival Research Center, Tufts University.

processes in both living systems and technology. As a true founder, he not only developed this new discipline but also gave it its name.

The term "cybernetics" comes from the Greek word for "helmsman," one who steers a ship. A helmsman takes in information from his eyes and instruments and provides outputs by adjusting the ship's course, an example of a system governed by *negative feedback* (see, for instance, Gurvitch, 2023, Vol. 1, Chapter 5). Wiener's insights into feedback, control, and communication laid the groundwork for many technological advances, influencing fields as diverse as automation, neuroscience, and, ultimately, AI.

Norbert Wiener was among the first to suggest that all intelligent behavior arises from feedback mechanisms — processes that machines could, in principle, emulate. For a time, research on neural networks was considered part of cybernetics, and in this sense, AI is an offshoot of that field. Moreover, Wiener played a pivotal role in the lives of Warren McCulloch and Walter Pitts.

The discipline Wiener established in the 1940s had far-reaching implications for industrial automation, robotics (then as distant and fantastical as AI), systems control, biology, neuroscience, computer science, neural network modeling, philosophy, and even the organization of society. The wide-ranging relevance and clarity of Wiener's presentation made his 1948 book *Cybernetics* widely influential. For a time, *cybernetics* served as the primary label for what we now recognize as various branches of engineering and computer science concerned with control and communication.

Norbert Wiener and John von Neumann were mathematical geniuses who, in addition to their fundamental contributions to mathematics, made significant advancements in applied fields, including computer science (both of them), economics (von Neumann), and engineering (again, both of them). However, they cannot be considered the founders of AI; they did not work on *thinking machines* per se, theoretically or otherwise. However, they have done so much for the relevant disciplines that AI, as we know it, would not have appeared without them. Still, Turing's, McCulloch's, and Pitts' contributions to AI were more direct.

Wiener's role in the lives of our neuron heroes was not all positive. Here is a brief account of their engagement.

By some accounts, Wiener met McCulloch at the neurophysiological meeting in New York in 1942, if not earlier. One year later, in the summer of 1943, he met Walter Pitts. The ubiquitous Jerome Lettvin described what happened:

> In 1943, I was working part-time in neurology and part-time in medical wards because I had to have general medical experience. My co-intern had a relative, a second uncle or something of that sort, by the name of Norbert Wiener, and said we should go over and visit him sometime. OK, so we go over and visit, and I meet Norbert Wiener for the first time. Wiener had just lost his right-hand man to a skiing accident and was desolate; he couldn't find anybody else of that quality.
>
> So I said, "I happen to know a young man from Chicago, a mathematician who is extraordinary," and so forth, and I described Walter Pitts.
>
> Winner said, "*There does not exist such a person.*"
>
> So I called McCulloch, and I remember the two of us chipped in together and got Walter a round trip train ticket to Boston.
>
> It was very funny...I remember Walter walking into Norbert's office, and Norbert says, "Hello," and no introductions, "Come, I want to show you something," and takes Walter next door where there is a blackboard on two walls, and Wiener says,..."I'm going to prove the ergodic theorem.[5] He starts out, and a short distance down the board, Walter begins some critical questions and asking for clarifications, and by the time they'd finished the first board, Walter going along making commentary, it's sort of obvious what's going to happen. By the end of the second board it's definite. Walter is going to work with Weiner. (Anderson and Rosenfeld, 1998, 4)

On Wiener's urging, at the end of 1943, Pitts left Chicago and moved to Cambridge, Massachusetts. When Wiener invited Pitts to MIT, he promised him a Ph.D. in mathematics (Conway and Siegelman, 2006, pp. 141–142). Knowing that Pitts has never obtained a high school diploma, Wiener planned to do it unconventionally — by first enrolling him as a "special student" under his guidance. He devised a special doctoral program for Pitts, which was supposed to fill the blanks in his education. Wiener, who immediately recognized Pitts' unique talent, took him

[5]The ergodic theorem is a mathematically subtle concept from statistical physics. Its rigorous proof, first provided by George Birkhoff in 1931 and later refined, requires advanced tools from measure theory and the theory of dynamical systems.

"under his wing." He treated him as his scientific assistant and, perhaps, even as a colleague.

In the winter of 1943–1944, Wiener brought Pitts to a conference he organized in Princeton, where he introduced him to another towering figure of the era: John von Neumann. Von Neumann, too, was impressed by Pitts' intellect. We have already noted von Neumann's acknowledgment of the 1943 McCulloch-Pitts paper on neural networks.

Wiener rarely worked alone; he thrived on the presence of capable listeners who could engage with his stream of ideas. Because cybernetics aimed to provide a unified mathematical framework for processes in both technology and living organisms, many of those around Wiener were physiologists, neurologists, or psychologists. This loosely associated group has sometimes been called the "Wiener Circle" — a clever pun, since "Wiener" in German means "someone from Vienna." The name evokes the earlier "Vienna Circle," a group of European philosophers in the 1920s and 1930s that included Rudolf Carnap and other leading figures of logical positivism. Some of the permanent members of the *Wiener Circle* in the 1940s included Arturo Rosenblueth, a Mexican physiologist and Wiener's collaborator; Warren McCulloch and Walter Pitts — we know who they were — Julian Bigelow, an engineer and mathematician; and, later in the decade, Jerome Lettvin, who, after graduating from medical school, became a competent experimental neurologist.

From 1943 until the end of the decade, Wiener and this group collaborated relatively amicably on problems in mathematical biology, communication theory, and control systems — topics that would form the foundation of cybernetics. However, toward the decade's end, Wiener grew increasingly dissatisfied with the course of these discussions, particularly with McCulloch's direction, which leaned toward psychiatry and speculative group psychology. One sign of this estrangement was Wiener's decision to abstain from the final three Macy Conferences, which had begun in the spring of 1946 with his enthusiastic participation.

The Macy Conferences on Cybernetics — initially titled "*Conferences on Circular Causal and Feedback Mechanisms in Biological and Social Systems*" — were a series of ten interdisciplinary meetings held in New York City between 1946 and 1953. These conferences were pivotal in shaping the field of cybernetics. Warren McCulloch served as chairman for all ten. Participation was by invitation only and included a small, elite group of world-leading figures in mathematics, physiology, psychiatry, anthropology, and emerging cybernetics. Norbert Wiener played a central role in the early meetings, attending the first seven.

Walter Pitts — Wiener's "special student" with a couple of foundational papers on neural networks under his belt — accompanied him to all seven of those conferences. He was the youngest participant and the only one without a doctoral degree or a formal academic position. True to his reserved nature, Pitts rarely spoke, but he did make a brief technical intervention during the second conference in the fall of 1946. According to the transcript, Pitts stated that he was preparing a doctoral dissertation under Wiener's supervision, aiming to extend the geometry of interconnected neural networks into three dimensions to better match the structure of the human brain. Moreover, he and Wiener envisioned modeling these three-dimensional neural networks as stochastic, probabilistic systems — a visionary concept that would become central to neural network theory four decades later. The idea was far ahead of its time, and the perceived mathematical difficulties seemed insurmountable. The chairman of the Macy Conference and the man who rarely missed a chance to put in his 10 cents of wisdom, Warren McCulloch, took this opportunity to remark that the mathematics required for such a project "has yet to be evolved" (Conway and Siegelman, 2006, p. 162). We will soon return to the fate of Pitts' dissertation, which, alas, was never finished.

Let us fast-forward five years. The year is now 1951. Walter Pitts has not completed his doctoral dissertation at MIT, in part because of his "incorrigible habit of procrastination and a terror of being judged" and in part because of his habit of taking long trips with his pals, principally Lettvin and Selfridge, another of "McCulloch's boys." As Wiener complained to McCulloch, "Walter is such a migratory person that we are not quite certain how to reach him" (Conway and Siegelman, 2006, p. 201).

In 1951, a new research group focusing on the brain's processing mechanisms in vision and sensory perception was established within MIT's *Research Laboratory of Electronics* (RLE). The initiative to form this group came from Jerome Wiesner, then a prominent figure at MIT and later its president. Wiesner facilitated the recruitment of Warren McCulloch, who left his tenured professorship at the University of Illinois to lead the group. The other group members were the inseparable Pitts and Lettvin, the latter an experimental neurologist. Another neurophysiologist in the group was Pat Wall.

It should be noted that Norbert Wiener was not involved in the formation of this group, and for him, its establishment at RLE was a source of anger and feelings of betrayal.

Wiener had his oddities and his mental health issues. In the summer of 1951, after he gave lectures on cybernetics in Paris and Madrid (during his European sabbatical), Wiener, instead of relaxing with his family on the French side of Lake Geneva, feverishly wrote memoirs of his childhood and youth, which he called "The Bent Twig." This title probably tells us all we need to know about this work. It was recommended to Wiener by his psychiatrist, who prescribed writing a biography as a remedy for Wiener's severe depression. Instead, it backfired, making Wiener unhappier than before.

The stress he experienced caused occasional intense headaches; in the summer of 1951, they caused him to be hospitalized in Geneva, Switzerland. Then, upon returning home to Massachusetts, still unwell, Wiener flew to Mexico City, where he was invited to participate in the celebrations marking the 400th anniversary of the University of Mexico.

According to a psychiatrist who saw him in the fall of 1951 in Mexico, Wiener was acutely unhappy, and the depression he was generally prone to reached new heights (Conway and Siegelman, 2006, p. 218). One reason for his unhappiness was related to that memoir he had just finished writing. It was about his youth and early career, and in it, Wiener did not "spare" his mother, his early mentors, or himself. What he had written made such a bad impression on the publishers to whom he sent his manuscript that, despite Wiener's fame, no one agreed to publish it.

Just at that time — the case of quintessential bad timing — Jerome Lettvin and Walter Pitts sent Wiener a letter about the good time they were having establishing that new group at MIT's Research Laboratory of Electronics. The letter was written in a mock medieval style, addressed to "Dear Arturo [Resenblueth] and Norbert." It opened with "*Know, o most noble, magnanimous and puissant lords, that we be now established in a winsome laboratory — a savage place...,*" etc. The letter ended with a humorous invitation for Arturo Rosenblueth to come to Cambridge: "*If you (Arturo) don't come, so that we can all work together, we have every intention of stealing Norbert from you.*"

All in all, it was a playful, harmless, and friendly letter. However, the letter's addressee saw it differently. A half-crazy, tired, depressed, and unhappy Wiener, who resented the idea of that lab being established "behind his back," snapped and sent an angry telegram to Wiesner, which said, "IMPERTINENT LETTER RECEIVED FROM PITTS AND LETTVIN. PLEASE INFORM THEM THAT ALL CONNECTION BETWEEN ME AND YOUR PROJECTS IS PERMANENTLY ABOLISHED. THEY ARE YOUR PROBLEM. WIENER." (Conway and Siegelman, 2006, pp. 218–219).

Wiener was not satisfied with that curt telegram to Wiesner, and on the same day, he wrote a letter to James R. Killian, Jr., the president of MIT. The letter was worse than the telegram. In it, Wiener attacked the entire group, including McCulloch, Wieser, and the "boys," accusing them of all imaginable sins, including the cardinal one: being dishonest with the money. At the end of his letter, Wiener wrote, "*I can definitely say that the present atmosphere is not one in which I can continue my work in cybernetics.*" (*ibid.*, 220–221).

In his reply, Killian tried to calm Wiener down, arguing that the activities at RLE were, in fact, "*the best tributes to the importance and validity of your work.*" Killian ended his reply with, "*I can assure you of the undiminished support of your friends and colleagues here at Tech [Massachusetts Institute of Technology] and of our desire to further your own program and recognition in every way available to us*" (*ibid.*, 221–222).

This was not the first time Wiener threw such tantrums, and there was a good chance that, in a little while, he would come to his senses and resume normal interaction with the McCulloch's group.

This thought — of a possible reconciliation — probably crossed the minds of several people: Killian, Wieser, McCulloch, perhaps even Wiener himself, and, significantly, one more person — Wiener's wife, Margaret. But she categorically did not want reconciliation to happen! She had been waiting for an occasion like that for a long time, and she was not going to miss it. She wanted to separate her husband from that group of rascals and drunkards, as she saw McCulloch and his entourage, for good. As Lettvin said, "*She hated Warren and she hated everybody associated with Warren*" (*ibid.*, 223).

What Margaret told her husband in December 1951 became known only years later. And even half a century later, eighty-year-old Jerome Lettvin was fuming when talking about it. In December 1951, in Mexico, Mrs. Wiener told her husband that four years ago, when their then nineteen-year-old daughter, Barbara, was visiting Chicago, she had been seduced by the corrupt "boys."

Even a cursory look at the facts would show that it was a false, trumped-up charge, which was new even to Barbara. "She made it all up! There was nothing to it," as Lettvin described it (*ibid.*, 226). But the truth did not matter; the damage was done, and Wiener never changed his mind.

So that was the end of McCulloch and Wiener's friendship and scientific collaboration, as well as the end of Pitts' apprenticeship under Wiener. This proved devastating for the vulnerable and already mentally fragile Pitts. According to multiple accounts, Pitts stopped working altogether and burned his unfinished manuscript on three-dimensional neural networks. "...For Walter it was fatal.... He burned everything he'd ever written... the whole manuscript he was working on for his dissertation and everything else. And from that point on, he did nothing" (*ibid.*, 229).

Pitts cut off contact with former colleagues and friends and began drinking heavily in local Cambridge pubs. When Michael Arbib, a mathematician who worked on neural networks and was a former graduate student of Wiener, met Pitts at MIT in the early 1960s and attempted to discuss the 1942–1943 papers with him, the conversation could not continue: Pitts began shaking uncontrollably due to *delirium tremens*.

Walter Pitts died in 1969, alone in a boarding house in Cambridge at the age of 46, from complications related to alcoholism. That same year, Warren McCulloch also passed away at the age of 71. Norbert Wiener had died five years earlier in 1964 at the age of 69 from a heart attack.

AI From the Late 1940s to the End of the 1950s

Alan Turing and AI in 1948–1950

In late 1947, Alan Turing returned to Cambridge from the National Physical Laboratory (NPL) in Teddington, London, where he had been developing one of Britain's first programmable electronic computers: the *Automatic Computing Engine* (ACE). In Cambridge, in 1948, Turing wrote *Intelligent Machinery* (Turing, 1948), the first-ever report explicitly focused on *artificial intelligence*.[6]

In this report, Turing envisioned a network of randomly connected electronic elements resembling binary McCulloch–Pitts neurons. Turing cited in his report the McCulloch–Pitts paper, where it was shown that such networks could carry out all conceivable logical and computational operations.

But Turing's focus was distinct from that of his predecessors: He wanted to know whether such a system could *learn*. He began by observing that a network with fixed inter-neural connections could not acquire new knowledge. He then proposed a more sophisticated architecture in which switches could be installed in the synapses — switches that could

[6]The term "artificial intelligence," or AI, appeared only in 1956.

either block or allow signals to pass. Crucially, these switches would be trainable and capable of adapting based on the system's prior experiences. From this, Turing concluded that such a system could, in principle, learn.

Once again, Turing demonstrated his remarkable ability to identify the simplest non-trivial solution, just as he had with the one-dimensional computing machine described in Chapter 2. That learning perspective is now a hallmark of modern AI.

Let us note that in the same work, Intelligent Machinery, Turing, as an example, considered the automation of chess and even proposed his own chess-playing algorithm (later known as Turochamp). Since no machine capable of executing it yet existed, he followed his own algorithm and "manually" calculated several chess games. Thus, overall, this report, prepared by Turing in 1948, was extraordinarily rich and insightful.

All the more striking is the fact that it was never published during the author's lifetime.

Turing's supervisor, Charles Galton Darwin, director of the NPL and grandson of the famed naturalist, dismissed the manuscript as a "schoolboy essay." In other words, he did not understand it. As a result, *Intelligent Machinery* remained unpublished until 1968, twenty years after it was written and fourteen years after Turing's death.

In 1950, Turing published a follow-up in the Oxford philosophy journal *Mind*: the now-classic *Computing Machinery and Intelligence* (Turing, 1950), often considered the first serious academic paper on AI (though it would have been the second if *Intelligent Machinery* had appeared in 1948 — and perhaps the fourth if the 1942–1943 papers by Pitts and McCulloch were recognized as AI milestones, as we believe they should be[7]).

Turing opens the *Mind* paper with the now-famous line: "*I propose to consider the question, 'Can machines think?'*." He examines whether the question is even meaningful, explores various types of machines, and considers how their operation might resemble human cognition. Significantly, he argues that machines could *learn*, much like a child learns through interaction with the world — interactions he referred to as

[7] It would be unfair to say that all the authors writing about AI history ignored McCulloch and Pitts' contributions. For example, their work is fully acknowledged in a comprehensive book about AI written by Stuart Russel and Peter Norvig (2010).

interferences, an intriguing word choice that perhaps reflects something of Turing's own peculiar sensibility.

He also draws upon the work of Gödel, Church, and his own earlier results to emphasize the inherent limitations of algorithmic procedures. Yet, in a profound and forward-looking comment, Turing noted that while we might prefer machines that never make mistakes, intelligence itself is not defined by infallibility. This observation resonates strongly with today's "hallucinating" large language models (LLMs) and chatbots — but more on that later.

Turing does not forget about chess, but in his 1950 paper chess appears only in passing. He notes that he has already devised a program for playing chess, although no suitable machine yet exists to run it [Turing, 1950, p. 446]. We will return to this theme when we discuss Claude Shannon.

In another essay, which became public only after his death, Turing suggested that thinking machines would need to possess an element of *randomness* — that they would be, in some respects, unpredictable. This insight was likely influenced by quantum mechanics, in which nature is indeterministic at the microscopic level. Turing suspected that the human brain might reflect this quantum uncertainty, potentially giving rise to free will. In expressing this idea, he anticipated the arguments made by Roger Penrose in *The Emperor's New Mind* (1989) by decades.

Turing's most well-known contribution to AI, also presented in his 1950 *Mind* paper, is the idea now known as the Turing test. We touched on it in the Introduction. It stems from the "Imitation Game," a Victorian parlor game involving a man, a woman, and an interrogator. The interrogator, unable to see the players, must determine who is who based solely on their written responses. The man tries to imitate the woman, and the woman tries to be herself. Turing proposed replacing the man with a machine (at which point the gender of the second player becomes irrelevant). If the interrogator cannot reliably distinguish between the machine and the human, the machine is said to have passed the test.

The Turing test is disarmingly simple — typical of Turing's genius for minimalism — but it remains, to this day, one of the most widely discussed (and contested) ideas in AI. While many AI researchers have historically dismissed the test as too narrow or superficial to define intelligence, attitudes may be shifting now that AI systems are finally beginning to pass it.

A 2024 study (Mei *et al.*, 2024) examined AI chatbots' behavior in several "behavioral" games (such as Dictator, Ultimatum, and Trust) designed to elicit characteristics such as trust, fairness, risk aversion, and cooperation. The study found that ChatGPT-4 exhibited behavioral and personality traits statistically indistinguishable from those of humans across various cultures.

In March 2025, researchers Cameron R. Jones and Benjamin K. Bergen from the University of California, San Diego, published a study titled *"Large Language Models Pass the Turing Test"* (Jones and Bergen, 2025). In their study, participants engaged in simultaneous five-minute text conversations with a human and an AI model and then judged which conversational partner they believed to be human. When prompted to adopt a humanlike persona, OpenAI's GPT-4.5 was deemed to be human 73% of the time, significantly more often than actual human participants.

Donald Hebb's theory of learning and memory

We now understand the basic function of a neuron in the brain: it receives a multitude of inputs from other neurons and either *fires,* sending an electrical pulse down its axon, or remains inactive, depending on the combined effect of those inputs. But how does this lead to learning and memory?

Canadian neuropsychologist Donald Hebb (1904–1985) was the first to offer a meaningful theory of how learning might arise from neural activity. In his influential 1949 book *The Organization of Behavior* (Hebb, 1949), he proposed that synapses — the junctions where neurons connect — undergo physical changes through repeated use. These changes increase the likelihood that a given synapse will trigger the firing of the downstream neuron.

This ability of synapses to change in response to experience is known as *neuroplasticity.* While the underlying electrochemical processes are complex, Hebb's insight can be expressed simply: With repeated activation, a synapse becomes more *electrically conductive*. In other words, learning and memory emerge from the strengthening of specific neural pathways. The repeated co-activation of certain neurons reinforces their connections. As the often-quoted summary puts it, "*Neurons that fire together, wire together.*"

When we place Hebb's concept of learning alongside Turing's 1948 *Intelligent Machinery* report, a striking correspondence emerges. Turing proposed a neural network model in which synaptic switches could be trained by experience — essentially, a primitive form of neuroplasticity. In his system, synapses were either ON or OFF, meaning they were either active or inactive, passing a signal or blocking it. This was Hebbian learning rendered in binary form, where synapses became more electrically conductive not gradually but through discrete, all-or-nothing changes.

This parallel between Turing and Hebb appears to have largely been overlooked in the literature. But it deserves recognition. Once again, Turing's genius reveals itself, this time as a visionary of learning systems.

Of course, neither man can be faulted for the lack of cross-referencing. Turing could not have known about Hebb's yet-to-be-published book, and Hebb had no access to the unpublished *Intelligent Machinery* manuscript, which remained hidden for decades in a Cambridge archive.

Since the 1970s, a substantial body of experimental evidence has confirmed and expanded upon Hebb's theory of synaptic plasticity. What began as a plausible hypothesis is now firmly grounded in neuroscience. While Hebb's original ideas have been refined over time, his core insight — that learning and memory arise from activity-dependent changes in synaptic strength — has stood the test of time.

Notably, Eric Kandel of Columbia University demonstrated this principle through studies of the sea slug's nervous system, showing how synaptic changes underlie learning and memory. Kandel was awarded the 2000 Nobel Prize in Physiology or Medicine for this work.

Claude Shannon and AI

We introduced the great American scientist Claude Shannon in Chapter 2, where we noted that in 1937, while still a graduate student at MIT, he demonstrated that an appropriately constructed electrical circuit built entirely from wires and on-off switching elements, such as relays, could implement the complete system of Boolean logic. Such a circuit, he showed, would be capable of solving any mathematically solvable problem. Shannon was just 21 years old at the time, and this foundational result, which underpins all digital computing, was the subject of his master's thesis, not even a Ph.D.

But that thesis marked only the beginning of one of the most remarkable scientific careers of the 20th century.

In 1948, while working at Bell Labs, Shannon published his landmark paper, "*A Mathematical Theory of Communication*," which laid the foundations of Information Theory. In it, he demonstrated that any form of information — text, images, sound, or any other kind of signal — can be represented as a sequence of binary digits, the familiar "1s" and "0s" of digital code.

While this idea may seem obvious to us today, it was far from self-evident in the analog world of 1948. Shannon's insight would eventually become the conceptual bedrock of all modern digital communication systems.

Perhaps most profoundly, Shannon made information *quantifiable*. As noted in Chapter 2, he coined and defined the basic unit of information — the *bit* (short for *binary digit*). But Shannon went further: He proposed a method for measuring the *amount* of information in a message. To do so, he borrowed a concept from physics: *entropy*.[8]

Traditionally, entropy has been used to describe heat exchange in thermodynamics and to quantify randomness in statistical physics and physical chemistry. Shannon ingeniously adapted this idea to the domain of communication, redefining entropy as a measure of uncertainty or surprise in a message. In doing so, he bridged physics and information, giving rise to a new scientific framework.

> Shannon's concept of entropy is fascinating: At times, it feels intuitively clear, and at other times, quite subtle. Broadly, if we compare two messages, the one with *higher entropy* is harder to predict — it may carry more informational content, but that isn't necessarily the case.
>
> For example, consider a source that produces a single symbol repeated endlessly: XXXXXXXXX.... It's easy to guess the next symbol: It will almost certainly be another X. According to Shannon, this message has extremely *low* entropy because there is no uncertainty.
>
> At the other extreme, suppose the source generates symbols entirely at random, with each symbol having an equal likelihood. In this case, the next symbol is entirely unpredictable, and the entropy is maximized.
>
> However, high entropy does *not* necessarily imply high informational content. A stream of pure randomness may be unpredictable, but it can also be meaningless — just noise. In that sense, entropy measures the *potential* for information, not its presence. It quantifies uncertainty, not meaning.

[8]Both Norbert Wiener and John von Neumann expressed ideas similar to Shannon's concerning the application of the concept of entropy to information theory (Conway and Siegelman, 2006, p. 166).

The reader may recall that in Chapter 2 we already mentioned Shannon's entropy when discussing the fundamental limits on the energy required to perform a calculation. That was only one manifestation of the ubiquitous role of entropy in information theory. Shannon's entropy is a versatile tool for quantifying uncertainty and information across many domains. It has found applications in information compression, cryptography, image and video processing, music analysis, and, increasingly, AI. More specifically, it underlies key techniques in machine learning, data compression, neural network training, natural language processing, reinforcement learning, and data privacy.

Shannon's generalization of physical entropy is arguably among the most profound scientific insights of the 20th century. It created a conceptual bridge between seemingly unrelated phenomena. After all, the original idea of entropy arose in the 19th century in the analysis of heat engines — systems concerned with steam, pistons, and thermodynamic efficiency. What could that possibly have to do with the informational content of a message?

The more we contemplate information, the more fundamental it appears to us. The famous theoretical physicist John Wheeler coined the slogan "*It from bit,*" suggesting that the fabric of the Universe itself is fundamentally informational.

But let us return to Claude Shannon and his many contributions.

In his 1948 paper, "A Mathematical Theory of Communication," Shannon addressed the problem of transmitting digital information in the presence of noise. He proposed what became known as the *Shannon limit* — the theoretical maximum data rate at which information can be transmitted over a noisy communication channel without error. This result still defines the boundaries of modern communication engineering.

During the Second World War, Shannon worked on military cryptography, and his interest in secure communication continued after the war. In 1949, he published "Communication Theory of Secrecy Systems," applying information theory to cryptography. This paper offered a rigorous mathematical framework for understanding and designing secure communication systems. It also became a foundational work in modern cryptography.

Writing about Claude Shannon is a bit like recounting the Labors of Hercules; though, come to think of it, the comparison flatters Hercules. While he slayed the Nemean Lion and the Lernaean Hydra sequentially, Shannon took on multiple monumental projects simultaneously, accomplishing more than seemed humanly possible.

We now turn to Shannon's direct contributions to the early development of AI.

As noted in Chapter 2, in addition to his foundational work in digital logic, information theory, and cryptography, Shannon was also one of the founding fathers of AI. In early 1950, he, almost simultaneously, published two landmark articles on machine intelligence. The first, "A Chess-Playing Machine," appeared in *Scientific American* and was aimed at a general audience. The second, more technical paper, "Programming a Computer to Play Chess," appeared in *The Philosophical Magazine* and was directed at experts in computing and logic.

In 1953, Shannon followed up with a third article, further exploring strategies for enabling programmable machines to play and win at chess. These papers marked some of the earliest serious discussions of game-playing as a model for machine intelligence — an idea that would prove profoundly influential in the trajectory of AI research.

Shannon started his 1950 *Scientific American* article thus:

> For centuries philosophers and scientists have speculated about whether or not the human brain is essentially a machine. Could a machine be designed that would be capable of "thinking"? During the past decade several large-scale electronic computing machines have been constructed which are capable of something very close to the reasoning process. These new computers were designed primarily to carry out purely numerical calculations. They perform automatically a long sequence of additions, multiplications and other arithmetic operations at a rate of thousands per second. The basic design of these machines is so general and flexible, however, that they can be adapted to work symbolically with elements representing words, propositions or other conceptual entities.[9] (Shannon, 1950)

In these articles, Shannon proposed the structure of a chess-playing computer program. As a strong chess player, he used his understanding of the game to formalize strategies for the machine. His *evaluation function* considered key factors such as material balance (the number and value of pieces), control of the center, king safety, and pawn structure.

Shannon also introduced the *minimax algorithm*, which seeks to minimize potential losses in the worst-case scenario, assuming the opponent

[9]Note that this is the thought expressed by Ada Lovelace in her Notes a hundred years earlier (see Chapter 2).

plays optimally. This was a foundational contribution to the field of game-playing AI. A few years later, when Arthur Samuel developed his evaluation function for checkers (see later), he acknowledged Shannon's influence.

Since we mentioned Alan Turing's concurrent interest in chess algorithms, we should briefly compare his contribution with Shannon's. Turing's approach, developed earlier, was more exploratory and algorithmic: he outlined a concrete procedure for move selection (Turochamp) and even executed it "by hand," but without the systematic formalization that would later become standard. Shannon, by contrast, placed the problem on a more general and rigorous footing: he clearly distinguished between different search strategies (now often called Type A and Type B), introduced an explicit evaluation function, and framed game playing in terms of minimax search. In this sense, Turing anticipated the idea, while Shannon gave it its canonical mathematical formulation.

Notably, Shannon also considered the possibility of improving the machine's skills through experience, an early reference to what is now known as *machine learning*. In his *Scientific American* article, he wrote:

> Some thought has been given to designing a program that would develop its own improvements in strategy with increasing experience at play... One possibility is to devise a program that would change the terms and coefficients involved in the evaluation function on the basis of the results of the games the machine had already played. (Shannon, 1950)

However, the learning aspect remained only a passing mention (as confirmed in his technical article); it fell to Arthur Samuel to design and implement an actual self-learning program.

Shannon's chess-playing program was not implemented in software; it was not a line-by-line coded algorithm but a conceptual framework: a thoughtful plan of attack and an exploration of possibilities. He never intended to create a mechanized world chess champion. In the early 1950s, that goal would have seemed wildly overambitious. Instead, Shannon hoped to build a system capable of playing "a fairly strong game" of chess — a platform for experimenting with the broader idea of *thinking machines* and exploring how such machines might eventually be applied to more practical problems.

Although a theorist of immense stature, Shannon was also a hands-on engineer who took great pleasure in building things. He constructed a small prototype of his chess machine — not at Bell Labs, where he had access to superb workshops and equipment, but at home, for the sheer fun of it. His wife, Betty, a Bell Labs technician, helped him with the construction. The result was the *Endgame Machine* — a physical device that could not play a full game of chess but could handle simple endgames involving only a few pieces.

While working on his chess-related projects, Shannon built another remarkable AI "toy" at home in 1950: a maze-solving machine. It consisted of a maze with adjustable metal walls and a small electromechanical mouse Theseus (which can be seen in Shannon's hands on Fig. 6 in Chapter 2), named so after the hero of Greek mythology who navigated the labyrinth of the Minotaur.

Theseus could solve the maze by trial and error, bumping into walls and gradually learning to reorient itself at each junction. Once it completed the labyrinth, it "remembered" the correct path. If placed back at the same starting point, Theseus would complete the run flawlessly. If the walls were moved or the starting position changed, it would relearn the new geometry on the first trial and then navigate it perfectly on the next. This intriguing machine demonstrated Shannon's early exploration of adaptive behavior — a cornerstone of what would become machine learning and robotics.

The mouse was equipped with a small bar magnet and metallic whiskers. When the whiskers touched the metallic walls of the maze, they completed an electrical circuit. Beneath the maze's floor was the machine's "brain" — a relay-based control system composed of 90 telephone relays. Also hidden beneath the surface was a motorized cart that carried another magnet; the cart was driven by X and Y motors. The magnetic attraction between the cart's magnet and the one embedded in the mouse caused Theseus to move as the cart shifted positions.

The mouse could be placed at any arbitrary location within the maze. Its task was to find its way out. On its first attempt, Theseus moved in a random direction. If it encountered a wall, the metallic whiskers made contact, triggering the relay circuit. The mouse would then rotate 90 degrees and try a new direction. When pointed down an open corridor, it would proceed until it reached a turn, where, once again, it would bump into a wall and adjust its direction.

Each wall encounter was memorized by the relay system beneath the floor. Through these experiences, the system built an internal map that allowed Theseus to escape the maze without error on subsequent runs. Once the correct path had been learned, the mouse would glide smoothly from start to finish without touching any walls.

To ensure that Theseus didn't get stuck in repetitive, unproductive behavior, Shannon added what he called an "anti-neurotic circuit." If the mouse repeated the same unsuccessful sequence six times in a row, the system would override the behavior and force it to try a new strategy.

Shannon's mouse learned — and that is what made *Theseus* a milestone in the history of AI. In 1952, Bell Labs produced a short film in which Claude Shannon introduced the maze-solving mouse and briefly explained how it worked. The demonstration captured the public's imagination. *Time* magazine published an article titled "Mouse with a Memory," while *Popular Science* ran a piece under the headline "This Mouse Is Smarter Than You Are."

Shannon's presentation of *Theseus* made a big splash at the 1951 *Macy Conference on Cybernetics*, which was devoted to what, in a few years, would be called AI. The idea that a machine could learn from experience — even in a toy-like context — was technically provocative and philosophically thrilling. It symbolized the shift from hardwired automation to adaptive intelligence, laying conceptual groundwork for future robotics and machine learning developments.

Marvin Minsky and the first neural network machine

Like many pioneers in AI, Marvin Minsky (Fig. 5) was a mathematician. He earned a B.A. in mathematics from Harvard in 1950 and a Ph.D. from Princeton in 1954. As an undergraduate, Minsky came across the seminal 1943 McCulloch–Pitts paper on neural networks and decided to build such a machine himself. But he wanted to go further; he wanted his machine to learn from experience. By 1949, Hebbian learning had been formulated, and Minsky saw a path forward.

He found the opportunity during his graduate studies at Princeton, where he proposed a dissertation focused entirely on neural networks and

Fig. 5. Marvin Minsky (1927–2016), American computer scientist and pioneer of AI. He built the world's first neural network machine in 1951 and attacked neural network research in 1969.

Source: Image reproduced under license CC-BY-3.0.

machine learning. His advisor, Professor Albert Tucker (better known for his work in game theory), had different research interests but allowed Minsky to pursue the topic. It probably didn't hurt that John von Neumann was keenly interested in the project.

While most mathematics dissertations were purely theoretical, Minsky took a different route. In 1951, with the help of a fellow physics graduate student, he built the first neural network machine capable of learning: the *Stochastic Neural Analog Reinforcement Calculator*, or *SNARC*. The cleverly named SNARC[10] was designed to simulate a network of neurons that could adapt through experience — a rudimentary form of what we now call *reinforcement learning*, where a system modifies its behavior in response to rewards or penalties.

SNARC was an analog computer built from 3,000 vacuum tubes and circuits simulating 40 interconnected McCulloch–Pitts neurons. The "synapses" between them could adjust their strengths based on outcomes from

[10] Snark is the fictional creature of Lewis Carroll's poem "Hunting for the Snark." Snark is elusive and very hard to track down.

previous runs — a physical realization of Hebb's maxim: "*Neurons that fire together wire together.*" SNARC's task was to guide a virtual mouse through a simulated maze.

The *synaptic strengths* were embodied in *weights*, i.e., coefficients multiplying the values of the signals exchanged by artificial neurons. These weights could be modified via control knobs. Initially, the network was randomly connected. As the virtual mouse stumbled upon the correct sequence to escape the maze, relevant synaptic knobs were manually adjusted to reinforce that pathway. A human operator monitored the results and tweaked the connections accordingly. Minsky was the first to introduce continually adjustable weights, which became the staple of machine learning. Today, three-quarters of a century later, LLMs are still trained by adjusting the weights.

Minsky's work shared some similarities with Claude Shannon's *Theseus* mouse, described earlier, but there were significant differences. Shannon used logic relay circuits and a relay-based memory to support learning behaviors. Minsky used vacuum tubes to simulate the McCulloch–Pitts neurons. Shannon's system was hardwired, maze-specific, and not designed to scale; expanding it required adding relays and rewiring the entire circuit. By contrast, Minsky's neural network was modular and expandable. And, as we said, the continually changing SNARC weights replaced the on-off switches employed in Theseus.

Building SNARC alone would have been enough to earn Minsky a place in the AI pantheon. But he did much more. He was only 24 when he built his electronic "school for rats," and he went on to have a long and distinguished career, remaining active until his death in 2016 at the age of 88.

In 1955–1956, Minsky was one of the key organizers and participants in the famous Dartmouth Conference, which gave the field its name: *Artificial Intelligence*. In 1959, he co-founded the MIT Artificial Intelligence Laboratory (AI Lab) with John McCarthy. Later, in 1985, he co-founded the MIT Media Lab. In 1969, he received the Turing Award, the highest honor in the field of computer science.

The next chapter in Minsky's scientific career was less admirable. That same year, when he received the Turing Award, Minsky and his MIT colleague Seymour Papert published *Perceptrons: An Introduction to Computational Geometry* (Minsky and Papert, 1969). In that book, they analyzed and criticized the limitations of a neural network model developed by Frank Rosenblatt, whom we will discuss later in this chapter.

Although their analysis was mathematically correct, Minsky and Papert cast doubt on neural networks in general, missing the promise and future potential of more sophisticated multilayered models. The authors intended to sideline neural network research in the U.S. and succeeded, setting the field back by at least a dozen years. We will revisit this story later.

Minsky is also remembered for one of the most spectacularly off-target predictions in the history of AI. In 1970, he told *Life* magazine: "*...from three to eight years, we will have a machine with the general intelligence of an average human being.*" Fifty-five years later, we still expect AGI to appear "from three to eight years" in the future — but this time with some confidence.

And finally, and most importantly, never mind the SNARC or the founding of two MIT labs: Marvin Minsky was also a technical advisor on Stanley Kubrick's *2001: A Space Odyssey*!

AI and the first computers

As described in Chapter 2, the first electronic programmable computers emerged in the U.S. and Great Britain, including ENIAC in 1946, the Manchester Mark I in 1948, EDVAC in 1949, UNIVAC I in 1951, the Ferranti Mark I in 1951, and IAS machine in 1952. By the mid-1950s, there were so many that they could not be listed in one paragraph. Each subsequent machine was more powerful than the previous one, encompassing all aspects of technology, from the speed of operations and memory to improved architecture and programs. It was the dawn of information technology, a time of daring projects — big expectations, big dreams, and, at least in the U.S. and Great Britain, big funding. The emerging computers supported new disciplines, such as *cybernetics,* which we discussed above. And, of course, they supported the first steps of *intelligent machinery*, which, at a minimum, required a machine to be present. The honor of being the first (somewhat) thinking machine fell on IBM's first commercial computer, IBM 701.

Arthur Samuel and the first self-learning program

In the 1950s, *intelligent machinery*, later known as AI, advanced from the mathematical models of McCulloch and Pitts, the musings of Alan Turing, and the computer program proposals of Claude Shannon to practical demonstrations on electronic computers. We have already discussed Marvin Minsky's SNARC, which was completed in 1951. By 1954, American engineer and computer scientist Arthur Samuel had developed a program to play checkers, which had *learned* the game and improved as it played more parties. Unlike Shannon's "toy" models based on relays and Minsky's small-scale SNARC, Samuel's work demonstrated machine learning on a state-of-the-art electronic computer. His learning scheme resembled Shannon's minimax algorithm.

Arthur Samuel was born in 1901 in Emporia, Kansas, to a family of Jewish immigrants (his father having immigrated from England). At age nine, he knew he wanted to be an electrical engineer. Perhaps his father, a schoolteacher, told him about one of the first electrical engineers from his old country, John Ambrose Fleming, who in 1901 made the transatlantic wireless telegraph transmitter for Marconi and, in 1904, invented the first vacuum tube. Much of Samuel's career was centered on electronic vacuum tubes.

Before Arthur Samuel became interested in *machine learning*, he had a distinguished career in electronic engineering and computers. Young Arthur first studied at a small college in Emporia, and at 23, he was admitted to the Department of Electrical Engineering at MIT. He was one of the students who helped Vannevar Bush in the early stages of building his electromechanical Differential Analyzer, which we discussed in Chapter 2. By Samuel's own admission, that got him hooked on digital computers for the rest of his life. After graduating from MIT (it was not a Ph.D. program; Samuel never got one), Samuel briefly worked at General Electric. Then, he joined the newly established Bell Telephone Laboratories, where he worked on various electronic engineering problems from 1928 to 1946. He often interacted with the legendary president of BTL, Mervin Kelly, whom we wrote about in previous chapters. Kelly repeatedly redirected Samuel's work, alternating between assigning him to management and returning him to research. Samuel's work at Bell was, by and large, of high caliber but would not make him famous. He applied for dozens of patents on various vacuum tubes and microwave devices. The renowned physicist from Columbia University, Professor Isidore Rabi, for whom Samuel built some unique instruments, mentioned his contributions in his 1944 Nobel Lecture.

Then, after 18 years at Bell Labs, Samuel felt restless. Despite his assured future at Bell Labs, where he was well respected and valued, he decided to change course. He accepted an invitation to become a tenured professor at the Department of Electrical Engineering at the University of Illinois. This invitation was made despite Samuel not having a Ph.D. Such offers to people without a doctorate were exceedingly rare (now non-existent); his reputation must have been brilliant for this to happen. Samuel spent three years in Illinois teaching EE courses and heading a vacuum tube laboratory. He also participated in the initial work on the university's own electronic computer, ILLIAC, which was finished after he left. By 1949, Samuel decided that he no longer wanted to be a *vacuum tube man* but a *computer man*. For that, he, then 48, was willing to sacrifice the full professorship with tenure — a unique case; nobody ever gives up a tenured professorship, except for Arthur Samuel!

A natural place for a future "computer man" was IBM, a company with significant resources that, in 1949, was ready to enter the electronic computer market. Samuel was interviewed and accepted to IBM Poughkeepsie as a senior engineer (soon to become Manager of Research and Advanced Development). He was offered an incredible salary of $20,000 (about $260,000 in today's dollars). His Bell Labs salary only three years earlier was four times lower! Presumably, IBM really wanted him.

Samuel's first assignment was to assist IBM engineers in building their first electronic (vacuum-tube-based) computer, the IBM 701. His knowledge of digital computers and vacuum tubes and his far-sighted view of the field helped that great company find its "feet" in the computer business, which would become IBM's main business in the years to come. Understanding that the era of vacuum tubes was ending and a new era of solid-state transistors was approaching, Samuel convinced IBM management to establish a solid state laboratory, a move that Bell Labs had made back in 1939 (as mentioned in Chapter 1).

The reader may recall that in Chapter 2, when we discussed computer memories, we mentioned an early computer memory based on the cathode ray tube (CRT) developed in 1946 in Manchester, England, called *Williams' tube memory*. In 1951, Samuel increased that memory fourfold, which was a big deal.

He made additional contributions to the IBM 701, but we will not dwell on them because a much more significant project — the development of a self-learning checkers program — was already in progress. Samuel chose checkers over chess, knowing that modern computers lack the memory needed to handle a more complex chess game. Checkers is not a trivial game, but it requires fewer resources than chess. It is ancient, dating back millennia. A skillful player "sees" a few moves ahead, envisioning and evaluating possible positions. In the late 1940s and early 1950s, computer memory was woefully insufficient to consider all possible moves. Like Shannon before him, Samuel needed to find ingenious ways to circumvent that, and he did.

Samuel worked on his program from 1948, when he was still in Illinois. It was a hobby, and he worked on it in his spare time. Once at IBM, he continued working, but now he was writing for a specific emerging machine, the IBM 701. Designing a new computer involves deciding on the instruction set — a collection of all the commands the computer uses in its operations. Samuel asked IBM engineers to include several new commands he considered helpful for his checker-playing program.

Samuel wrote a program that included the game's rules but did not teach the machine how to play the game. The program enables the machine to recognize when it achieves good results, relying on trial and error to learn the game. It was the first *self-learning* program, and Samuel's work was among the first attempts to work with computers on *non-computational* problems.

Similar to Shannon, Samuel designed an *evaluation function* that assessed the desirability of a given board position. This function helped the program determine which moves were more effective than others. Unlike Shannon's rigid program, however, Samuel's program continually adjusted and refined its evaluation function based on the outcomes of the games it played. Moves that led to wins were reinforced, while moves that led to losses were avoided in future games. This strategy was later known as *reinforcement learning*, and Samuel was the first to implement it, albeit in a rudimentary form. Reinforcement learning emerged as a distinct field and was formalized later, particularly in the 1980s, with contributions from Richard Sutton, Andrew Barto, and others.

A computer played the game numerous times against a human player, improving its skill. Eventually, Samuel's program achieved a "respectable amateur status"; it did not become the world champion in checkers, as that was not possible until much later, when computer memories improved many times over compared to the 1950s.

The first checker-playing program was completed in 1954; the second, with more effective learning capabilities, was completed at the end

Fig. 6. Arthur Samuel (1901–1990) playing checkers with the IBM computer he programmed.

Source: Reprint courtesy of IBM Corporation © (2026).

of 1955. At that point, IBM, which had not previously publicized Arthur Samuel's program, decided to go public. The program's creator was invited to participate in an interview on CBS's "Good Morning," hosted by Will Rogers, Jr., the son of the famous humorist Will Rogers. It was February 24, 1956; an hour-long "Good Morning" show featured Rogers interviewing Samuel remotely, with Rogers in his studio and Samuel at IBM, at the controls of the IBM 701 (Fig. 6). Then, Rogers invited a strong checkers player to have a match against IBM 701. I don't know who won,[11] but the machine performed impressively enough for IBM stock to jump 15 points. Around the same time, U.S. and Canadian film crews filmed matches between IBM computers and strong checker players. One match featured a leading player from Canada and another from the U.S., and this footage was combined and broadcast on American television, as well as on many TV programs worldwide.

All of this placed AI on the map of public opinion; it is one thing to muse about computers that might one day show human-like intelligence

[11] Apparently, the machine demonstrated strong play, but did not win; it may have ended the game in a draw. Otherwise, we would certainly have heard of a victory, and it would have entered the historical record. Later, as Samuel further improved his program, machines began to defeat even strong players. Thus, in 1962, Samuel's program defeated the well-known checkers master Robert Nealy.

and quite another to see on television a strong checkers player lose or tie to the machine.

In 1959, Arthur Samuel summarized his work in a paper, "Some Studies in Machine Learning Using the Game of Checkers" (Samuel, 1959). In it, he coined the term *Machine Learning*, which did not exist before. In the abstract of that paper, he wrote:

> Enough work has been done to verify the fact that a computer can be programmed so that *it will learn to play a better game of checkers than can be played by the person who wrote the program* [Italics ours; a wonderfully lucid definition of machine intelligence!]. Furthermore, it can learn to do this in a remarkably short period of time (8 or 10 hours of machine-playing time) when given only the rules of the game, a sense of direction, and a … list of parameters which are thought to have something to do with the game….

Samuel retired from IBM in 1966 and, for the remaining 24 years of his life, served as an adjunct professor of computer science and later as a professor emeritus at Stanford University. There, he worked on speech recognition and participated in the development of TeX, a typesetting system that became very popular in the field of computer science. He continued writing computer programs until he was over 80, a world record according to his younger associates (Weiss, 1992).

Nothing Arthur Samuel did before or after his work on the self-learning checkers program could match the significance of that milestone. He and Marvin Minsky will always be remembered as the people who initiated machine learning. While Shannon was planning how to program a computer to play chess and Minsky, using a neural network, was teaching virtual mice to escape from imaginary mazes, Samuel made the world aware that machines could learn to do certain things — not just computations — as well as, or even better than, humans.

Untimely thoughts

After writing all this, I paused and wondered: *Wasn't that the moment to cry wolf?* When a machine first learned to play a game on par with a human, should that not have been the moment to say, *Stop! Don't go any further.*

But of course, such thoughts were inconceivable in the 1950s. That wonderful man, Arthur Samuel, probably would not have understood my concern. And even today, they are not particularly popular. We are delighted by AI. ChatGPT has appeared only recently, and it already seems that life would be impossible without it.

No matter what any of us may think, the show goes on. We are destined to walk into this *gilded mousetrap*, which always looks so inviting.

The workshop that named AI

In 1955, John McCarthy, then an assistant professor of mathematics at Dartmouth College, proposed a summer research project on "thinking machines." He gave it the name *Artificial Intelligence*. To organize the workshop, McCarthy invited three other prominent scientists: Marvin Minsky, Claude Shannon, and Nathaniel Rochester. Together, they planned what became known as the *Dartmouth Summer Research Project on Artificial Intelligence,* held for several weeks in the summer of 1956 at Dartmouth College in New Hampshire.

Roughly 50 researchers participated, including short-term visitors. Among the names we now recognize were Claude Shannon, Marvin Minsky, Warren McCulloch, Julian Bigelow (a computer engineer who worked with Norbert Wiener on cybernetics and oversaw the construction of von Neumann's IAS computer; see Chapter 2), John Nash (the brilliant and troubled mathematician portrayed in the 2001 film *A Beautiful Mind*, who would go on to receive the 1994 Nobel Prize in Economics and the 2015 Abel Prize in Mathematics), and Arthur Samuel, whose work we've just discussed.

Despite the star-studded lineup, two relatively unknown scientists gave the most successful presentation: Allen Newell of the RAND Corporation and Herbert Simon of the Carnegie Institute of Technology (now Carnegie Mellon University). They demonstrated a computer program they had written with J. C. Shaw, called the *Logic Theorist* (LT) — a program that, according to Simon, could actually reason. As he put it, "We have invented a computer program capable of thinking non-numerically...."

The LT could prove 38 of the first 52 theorems in Russell and Whitehead's *Principia Mathematica* — the same work we referenced earlier when discussing Walter Pitts. Despite this remarkable achievement, the *Journal of Symbolic Logic* rejected a paper co-authored by Newell, Simon, and the LT itself.[12]

It may come as a surprise that programs capable of proving complex mathematical theorems existed as early as the 1950s, but they did. These early systems were narrow in scope but could tackle logical, algebraic, and geometrical problems with some success.

Following the Dartmouth Workshop, between 1957 and 1959, Newell and Simon, collaborating again with J.C. Shaw, developed a more ambitious program: the *General Problem Solver* (GPS). It was designed to model human problem-solving behavior across a wide range of domains. GPS broke down problems into subproblems and used *heuristics* — rules of thumb — to guide the search for solutions.

One of Newell and Simon's most lasting theoretical contributions was made in 1975 when they formulated the *Physical Symbol System Hypothesis*. According to this hypothesis, intelligent behavior arises from manipulating symbols according to rules. In their view, symbol processing was not only necessary but also sufficient for general intelligence — a view with deep historical roots in the works of Ramon Llull and Gottfried Leibniz, which we discussed earlier. This idea became the cornerstone of *symbolic AI* and shaped the emerging field of cognitive science, which sought to explain human thinking in terms of rule-based systems.

Newell and Simon's pioneering role in the birth of symbolic AI, along with their broader theoretical insights, earned them the *1975 Turing Award*, the highest honor in computer science.

But to return to the Dartmouth Workshop itself, whether or not it produced lasting technical results, it unquestionably focused the scientific community's attention on the new and exciting idea of machine intelligence. It helped define many of the field's recurring themes — language processing, neural networks, machine learning, and symbolic reasoning — and, perhaps most enduringly, it *gave the field its name*. That name stuck.

[12]This was the first known instance of a machine being listed as a co-author of a scientific paper. Until then, science had occasionally seen fictional or anonymous authors but never a truly non-human one. The editors of the *Journal of Symbolic Logic* declined to publish the paper.

Machine Translation (MT)

So far, we have described early attempts at making machines think and learn. No one knew how to distinguish true thinking from mere imitation, and many AI practitioners, then and now, suspect there may be no real difference. By the mid-1950s, machines played popular games and guided their virtual surrogates through labyrinths, learning through trial and error how to do these things better.

However, there was another line of attack: Researchers sought to develop machines that could automatically translate from one language to another. Unlike playing games, machine translation (MT) has broad and immediate human value. Language has always been seen as a quintessential human prerogative — "the mirror of the intellect," as Leibniz called it. That, combined with the onset of the Cold War in the late 1940s and the need to translate Russian, made MT a coveted target of scientific effort and government funding.

I asked ChatGPT-4o about the history of MT, and here is what I got (slightly edited and rephrased):

1954: The Georgetown-IBM experiment marked the first successful demonstration of machine translation. It translated 60 Russian sentences into English using a rule-based approach, showing the potential of automated translation. This led to substantial government funding of MT research in the U.S.

1957: The launch of *Sputnik* in 1957 intensified U.S. fears of technological inferiority and triggered a surge in government funding for science and defense, including machine translation. MT was seen as strategically vital, especially for translating Russian scientific and technical literature. Several research groups, including those at Georgetown, IBM, MIT, and RAND, expanded their rule-based systems with increased federal support. Despite this momentum, progress remained limited. Translations were often literal and unreliable, contributing to the growing skepticism in the years that followed.

1960s: Researchers focused on rule-based machine translation (RBMT), which used hand-crafted rules and bilingual dictionaries. But by the mid-1960s, progress had stalled. The *Automatic Language Processing Advisory Committee* (ALPAC), commissioned by the U.S. government, published a devastating 1966 report recommending reduced funding; as a result, MT research in the U.S. nearly ground to a halt.

1980s: Advances in computing power and access to large bilingual datasets enabled the rise of *statistical machine translation* (SMT), which used probabilistic models to predict word sequences.

1993: IBM's *Candide* project applied SMT methods to practical translation tasks, improving fluency and accuracy.

2000s: With the rise of the internet and the increasing availability of large-scale text data, SMT systems became more sophisticated and widespread.

2006: *Google Translate* launched using SMT. It leveraged web-scale (*extremely large*) bilingual datasets to improve translation quality and made MT accessible to the general public.

2014: The introduction of neural networks brought about a major shift. Researchers developed neural machine translation (NMT) systems using deep learning to model the translation process.

2016: Google Translate transitioned from SMT to NMT with the launch of the *Google Neural Machine Translation* (GNMT) system. Translation quality has significantly improved, particularly in terms of fluency and contextual accuracy.

2017: The *Transformer model*, introduced by researchers at Google, became the foundation for most modern NMT systems due to its ability to model long-range dependencies and parallelize training (Vaswani *et al.*, 2017).

Current state of machine translation (as of early-2026)

The emergence of *large language models* such as ChatGPT-3, ChatGPT-4, and more recently, GPT-4o and GPT-5 has transformed machine translation far beyond the capabilities of earlier NMT systems:

1. *Context-Aware and Nuanced Translation*: Earlier NMT systems, such as Google Translate or DeepL, often struggled with idioms, nuance,

and long-range coherence. LLMs maintain consistent context across entire paragraphs and documents, producing translations that better preserve tone, meaning, and cultural subtleties.

2. *Multilingual Flexibility and Low-Resource Gains*: Traditional systems, which were trained on specific language pairs, required extensive bilingual data to achieve their goals. LLMs are inherently multilingual and can generalize from limited examples, significantly improving translation quality in low-resource languages — a longstanding weakness for MT.
3. *Idiomatic and Stylistic Control*: Modern LLMs translate idioms, slang, and proverbs with surprising fluency. They also let users specify the tone and style — formal, casual, technical, poetic, etc. — making translations more flexible and audience-appropriate.
4. *Real-Time and Interactive Translation*: With GPT-4o, real-time translation became a practical reality. These models now support live, multilingual interaction, transforming customer support, global meetings, and cross-border collaboration.
5. *Multimodal Capabilities*: GPT-4o introduced multimodal translation, enabling the model to handle spoken language, images containing text, and traditional written input, all within one interface. This innovation is particularly valuable for accessibility, travel, education, and media consumption.
6. *Adaptive and Personalized Learning*: Some MT systems now improve over time by learning from user feedback. This adaptive translation supports specialized domains, such as legal, medical, and technical writing, without requiring retraining of the entire model.
7. *Seamless Task Integration*: LLMs integrate translation with summarization, sentiment analysis, rewriting, and Q&A tasks, making them powerful for content localization, multilingual search, and *real-time communication across languages.*[13]

In short, modern MT is no longer simply about converting words between languages. It is about reexpressing meaning with an awareness of context, style, and purpose. LLMs have increasingly blurred the line between translation and conversation.

---

[13] *Real-time communication across languages* refers to live, two-way interaction between speakers of different languages, enabled by instant translation of speech or text as the conversation unfolds. Advanced systems can support spoken or written input, allowing participants to communicate naturally in their own language.

I used ChatGPT-4o to help write the text above. I was curious to see what it could do, and perhaps you'll agree that it did a most capable job.

To conclude this section in a "human voice" and to hear a professional opinion, I asked my longtime friend Rosa Kavenoki, Professor Emerita at the Middlebury Institute of International Studies at Monterey — where she headed the Russian–English translation program for many years — to add a few words:

> There are two domains for transferring thoughts between languages: translation, which is written, and interpretation, which is oral.
>
> Machine translation applies specifically to written texts. Today, AI tools can translate nearly anything in seconds — technical documents, emails, news articles, even some literary works. As a result, many freelance and staff translators have shifted roles. Instead of translating from scratch, they now work as editors, revisers, and content formatters, refining machine output for clarity, tone, and accuracy.
>
> That said, prose and poetry, with their layered meanings, rhythm, and nuance, still rely on human translators. Machines struggle with metaphor, cultural references, and artistic intent. Here, human creativity remains essential.
>
> Interpretation is still a different world and remains mostly in human hands. But that grip is loosening. It's now under aggressive pressure from emerging speech and voice generation technologies. These tools are advancing fast, mimicking tone, timing, and even emotion. At this pace, machine-driven interpretation could take over within a few years, especially in standardized or lower-stakes settings. High-level diplomacy may hold out longer, but the shift is already coming.

Frank Rosenblatt and the Perceptron

Frank Rosenblatt (1928–1971) was an American psychologist and neuroscientist who was fascinated by Hebbian learning and the neural network model of McCulloch and Pitts. Putting "two and two together," Rosenblatt, in 1957, developed a mathematical model for how a brain might process visual input. Based on this model, he designed and built a machine, *Mark I Perceptron*, in 1958 (Fig. 7). The artificial neurons were implemented using analog summation circuits and mechanical

Fig. 7. Frank Rosenblatt with his Perceptron photographed in 1958–1960.
Source: Image in the public domain.

potentiometers (variable resistors). They produced binary outputs based on whether the summed input exceeded a threshold.

The machine's name — *Perceptron* — evoked a machine's ability to perceive the world.

Although not trained in mathematics or computer science, the 29-year-old Rosenblatt proposed a simple yet effective binary classification algorithm for assigning objects to one of two categories. This was a wise choice: Classification is one of the simplest cognitive tasks, requiring intelligence. Rosenblatt believed that future versions of his machine would become vastly more capable. That future never arrived. In 1971, amid Perceptron experiments, Frank Rosenblatt tragically died in a boating accident at the age of just 43.

What captured the public's imagination was that the Perceptron had "eyes" — it could perceive objects in its visual field. That, combined with Rosenblatt's learning algorithm and its novel ANN-based hardware implementation, made the Perceptron a landmark in the early history of AI.

The first Perceptron, *Mark I*, was an analog electromechanical machine built at Cornell University's Aeronautical Laboratory in Buffalo, New York. It was equipped with a 20 × 20 grid of photoelectric cells that acted as the machine's eyes. After Mark I, several others were developed, but subsequent Perceptrons were implemented in software, running on general-purpose digital computers, such as the IBM 704 and later machines.

The Perceptron was shown simple visual objects, one at a time: squares and circles, squares and diamonds, the letters X and E, or other basic shapes belonging to two distinct categories. The size of the image sets varied. For example, the X-and-E set contained 10 examples of each letter (a total of 20), while the circle-and-square set included 500 shapes. A typical training session involved dozens of runs — sometimes many more — during which each object in the binary set was repeatedly presented to the machine. The Perceptron was gradually "taught" how to classify them (see the following special section for details on the learning process).

How the Perceptron Learned

The "eyes" of the Perceptron (Fig. 8) — photoelectric cells — produced 400 electrical signals when an object was placed before them. Each cell generated a voltage proportional to the amount of light it received. The collection of these 400 voltages — today, we would call them 400 pixels — effectively captured the shape of the object placed within the grid's field of

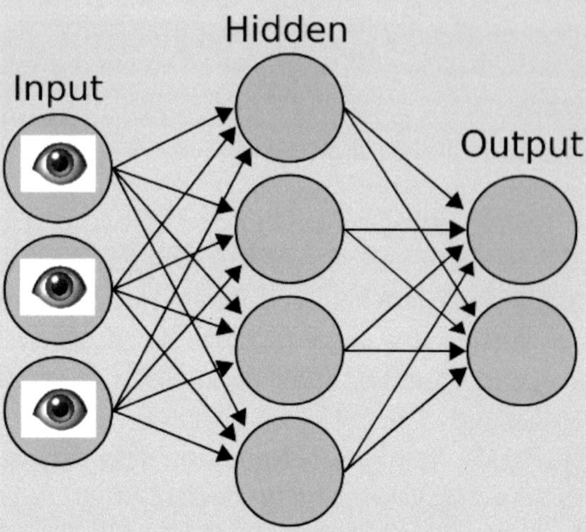

Fig. 8. A schematic diagram showing a neural network with an input layer ("eyes" or photocells), a hidden layer (neurons engaged in processing the signals they receive from the input layer), and a two-neuron output layer making the classification decision. Only the number of output neurons is realistic; in all the other layers the number of neurons is reduced compared to the real Perceptron.

Source: An original image created by Colin Burnett, Lic. CC BY-SA3.0; "eyes" added by the author.

vision. Each of these input voltages was multiplied by a coefficient called a *weight*. These weights represented the importance of each input in determining the output. Initially, all weights were set to random values but adjusted during training based on feedback from the classification results. We first met weights in a section about Marvin Minsky's SNARC, where they were used to train a neural network-guided virtual mouse to escape a virtual labyrinth. Rosenblatt, for the first time, used weights in training the system to recognize images.

The inner circuitry — what we would now call the *hidden layer* — consisted of approximately 500 McCulloch–Pitts neurons, now commonly referred to as *nodes*. Each node received a weighted combination of signals from many input cells and performed a simple summing operation followed by a threshold test: If the sum exceeded a threshold value, the neuron "fired"; if the sum was below the threshold, it remained silent. These hidden-layer neurons acted as feature detectors, responding to different patterns in the input image. Their outputs were passed to the output layer.

The output layer, which made the final decision, usually consisted of one or two neurons. In the simplest case, a single output neuron handled the classification: If it fired, the Perceptron recognized a circle; if it did not, it recognized a square. In other versions, two output neurons were used, each representing one of the two possible choices. The decision was then made by comparing the amount of input each received — the neuron with the stronger total signal determined the outcome. Although the Perceptron's neurons were inspired by the McCulloch–Pitts model, where a binary neuron fires or not, the actual machine worked with continuous *analog* signals. This allowed it to compare signal strengths before applying the final yes-or-no decision.

If the classification was correct — for example, if the machine correctly identified a circle — the potentiometers controlling the weights remained unchanged, and the next image was presented. If the classification was incorrect — say, if the machine reported a square instead of a circle — the operator adjusted the relevant weights by turning the knobs. The same image was then shown again. This process was repeated until the machine provided the correct response, after which the operators proceeded to the next object.

Rosenblatt also introduced a concept that would later become central to AI: the *loss function*. In the original Perceptron, the "loss" was represented by simple binary feedback: either a thumbs-up or thumbs-down. A correct answer meant no changes; an incorrect one triggered a minor uniform weight adjustment. This primitive loss function did not quantify *how wrong* the guess was — only that it was wrong. Later, neural networks would adopt far more nuanced loss functions that measured the degree of error and provided more specific guidance for improvement.

In Rosenblatt's Perceptron, weight adjustments following an error were fixed in size — each weight was incremented or decremented by a constant amount, called the *learning rate*. The direction of the adjustment depended on whether the internal computation had overshot or undershot the decision threshold.

Although the final decision was binary, Perceptron's internal computation was not. It calculated a continuous, weighted sum of all inputs and then applied a threshold function: If the sum was positive, it "saw" a circle; if negative, it saw a square. The weight adjustments aimed to nudge this internal sum toward the correct side of the threshold.

Strictly, the original Perceptron did not require a formal mathematical loss function. But the core idea — that a system must evaluate its performance and update its internal parameters accordingly — was already present. This principle became the foundation of learning in all future ANNs.

In today's deep learning systems, this idea has evolved into a precise science. Modern neural networks use continuous loss functions that not only indicate *how* wrong a prediction was but also provide a direction and magnitude of weight adjustments for improving it. We will revisit this in more detail when we discuss the *backpropagation* algorithm.

After training, the Perceptron generally succeeded at binary classification. Its effectiveness depended on the types of shapes and conditions used. For example, it could distinguish between circles and squares with 99.8% accuracy.

Concerning this ability to learn from data and produce correct outputs (such as accurate classifications), Rosenblatt proved the following result, now known as the *Perceptron Convergence Theorem*:

If the training data are linearly separable, then the Perceptron learning algorithm is guaranteed to converge in a finite number of steps to a set of weights that correctly classifies all training examples.

The term, *linearly separable data,* is hard to explain without using math. In rough terms, it means that the categories you are trying to separate — say, circles versus squares or the letters X versus E — have clearly distinguishable features that the machine can learn to tell apart. By contrast, categories such as cats versus dogs are more subtle and overlapping and cannot be separated so easily. Note that this theorem does not say that nonlinearly separable problems could never be solved with the Perceptron; it only says that this result isn't guaranteed.

This theorem and other insights were presented in Rosenblatt's 1962 book, *Principles of Neurodynamics: Perceptrons and the Theory of Brain Mechanisms*.[14]

Later, we will see that Rosenblatt's Perceptron received a negative review from Marvin Minsky and others in mainstream AI research. We will also describe a powerful method that allowed AI scientists to overcome the Perceptron's limitations, known as "backpropagation." The technique had many "fathers," one of whom, arguably, was Frank Rosenblatt himself. His 1962 book already contained the seeds of that future revival: It contained the concept of backpropagation, although not in exact mathematical terms. Here is the excerpt from an interview with Professor Jack D. Cowan,[15] recorded in 1993:

> Backpropagation was the key to it; I think so. If you read Rosenblatt's book [*Principles of Neurodynamics*], he talks about backpropagation in there — the architecture for solving the problem and everything, even

[14] One can't help but be impressed by Frank Rosenblatt — a psychologist and neuroscientist who not only built one of the earliest "thinking machines" but also proved foundational theorems in computer science. Not bad for someone without formal training in the exact sciences.

[15] Jack D. Covan, a professor in the Department of Mathematics at the University of Chicago, significantly contributed to the statistical theory of neural networks and formulated the mean-field theory for large-scale brain activity.

> including the term "backpropagation." The only thing that's not there is the actual algorithm, the details of it, but the structure and everything else is there. Poor old Frank was vastly undervalued in his day. (in Anderson and Rosenfeld, 1998, 115)

We will describe backpropagation in more detail later.

A newcomer to computer science and engineering, Rosenblatt welcomed public interest and recognition. He was not averse to favorable publicity, even when it was fueled by somewhat inflated promises. When the *Perceptron Mark I* was unveiled in July 1958, journalists were invited to the demonstration, and *The New York Times* published an article entitled *"Electronic Brain Teaches Itself."* In it, we find some rather colorful statements:

> Navy officers demonstrating a preliminary form of the device in Washington said they hesitated to call it a machine because it is so much like a "human being without life." Dr. Frank Rosenblatt…designer of the Perceptron, conducted the demonstration. The machine, he said, would be the first electronic device to think like a human brain. Like humans, Perceptron will make mistakes at first, but "it will grow wiser as it gains experience."…Later Perceptrons, Dr. Rosenblatt said, will be able to recognize people and call out their names. Printed pages, longhand letters and even speech commands are within its reach. Only one more step of development…is needed for the device to hear speech in one language and instantly translate it to speech or writing in another language…In principle, Dr. Rosenblatt said, it would be possible to build Perceptrons that could reproduce themselves on an assembly line and which would be "conscious" of their existence.

I do not wish to criticize the gifted Frank Rosenblatt; however, it is clear that such exuberant predictions were bound to backfire — and indeed, they did, though not immediately, but about a decade later.

Rosenblatt's work was supported by the Office of Naval Research (ONR), which is why the *New York Times* article mentioned Navy officers. Rosenblatt built second-generation Perceptrons (mostly in software) and attempted to teach them more complex tricks. However, the Perceptron's abilities were limited. It never learned to classify anything but the simplest shapes, and even then it was lost when the shape sizes or orientations varied. Rosenblatt hoped to overcome these limitations and continued improving his Perceptrons until his untimely death in 1971.

The AI Winter That Lasted for a Decade

Critique of Perceptrons

In 1969, Marvin Minsky and Seymour Papert published *Perceptrons: An Introduction to Computational Geometry* (Minsky and Papert, 1969),[16] a book that critically examined the capabilities and limitations of Rosenblatt's Perceptron. With mathematical rigor, the authors demonstrated why there was little reason to expect such a device to succeed beyond a narrow class of problems.

Rosenblatt understood the limitations of his creation better than he is sometimes given credit for. His *Perceptron Convergence Theorem*, which we stated in the previous section, showed that when the data can be separated by a straight line — that is, when it is *linearly separable* — the Perceptron will always find the correct solution in a finite number of steps. For more complex, nonlinear problems, he was less certain: They might be solvable, he thought, or they might not.

Seven years later, Marvin Minsky and Seymour Papert settled the question mathematically. They proved that a single-layer network like Rosenblatt's could *never* solve such nonlinear problems. It was not a refutation of Rosenblatt's theorem but a clarification of its boundaries, showing precisely where the line between possibility and impossibility lay.

One example they emphasized, which would become infamous in the history of AI, was the *exclusive-or* (XOR) problem. XOR is one of the basic operations in Boolean algebra. The XOR function takes two binary inputs and returns true (logical 1) if *exactly one* of them is true; in all other combinations, it returns false (logical 0). While this logic is trivially easy for humans to grasp, Minsky and Papert demonstrated that a single-layer Perceptron cannot compute XOR. The reason is that the data which provides an XOR input (those 1s and 0s) is not linearly separable: You can't draw a straight line that cleanly divides the "true" and "false" outputs in a two-dimensional input space. Since the Perceptron had no hidden layers

[16]At the time of the 1969 publication of *Perceptrons*, Marvin Minsky was a professor of electrical engineering at MIT and co-founder of the MIT Artificial Intelligence Group. Seymour Papert was an associate professor of applied mathematics at MIT and an active member of the AI research community there.

and relied only on linear decision boundaries, it was fundamentally incapable of solving such problems.

The XOR case became both a symbolic and technical demonstration of the Perceptron's limitations, and it was central to Minsky and Papert's argument that neural networks — at least in their then-current form — had hit a conceptual dead end.

Minsky and Papert extended their critique beyond single-layer models, casting doubt on the promise of neural networks in general. The *multilayer* Perceptrons were not criticized as inherently flawed but were dismissed mainly because no practical way to train them existed at the time. Minsky and Papert's conclusions were widely interpreted as a broader rejection of neural networks as a viable path toward AI.

Two aspects of the Minsky–Papert book warrant careful examination: the validity of its scientific conclusions and the underlying motivations of its authors. On both fronts, contemporary members of the neural network community expressed strong reactions. First, about the validity of the Minsky–Papert conclusions. Here is the opinion of Jack D. Cowan, whom we quoted previously:

> There was a very heated discussion between Dave Block and Marvin Minsky. Block reviewed Minsky and Papert's book (Block was a physicist at Cornell who worked with Rosenblatt in the early 1960s.) The review was critical, and then there was the rejoinder [*a sharp or witty reply*]. Block claimed that the multilayer Perceptron could do everything that Minsky and Papert claimed couldn't be done by a Perceptron. With hindsight, Dave was correct, although he didn't know how to do it. He was right that the multilayered Perceptron is capable of doing all the things that they said couldn't be done. (Anderson and Rosenfeld, 1998, pp. 115–116)

Of course, we know that multilayered ANNs, when adequately trained, could do everything Mansky and Papert requested of them in 1969 and much more. What interests us today is that the anti-Perceptron's conclusions were strongly disputed as soon as they appeared — something you do not often hear when reading about the history of ANNs.

And now, to the second question: What was Minsky and Papert's motivation? I always thought that the blow to neural networks they delivered in 1969 was unintentional, that it was collateral damage from their serious mathematical analysis of the Perceptron's limitations. But apparently, it was not so innocent. Here is a devastating account of that saga, given by ANN scientist Robert Hecht-Nelson[17] in 1993:

> ...at least some of these researchers *on the outs* [*out of fashion in the 1960s MG*] found that emotionally intolerable that some of what they saw as "lightweight" research on neural networks was receiving a lot of press attention. A particularly dramatic case of this was Marvin Minsky. Minsky had gone to the same New York "science" high school [*Bronx High School of Science*] as Frank Rosenblatt, a Cornell psychology Ph.D., whose Perceptron neural network pattern recognition machine was receiving significant media attention. The wall-to-wall media coverage of Rosenblatt and his machine irked Minsky. One reason was that although Rosenblatt's training was in "soft science," his perceptron work was quite mathematical and quite sound — turf that Minsky, with his "hard science" Princeton mathematics Ph.D., didn't feel Rosenblatt belonged on. Perhaps an even greater problem was the fact that the heart of the Perceptron machine was a clever motor-driven potentiometer adaptive element that had been pioneered in the world's first neural computer — the Snark — which had been designed and built by Minsky several years earlier. In some ways, Minsky's early career was like that of Darth Vader. He started out as one of the earliest pioneers in neural networks but was then turned to the "dark side of the force" and became the strongest and most effective foe of his original community. This view of his career history is not unknown to him. When he was invited to give the keynote address at a large neural network conference in the late 1980s to an absolutely rapt audience, he began with the words, "I'm not the Devil."
>
> By the mid-1970s [*This is the typo; it should have been "mid-1960s," MG*], Minsky and his colleagues (notably Seymour Papert) began to take actions designed to root out neural networks and ensure

[17]Robert Hecht-Nielsen (1947–2019) was an early advocate of artificial neural networks and a pioneer in the field of neurocomputing. He co-founded HNC Software, developed neural network models for commercial applications, and was known for his work on the theoretical foundations of neural learning.

> large and, in their view, richly deserved funding for AI research by getting some money currently being wasted on neural networks and more to boot, redirected. They did two things. First, Minsky and Papert began work on a manuscript designed to discredit neural network research. Second, they attended neural network and bionics conferences and presented their ever-growing body of mathematical results being compiled in their manuscript to what they later referred to as "the doleful responses" of members of their audiences…
>
> Of course, in reality, the whole thing was intended from the outset as a book-length damnation of Rosenblatt's work and many of its variants in particular, and, by implication, all other neural network research in general.
>
> Minsky and Papert's book *Perceptrons* worked. The field of neural networks was discredited and destroyed. The book and the associated conference presentations created new conventional wisdom at DARPA and almost all other research sponsorship organizations that some MIT professors have proven mathematically that neural networks cannot ever do anything interesting. The chilling effect of this episode on neural network research lasted almost 20 years. (Anderson and Rosenfeld, 1998, pp. 303–305)

In a peculiar twist of history — given that Minsky's own 1956 SNARC machine was based on a neural network — the publication of *Perceptrons* caused the collapse of the neural network research funding across the U.S. and, soon after, in much of Europe. Rosenblatt's ONR funding was the first casualty. In addition to believing Minsky and Papert, perhaps the Navy program managers were disappointed by the Perceptron's inability to *recognize them* and *call out their names*, as Rosenblatt promised them in 1958.

The Lighthill report

Sir James Lighthill was a distinguished British scientist and Lucasian Professor of Applied Mathematics at the University of Cambridge. He was best known for his foundational work in aeroacoustics — the theoretical study of noise generated by turbulent flows, such as those produced in jet engines. For his scientific achievements, Lighthill was elected a Fellow of the Royal Society and the Royal Aeronautical Society. In 1953, he was knighted by Queen Elizabeth II.

Lighthill was a first-rate mathematician with impeccable credentials and no direct ties to the field of AI. That is why, in 1972, the British *Science Research Council* (SRC) — the body funding AI research in the UK — asked him to evaluate the state of British AI and, by extension, global AI research.

Lighthill delivered his report in the summer of 1972, which was followed by a widely discussed BBC television debate in July 1973, titled "*The Prospects for Artificial Intelligence*."

In his brief (21-page) report, Lighthill reviewed AI's major themes and achievements over the prior 25 years. There was no bibliography or list of references; it contained only the names of the individuals he interviewed. Thus, it is difficult to determine what, if any, AI literature Lighthill reviewed — his investigations appear to have relied almost entirely on interviews with prominent AI researchers in the U.K. and the U.S. Among them were Marvin Minsky and John McCarthy, with the latter being the same McCarthy who had organized the seminal Dartmouth Conference in 1956.

Lighthill concluded that AI failed to live up to its early promises, except for narrow areas such as data storage and retrieval, and even then, only when the data were "confined to very precisely defined and analyzed types." He summed up the prevailing mood among the researchers he interviewed with the following observation:

> Most workers in AI research and in related fields confess to a pronounced feeling of disappointment in what has been achieved in the past twenty-five years. Workers entered the field around 1950, and even around 1960, with high hopes that are very far from having been realised in 1972. In no part of the field have the discoveries made so far produced the major impact that was then promised. (Lighthill, 1973)

This was precisely what the British Science Research Council was predisposed to hear. It promptly responded by canceling nearly all U.K. government funding for AI research, leading to the closure of multiple AI labs, most notably at the University of Edinburgh, which had been a global leader in the field. The consequences of this decision were felt across the U.K. from 1973 through the mid-1980s. In short, Sir James single-handedly triggered Britain's national "AI winter."

Ever since 1776, what has been good for Britain has often been good for America — and vice versa. The U.S. experienced its own triggers for the AI winter, some of which we have already mentioned: the 1966 ALPAC report, which criticized the lack of progress in machine translation, and the 1969 Minsky–Papert book, which exposed the limitations of the Perceptron. It's also likely that there were transatlantic ripple effects — skepticism and funding reductions echoing back and forth across the ocean.

As a result, the field of AI entered a period of stagnation worldwide, except in a few countries, notably the Soviet Union and Japan. In the West, the *AI winter* lasted from the 1970s through the early 1980s. The freeze lasted even longer for machine translation, which did not see serious renewal until the 1990s.

Symbolic AI vs. Artificial Neural Networks

We may envision the history of AI as a plant with long underground roots. The part of the plant that emerged above ground — what we call AI proper — appeared sometime in the 1950s. But its roots reach much further back, spanning decades, even centuries, and millennia. Some of these roots were explored in Chapter 2 and at the beginning of this chapter.

When we shift our gaze to the visible part of the plant — the part representing modern AI — we see that it does not resemble a tree with a thick central trunk and smaller branches sprouting from it at some height. Instead, it resembles a bush, with several stems growing side by side from the very beginning. Initially, these stems were not sharply distinguished from one another. But by the end of the 1950s, two main stems had become visible: one larger, representing classical or so-called *symbolic AI*, and a smaller one — AI based on *artificial neural networks*.

The distinction between symbolic AI and ANNs came into sharp focus with the controversy surrounding the Perceptron and the publication of Minsky and Papert's book, which attempted to prune the field by cutting away what was seen as an unpromising branch of ANNs. As we know today, that judgment was mistaken. Paradoxically, the ANN stem turned out to be viable and ultimately dominant, while the symbolic AI stem withered and has nearly disappeared.

While the initial impulse was toward developing AI that would simulate the human brain (recall Alan Turing's early AI papers), most of the AI work done in the 1950s, 1960s, and 1970s was carried out within the framework of symbolic AI. This approach was most clearly articulated in the 1976 paper "Computer Science as Empirical Inquiry: Symbols and Search" by the Dartmouth Workshop participants Allen Newell and Herbert Simon. This paper was presented as their joint lecture after they received the 1975 Turing Award.

The roots of symbolic AI can be traced back to the ideas of Ramon Llull and Gottfried Leibniz. Llull famously inscribed letters and symbols around rotating paper disks, attempting to generate truths through mechanical combinations. Leibniz, inspired by Llull, envisioned a universal symbolic language governed by the rules of logic — a formal calculus capable of solving any intellectual problem. While that universal language was never realized, Leibniz's pursuit gave rise to what we now call mathematical logic. We discussed all of that at the opening of this chapter.

Symbolic AI, the main branch of the AI tree during those early decades, followed a similar approach: modeling intelligence by explicitly encoding human knowledge and manipulating it according to formal logical rules. When the limitations of this approach became evident in the 1980s, it was retrospectively labeled "symbolic AI" — or, somewhat mockingly, "good old-fashioned artificial intelligence" (GOFAI). It was also called "classical AI."

It would be unfair to call symbolic AI barren — far from it. Much of the early progress in AI, including rule-based machine translation and game-playing programs, emerged from this tradition. Researchers also built successful expert systems, i.e., programs that performed well in narrow, well-defined domains such as medical diagnosis, chess, and mathematics. These were constrained environments in which logic-based approaches made sense and yielded valid, transparent, and interpretable results.

However, by the late 1980s, it had become increasingly clear that symbolic AI struggled in ill-defined contexts involving ambiguity, perception, or learning from raw data. Meanwhile, neural networks were gaining strength.

It would be naive to imagine, however, that as a result of the long competition between symbolic AI and AI based on neural networks and deep learning, the entire battlefield was left to ANNs and that symbolic AI disappeared from the face of the earth. This is not the case. Some AI scientists believe significant progress will be achieved by combining these two approaches in the future.

I would like to quote from a recent discussion between Alex Kantrowitz, host of the *Big Technology Podcast*, and Gary Marcus, professor emeritus of psychology and neural science at New York University, an AI expert, and author of several books on AI. Professor Marcus mainly spoke about the "scaling wall" — how LLMs lately (since the appearance of ChatGPT-4) are delivering diminishing returns on investment and that new approaches to building AGI are probably needed. At the end of that podcast, Marcus, at his host's request, gave a lively summary of symbolic AI versus neural networks, the subject of this section. His thoughts about combining symbolic AI with deep neural networks seem particularly intriguing:

> AK: Gary… I want you [*to say*]…about the path that you think needs to be taken to AGI…
> GM: A lot of people will have read Danny Kahneman's book "Thinking, Fast and Slow,"[18] and there he talked about *System 1* and *System 2* cognition.
>
> System 1 was fast and automatic. Reflexive System 2 was more deliberate, more like *reasoning*…I would argue that the neural networks that power generative AI are basically like System 1 cognition. They're fast, they're automatic, they're statistically driven, but they're also error-prone; they're not really deliberative. They can't sanity-check their own work…System 2 is more like classical AI, where you can explicitly represent knowledge…It looks more like computer programming.
>
> These two schools have both been around since the 1940s, but they've been very separate for what I think are sociological and economic reasons. Either you work on one, or you work on the other… there's been a great deal of hostility between the two, but the reality is — they kind of complement each other; neither of them has worked on its own. The classical AI failed…people built all these expert systems, but there were always these exceptions, and they weren't really robust. You'd pay graduate students to patch up the exceptions.
>
> Now, we have these new systems [*ANNs*]; they're not really robust either…The advantage of System 1 is [*that*] it learns very well from data. The disadvantage is — it's not very abstract…The large language models and that kind of approach — transformers — are very good at learning, but they're not very good at abstraction. You can give them billions of examples, and they still never really understand what

[18] Daniel Kahneman, *Thinking, Fast and Slow* (2011). Kahneman is the winner of the 2002 Nobel Prize in Economics.

multiplication is, and they certainly never get any other abstract concept. Well, the classical approach is great at things like multiplication; you [*make*] a calculator, and it never makes a mistake, but it doesn't have the same broad coverage, and it can't learn new things. So you can wire multiplication in, but how do you learn something new? The classical approaches have had trouble with that.

And so I think we need to bring them [*the two approaches*] together, and this is what I call *neuro-symbolic AI*. It's really what I've been lobbying for decades, and I think it was hard to raise money to do that in the last few years because everybody was obsessed with generative AI. But now that they're seeing the diminishing returns [*of scaling*], I think investors are more open to trying alternatives…*AlphaFold*[19] is actually a neuro-symbolic model, and it's probably the best thing that AI ever did — decoding protein folding, figuring out the three-dimensional structure of a protein from a list of its nucleotides [*Marcus probably means — amino acids, MG*]. ("Are We at the End of AI Progress With Gary Marcus," May 7, 2025, *Big Technology Podcast*)

The last great triumph of symbolic AI

In 1997, symbolic AI demonstrated its power one last time. When a machine defeated the reigning world chess champion, something broke — something snapped in our collective worldview. After 200,000 years[20] of undisputed intellectual dominance over the natural world, we suddenly found ourselves outmatched. At chess, for now, but other defeats felt imminent.

For a millennium and a half, chess has been revered as the most demanding and intellectually stimulating of all games. If language is "the mirror of the intellect," then chess could be seen as its magnifying mirror, intensifying and revealing the workings of the best human minds. The chess masters — grandmasters and world champions — were held in the highest esteem; the public admired them on par with the greatest scientists, writers, and philosophers — perhaps more. On second thought, definitely more than philosophers.

[19] We will talk about AlphaFold and its achievements later in the chapter.

[20] The approximate age of Homo sapiens.

Since 1886, following the establishment of the World Chess Championship, there have been 14 undisputed world champions. The last before the machine's victory, reigning from 1985 to 1997, was Garry Kasparov. He was widely considered the strongest human player of his era — and perhaps of all time — even after the title eventually passed to others.[21]

Since the inception of AI in the late 1940s and early 1950s, researchers have regarded chess as a "firing range" for AI. Alan Turing and Claude Shannon both worked on the earliest chess programs. Shannon even built the first chess-playing machine — an endgame solver — and argued that success in chess would be a meaningful demonstration of machine intelligence. At the time, however, few dared to dream that a computer might one day defeat the best human chess players. Shannon hoped, for example, only for a machine that could play "a fairly strong game" of chess. Anything more would have seemed arrogant — and perhaps even undesirable, the stuff of human-bashing science fiction.

But little by little, machines played better and better, thanks to the ever-increasing speed of computations, larger memories, and improved chess-playing algorithms. These symbolic, rule-based programs evaluated move sequences using strategies handcrafted by human chess masters. John McCarthy, who coined the term "artificial intelligence" in 1956, called chess "*the drosophila of AI*," referring to the small fruit fly on which geneticists conducted their experiments.

By 1977, chess-playing machines were already capable of beating 95% of all human players. Over the next decade, their strength advanced rapidly. By the late 1980s, they could defeat nearly everyone except for top-level grandmasters — a group that represented only a tiny fraction of all chess players, perhaps as little as 0.01%.

Direct Lineage of Deep Blue (written by ChatGPT-4o)

1. *ChipTest* (1985–1986) was developed at Carnegie Mellon University by Feng-Hsiung Hsu, Thomas Anantharaman, and Murray Campbell.

[21] In 1993, Kasparov broke away from FIDE, the official governing body of chess, and continued to defend his title under the rival Professional Chess Association (PCA). This led to a split in the world championship, with FIDE organizing its own title matches. The divide lasted until 2006, when the title was reunified. Kasparov remained the dominant figure in chess through the 1990s, even without FIDE's formal recognition.

It was the first chess machine using custom VLSI (*Very Large Scale Integration*) chips to accelerate move evaluation.

2. *ChipTest-M* (1987) was an enhanced version of ChipTest, offering greater power and improved search heuristics. This machine began beating strong human players.
3. *Deep Thought* (1988–1989), named after the fictional computer from *The Hitchhiker's Guide to the Galaxy,* was a major leap forward. It was the first computer to defeat a grandmaster, Bent Larsen, in a tournament game. However, it lost to Garry Kasparov in a 1989 two-game match (0.5–1.5), making a strong impression on him. The team behind Deep Thought was soon hired by IBM, forming the core of the Deep Blue project.
4. *Deep Blue Prototype* (1995–1996) played a six-game match against Kasparov in 1996, winning one game but losing the match 2–4. That was the first time a computer beat a reigning world champion in a single game under standard tournament conditions.
5. *Deep Blue* (1997) was significantly upgraded from the prototype. It was equipped with 30 processors and 480 special-purpose chess chips. It made history by defeating Kasparov in a 6-game rematch (score: 3.5–2.5).

The 20-year-long struggle culminated in IBM's specialized chess-playing supercomputer *Deep Blue* defeating world champion Garry Kasparov in May 1997. The impact on Kasparov, as well as on hundreds of millions of chess players worldwide and humanity itself, was profound. In a sense, Kasparov embodied the pinnacle of human intellectual potential, and when he fell, all of humanity felt the loss. *Newsweek* published a cover with the headline, *"The Brain's Last Stand."*

In 2017, Garry Kasparov published the book "Deep Thinking," in which he recounted his side of the story. His fights with predecessors of Deep Blue and later with Deep Blue itself were tense and often dramatic. In that book, he said, among other things:

> My matches against computers, which spanned nearly the entire twenty years I spent as the world's top-rated player, allowed me to think about chess as something other than a competition. Battling each new generation of chess machines meant participating in a hallowed scientific quest, sitting at the nexus of human and machine cognition, and holding up the banner for mankind. (Kasparov and Greengard, 2017, 23)

About the 1996 confrontation with Deep Blue Prototype, where Kasparov won the match but lost to the machine in one of the games, he wrote:

> A month after the match, I wrote in *TIME* that I felt I could sense "a new kind of intelligence across the table" that day [*the day of his loss*], and in a way, it was true. I wasn't suggesting any metaphysical interpretation, but could sheer speed really produce such impressive chess? Several of its moves were almost as if it were saying, "I bet you didn't think a computer could make a move like this!" For example, at one point in the middlegame, it sacrificed a pawn for *activity*, a very human-like idea not at all in keeping with the usual machine materialism. (*ibid.*, 141)

And then came the fateful rematch in May 1997. Kasparov writes,

> The new Deep Blue was operating on a new supercomputer model that was twice as fast as the old one, and it contained even more of Hsu's [*member of IBM team*] new and improved chess chips, 480 of them, and reached 200 million positions per second at its peak.[22] (*ibid.*, 165)

As we mentioned, in 1996, although the Deep Blue prototype (sometimes referred to as Deep Blue I) lost the match, it did win the first game. On that news — that, for the first time in history, a machine had beaten the World Chess Champion — IBM's stock rose so sharply that the next day the company was worth over three billion dollars more than the day before.

So, in the 1997 rematch, IBM was determined to win. They spent tens of millions organizing the match and likely three times more developing the new machine. Kasparov writes:

> I understand I underestimated that with so much on the line, IBM wasn't only building a chess machine to beat me at the board, but a machine to

[22] Kasparov means that, given the current state of the board, Deep Blue could evaluate up to 200 million possible future moves and resulting positions per second, selecting the one it deemed most advantageous. No wonder no human could compete with that! That Kasparov managed to win some games against such a machine is a testament to his extraordinary chess genius.

> beat me, period ... Gone was the friendly and open attitude that had been on display around the Philadelphia match [*of 1996*] run by ACM [*Association for Computing Machinery*]. With IBM in charge from top to bottom, this chumminess had been replaced by a policy of obstruction and even hostility... Bumping into several American Grandmasters who had been working with them was a bit of a shock. My team found out during the third game that despite IBM's statements that they hadn't been working with other GMs [*grandmasters*], several were staying at the hotel with the rest of the IBM team. The New York Times reporter would later confirm that they had been hired by IBM. (*ibid.*, 155, 161, 167, 168)

A significant portion of Kasparov's book is dedicated to analyzing the moves, both his own and the machine's. Since I know very little about the game of chess (the last time I played it was when I was about 15), I will not dwell on that. But I was curious to understand the feelings Kasparov experienced. He says himself that he is, by nature, very competitive and does not take losing lightly. And he was the first top-tier intellectual in history to suffer a devastating blow from AI. So, how did it feel? In his book, a few lines offer insights:

> When I sat across from Deep Blue twenty years ago [*in 1997*], I sensed something new, something unsettling. Perhaps you will experience a similar feeling the first time you ride in a driverless car, or the first time your new computer boss issues an order at work. (*ibid.*, 9)
>
> At the press conference, I was again frank about how impressed and surprised I was by some of Deep Blue's moves...When asked about remarks by Illescas [*one of the grandmasters participating in the event*] that I was afraid of Deep Blue, I was again candid. "I'm not afraid to admit I'm afraid! And I'm not afraid to say why I'm afraid. It definitely goes beyond any known program in the world." (*ibid.*, 212)

After Deep Blue won the 1997 match, IBM's stock valuation rose by $11.4 billion, which was *still a lot of money back then*.[23]

Deep Blue was a technological marvel, but it was narrow and specialized — brilliant at chess, yet incapable of generalizing to anything else. It

[23]The reference is to Daddy Warbucks, the fabulously wealthy industrialist in the comic strip and musical *Little Orphan Annie*. He remarked about his early fortune that "two hundred million was a lot of money back then."

couldn't learn new games, write a capable essay, drive a car, or recognize faces. It embodied the strengths and limitations of symbolic AI: powerful in narrowly defined domains with explicit rules but unable to function outside of them.

Two decades later, the strongest chess-playing machines were no longer based on symbolic AI but on self-learning neural networks. By the late 2010s, a new generation of chess engines had emerged that relied not on human-crafted rules and brute-force search, but on artificial neural networks trained largely through self-play. Against such a system, Deep Blue wouldn't stand a chance.

In 2017, two machine programs competed in chess: the computer world champion *Stockfish 8* and Google's program, called *AlphaZero*. Stockfish 8, like Deep Blue before it, was trained on the accumulated chess wisdom of the centuries. It held in its memory all the recorded games played by past chess players, and it also memorized all the games played by computers since the inception of computer chess. And it could calculate 70 million chess positions per second.

In contrast, AlphaZero could calculate a mere 80,000 positions per second, i.e., it calculated a thousand times slower. And it was not taught anything except for the game's rules. And yet AlphaZero won! The two machines played 100 games, of which AlphaZero won 28 games and tied the rest, not losing a single game. Next year, in 2018, AlphaZero increased its wins in a thousand-game match against Stockfish 8. How is that possible?

It was a good illustration of the advantages of ANNs over symbolic AI. Google's program emphasized *learning*. In preparing for the tournament, the machine, which had no prior knowledge of chess, played against itself. The shocking fact is that it took AlphaZero only *four hours* of preparation (Harari, 2019, 32). In four hours, it went from knowing only the game's rules to becoming the best chess player in the world — not among humans; that would be old news — but among chess-playing computers.

Artificial Neural Networks Rise From the Ashes

The criticism, which, we must admit, was well-founded, led to the withdrawal of nearly all government funding for neural networks in both the U.S. and the U.K. As a result, work on ANNs ceased in the early 1970s and did not resume meaningfully until the mid-1980s. Scientists know how such things happen. Thomas Kuhn famously described this phenomenon in *The Structure of Scientific Revolutions* as a *paradigm shift*, where the intellectual landscape changes overnight. What was once a lively and legitimate field of research is suddenly regarded as an academic graveyard — and not merely dead, but radioactive. Nobody in their right mind would go there. To do so is not only professionally risky but also intellectually isolating, with no peer support and no funding.

Yet, winter does not last forever if a field possesses true intellectual merit. As often happens in science, the seeds of its revival had already been planted, lying dormant for the time being and waiting for Kuhn's paradigm to shift back.

Backpropagation

A rather simple idea that would eventually trigger the revival of ANN research was already proposed in the 1960s. It was a variation of the powerful method of *negative feedback*. This insight, known as *backpropagation*, first emerged within the field of *optimal control theory*, a part of cybernetics.

In 1960, Henry J. Kelley proposed a method for efficiently computing *gradients* in dynamic systems, which is now recognized as an early precursor to the backpropagation algorithm. Kelley's technique was further developed in 1969 by Arthur E. Bryson and Yu-Chi (Larry) Ho.

In his 1962 book, *Principles of Neurodynamics*, Rosenblatt devised a similar idea for training multilayer Perceptrons. He talked about backpropagating errors to adjust the weights. However, he could not actually do it because his McCulloch–Pitts neurons had a stepwise response. Implementing backpropagation requires taking gradients, and Rosenblatt was unable to do so.

Paul Werbos (Fig. 9), in his 1974 graduate thesis, developed the mathematical structure for using it to train multilayered neural networks. Werbos' priority in developing the algorithm for backpropagation has been acknowledged by awarding him the 1995 *IEEE Neural Network Pioneer Award* and the *IEEE Frank Rosenblatt Award* in 2022, among other honors.

The story of his struggles with backpropagation is quite interesting. It illustrates the difficulty of undertaking anything, even the most valuable endeavor, during the ongoing "winter." Let me first describe backpropagation in simple terms in the special section and then tell the story of how Paul Werbos developed it.

Fig. 9. Paul Werbos (b. 1947), photographed in 1991; the inventor of the mathematical method of *backpropagation* for training multi-layered ANNs, photographed in 1991.

Source: Image by Rolf Kichuth, Lic. CC BY-SA 4.0.

What is Backpropagation?

A neural network consists of layers of artificial neurons (nodes) that pass signals from one to another. In Fig. 8, we illustrated a simple single-layer network similar to the one used in the original Perceptron. That simplicity came at a cost: The Perceptron was limited in its capabilities.

The idea of building *deeper* networks — with multiple layers — was evident to researchers from the beginning. Frank Rosenblatt himself understood that stacking layers might help overcome the limitations of a single layer.

Figure 10 shows a schematic of a deep neural network. We see the familiar input and output layers, but now there are several *hidden layers* between them. In such networks, each node in a given layer (say, layer n) receives input from all the nodes in the previous layer ($n - 1$). Each incoming signal is multiplied by a weight, and all these weighted inputs are summed. If this sum passes a threshold determined by the *activation function*, the node "fires," producing an output signal. That signal is sent to all

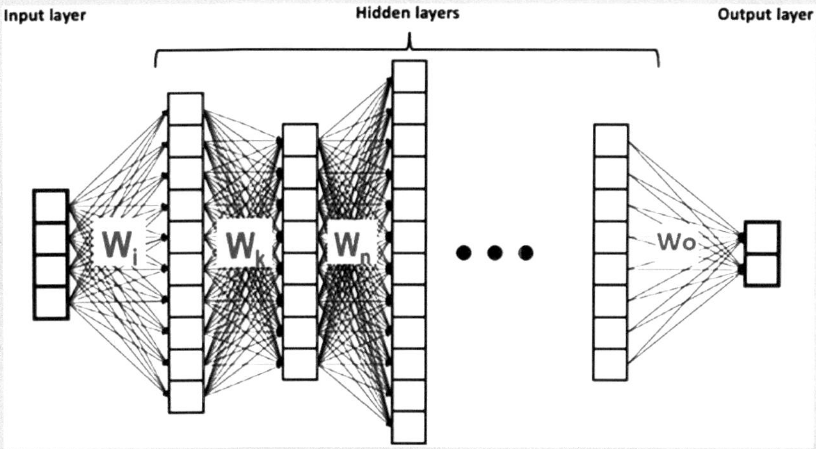

Fig. 10. A deep neural network with many layers. Each neuron in one layer is connected to every neuron in the next layer. Each connection has its own adjustable weight, symbolized by "W"s superimposed on the diagram.

Source: Base image by BrunelloN, Lic. CC BY-SA 4.0; weight labels added by the author.

the nodes in the next ($n + 1$) layer, each multiplied by its weight. Ultimately, the network produces a final result, for instance, classifying an image as a cat or a dog.[24]

Now, here comes the key question: How do we *train* such a system?

To Rosenblatt's credit, he understood and wrote in his 1962 book that the way to train a multilayered network would be to propagate information about an output error backward toward the network's inputs. However, no one knew how to do that. The difficulty was exacerbated by the fact that all neural networks, including the Perceptron, were based on the classical McCulloch–Pitts neurons, which exhibit a *stepwise* response: When such a neuron fires, it outputs a fixed voltage representing a logical 1; when it does

[24]As mentioned earlier, distinguishing between cats and dogs — animals with varied and overlapping features — is significantly more challenging than separating simple geometric shapes. The latter tend to have well-defined boundaries and consistent patterns, whereas cats and dogs can appear in many poses, colors, and contexts.

When discussing quantum thought experiments, scientists often name the two participating characters Alice and Bob. Following this tradition, I named an electron "Bob" in Chapter 3. It seems to me that in neural networks, Alice and Bob have been replaced by cats and dogs. Cats are especially prominent in ANN scientists' thinking, as we will see when discussing the views of Yann LeCun in the following chapter.

not fire, the output is zero. You cannot *differentiate* such a function — you cannot find its slope, which would be infinite — and Rosenblatt did not see a way out of that impasse.

Continuous quantities and their derivatives

The way to resolve the difficulty was to replace all *step functions* — those that jump between different values — with continuous ones, functions that have *slopes,* also called *gradients*. We must examine all functions we encounter in a network, going from the network's output to its input, smoothing everything out as we proceed: no steps anywhere, just smooth ramps. Mathematically, this means that all functions must be *differentiable*; a *derivative* is another word for a slope. In this sense, we are returning to the origins of calculus, which was developed in the 17th and 18th centuries, beginning with the foundational work of Leibniz and Newton. They, too, dealt with smooth, continuous functions and their derivatives.

You may wonder: if the output is just "cat" or "dog," how can we talk about making a "smaller" or "bigger" error? Isn't the output simply right or wrong? It turns out that although the final decision is binary, the network computes internal values continuously; for example, the strength of belief that an image is a cat might be 0.85 (85% confidence), while for a dog, it might be 0.15 (15% confidence). The system then selects the more confident option. If it guesses incorrectly with high confidence, that is a bigger error than guessing incorrectly with low confidence. So, the error can indeed be graded.

A good analogy is archery. You shoot an arrow at a target. If you hit the bullseye, the error is zero. But if you land in an outer ring, you are still on target — just less accurately. The farther from the center, the larger the error. It's not simply a matter of right or wrong, but of how close you came. That's what modern neural networks measure using a *loss function*: a continuous measure of error.

Neural networks also rely on *weights*, which are continuous variables. This means they can be adjusted in arbitrarily small steps. We compute how the loss changes *with respect to each weight* by taking a derivative. This indicates the direction to nudge each weight to minimize the error.

Finally, there are the *activation functions* of neurons. In the original McCulloch–Pitts model, neurons were binary: They either fired or they did not. That needs to change. Modern neural networks employ smooth activation functions that transition gradually between low and high values, effectively representing 0 and 1 as a continuous range. These functions are *differentiable*, which is essential for learning. During backpropagation, we calculate how each weight affects the final loss, and this calculation involves the activation function. If it were not differentiable, we would not be able to compute the gradients needed to adjust the weights and train the network. Typically, only two or three activation functions are used in practice, and we do not need to delve into their exact formulas. As long as they change smoothly between 0 and 1, they serve their purpose.

How will we train a network?

Each weight contributes, in some small way, to the final result. If the result is wrong, we want to *adjust* the weights in a way that reduces the error next time. However, to do so, we must determine how much each weight contributed to the error. This is where *derivatives* or *gradients* come in. A *gradient* and a *derivative* are just fancy words for a slope. In calculus, it tells us how much a function changes when we tweak its inputs. If a slight change in a variable causes a big change in the function, we say the gradient is steep. If the change is small, the gradient is shallow.

But what if one thing affects another indirectly — what if A affects B, and B affects C? Then, we use something called the *chain rule*, a basic rule of calculus.[25] The chain rule enables us to determine how A ultimately affects C by multiplying the slopes along the way.

In neural networks, we use the chain rule to trace how changes in each weight affect the final error. If changing a particular weight reduces the error, we adjust it in that direction; if it increases the error, we move in the opposite direction. This step-by-step adjustment process is called *gradient descent* — we move "downhill" to reduce the error.

Enter backpropagation

Backpropagation is the algorithm that uses the *chain rule* from classical calculus to calculate how much each weight in every layer contributed to the overall error. It does so in reverse, starting at the output layer and working step-by-step toward the input. That's why it's called *back-propagation*.

[25] The chain rule emerged from Leibniz's early differential notation, making him the originator in spirit. However, its full mathematical formulation and rigor were developed over the following century by others, notably Cauchy.

The process resembles *negative feedback*, in which a fraction of the output signal is *subtracted* from the input (Gurvitch, 2023, Vol. 1, pp. 221–225).

For each layer, the algorithm calculates the gradients, which tell us how much the error changes with respect to each weight. Once the gradients are known, the network uses them to update the weights in the direction that produces the maximum error reduction. This process is repeated millions or even billions of times, depending on the size of the training dataset and the complexity of the network. Each training step adjusts billions of weights, nudging every connection slightly. Over time, across many iterations, the network gradually refines itself, learning to perform its task with increasing accuracy. That's how a modern neural network learns, and that is also why this learning is so computationally demanding. This massive calculation, repeated billions of times, is where those megawatts of electricity go in training today's LLMs.

Paul Werbos began thinking about relevant subjects around 1967–1968 while he was a 20-year-old undergraduate at Harvard. He continued to develop the idea while pursuing his master of science (M.S.) degree at the London School of Economics (LSE) and later again at Harvard, where he completed his Ph.D. in 1974. However, he faced considerable resistance. His Harvard professors were skeptical of neural networks, believing them to be a dead end (as was the consensus during the "AI winter"). They were also somewhat wary of Paul Werbos, who was exceptionally bright but prone to juggling multiple ambitious projects simultaneously and seemed oblivious to academic procedures (such as securing a formal thesis advisor as a prerequisite for defending a Ph.D.).

Werbos later recounted his experience vividly in a 1993 interview with Edward Rosenfeld. Here are a few excerpts:

> I sent a paper to *Cybernetics*, which was published in 1968. It was not a coherent thing, but it had the germs of all the later ideas… I talk in there about the concept of translating Freud into mathematics. That is what took me to backpropagation…. (Anderson and Rosenfeld, 1998, p. 340)

Later, while enrolled in graduate courses at Harvard, Werbos sought to develop his method further. He approached the course committee, asking to base a required paper on his new technique:

> …I have this little method for adapting multilayer Perceptrons. It's really pretty trivial…And I'd like to do it for my paper for this computer course." They said, "Why don't you go talk about it to Larry Ho?" (*ibid.*, 341)

Professor Ho — the same Yu-Chi (Larry) Ho who, in 1969, computed gradients in dynamic systems — did not like that Werbos wanted to pursue a course problem different from the one initially assigned in his course. He recommended that Werbos take an "incomplete" in his course instead. One can feel Werbos' frustration with his Harvard professors in the interview he gave some 20 years later:

> So they wouldn't let me do it. One of the reasons that is amusing to me is that there are now some people who are saying backprop[agation] was invented by Bryson and Ho. They don't realize it was the same Larry Ho, who was on my committee and who said this wasn't going to work…The committee didn't like the neural network area generally, but they said, "Look, you find a patron [*thesis advisor*, MG] for your thesis, and we'll let you graduate in this area." (*ibid.*, 341, 343)

Looking for a *patron*, Werbos went to see Steve Grossberg, Marvin Minsky, and Jerome Lettvin, all of whom were at the neighboring MIT and had been involved, at one time or another, with neural networks. They all had different reasons for disliking Werbos' ideas.

Paul Werbos recounts a particularly wonderful exchange with Professor Grossberg, who said to him, "*The bottom line is, this stuff you've done, it's already been done before. Or else it's wrong. I'm not sure which of the two, but I know it's one of the two.*"

He also talked to Professor Minsky, who, as we know, was by that time a sworn enemy of neural networks. In particular, Werbos explained to Minsky that his proposed scheme would require smooth activation functions for individual neurons. This is how he described their conversation:

> I went to Minsky for help, but Minsky would not offer help. Minsky basically said, "Look, everybody knows a neuron is a 1-0 spike

> generator. That is the official model from the biologists. Now, you and I are not biologists. If you and I come out and say the biologists are wrong and this thing is not producing its 1s and 0s, nobody's going to believe us. It's totally crazy. I can't get involved in anything like this. (Anderson and Rosenfeld, 1998, p. 344)

Sometime in the early 1970s — no later than 1974 — Werbos, who at that time thought that backpropagation would not be a part of his Ph.D. thesis, wrote a detailed paper on its application in ANNs, but that paper was not accepted for publication, perhaps because he did not have the money for the journal's *page charges*. At that time, being cut off from financial support at Harvard, he had no money for food, let alone page charges.

Finally, Werbos managed to insert backpropagation into his 1974 Harvard Ph.D. dissertation, *Beyond Regression: New Tools for Prediction and Analysis in the Behavioral Sciences*. Backpropagation was described in the context of *prediction systems* and *control systems*. Applications in training multi-layered ANNs have not been placed upfront, presumably due to all the flak he was getting from his superiors. Nevertheless, the essential mathematical machinery — making neuron *activation functions* continuous and using the *chain rule* (see the special section above) to propagate gradients backward through layered systems — was adequately described.

Generally, a thesis is not considered a scientific publication. In that 1993 interview, Werbos stated that the first time backpropagation was described outside his thesis occurred in the *Time Series Processor* (TSP) computer manual, which was authored by Brode, Werbos, and Dunn and published internally at MIT and the DOD in 1981 — not precisely an open scientific publication either. In the 1990s, Werbos penned papers and books in which backpropagation was properly presented, but that came later. The way it appeared in the 1970s was not visible to the ANN community — whatever was left of it — and thus, it took a decade to be rediscovered and applied to real-world systems.

It is important to remember that even with the idea in hand, applying backpropagation to large, multilayered ANNs was computationally impractical. As someone aptly put it online, "*Training a neural network is akin to adjusting the knob on a radio to catch the right signal. Except there are billions of knobs, and each one needs millions of small adjustments.*" In the 1960s and 1970s, computers were not powerful enough. Only later, when computational capabilities caught up, could the true potential of Werbos's insight be realized. By then, this conceptually simple technique — remember how Werbos said, "*It's really pretty trivial*" — had been independently rediscovered by others.

How backpropagation was rediscovered and used in ANN training

Everyone agrees that the pivotal paper on backpropagation in ANNs, the one that revived the field and was picked up by a large number of researchers, effectively ending the long AI winter, was the 1986 *Nature* paper by David E. Rumelhart, Geoffrey Hinton, and Ronald J. Williams, *Learning Representations by Back-Propagating Errors* (Rumelhart, Hinton, and Williams, 1986).[26]

In 1985–1986, when the paper was written and published, Rumelhart (Fig. 11) was a professor of cognitive science at the University of California, San Diego (UCSD); Williams was his graduate student; and

Fig. 11. David Everett Rumelhart (1942–2011), an American mathematical psychologist who, with his colleagues and collaborators, introduced backpropagation into the mainstream ANN research.

Source: Photographed by Rolf Kickuth, Lic. CC BY-SA 4.0.

[26]The reader may recall what computer scientist Robert Rosin said about von Neumann's paper about stored program computer architecture: "*The most influential paper in the history of computer science, whether or not anyone expressed similar idea earlier...*" (in Chapter 2). Something similar could be said about the 1986 *Nature* paper of Rumelhart, Hinton, and Williams.

Hinton, who was at one time Rumelhart's post-doc, was at UCSD on sabbatical leave from Carnegie Mellon University (CMU) in Pittsburgh.

We will speak much more about Geoffrey Hinton, a recent Nobel laureate and one of the most famous figures in the field. We will discuss his scientific contributions and, in the next chapter, his views on the upcoming AGI.

Hinton, like many other prominent ANN scientists, was interviewed by Anderson and Rosenfeld for their collection of interviews, *Talking Nets*, where he discussed the backpropagation discovery and described how he initially opposed it:

> There's one other curious piece of history that I left out. I went back to San Diego in '82, I guess...David Rumelhart came up with this idea of backpropagation. He wasn't the first to come up with it, but...[*the last ellipses appearing in Hinton's interview; think of them as a summary of our whole story of backpropagation discovery*]. I first of all explained to him why it wouldn't work, based on an argument in Rosenblatt's book... it was an algorithm that couldn't break symmetry...Now, as you know, the standard way of learning is to have random initial weights...So Dave Rumelhart pointed out that we didn't need to worry about symmetry.
>
> The next argument I gave him was that it would get stuck in local minima.[27] There was no guarantee you'd find a global optimum. Since you're bound to get stuck in local minima, it wasn't really worth investigating.
>
> Nevertheless, Dave programmed it up and did a few things with it. I think he showed it could learn X-OR [*exclusive OR*].... (Anderson and Rosenfeld, 1998, 376)

Later, after he returned from San Diego to Pittsburgh, Hinton himself tried a few things with backpropagation and concluded that his initial hunch was correct: It was useless. And then he saw that it was not useless but wonderful! I love the way he described that in his *Talking Nets* interview:

> So I ran backpropagation on this problem and looked at the weights of the hidden units [*"units" or "nodes" are alternative names for artificial neurons in hidden layers, MG*]. They didn't have the right structure

---

[27] We will explain these terms — symmetry and local minima — in the following section.

> of weights. I decided it wasn't really working; it wasn't solving the problem. I never bothered to look at the error measure, which had gone to zero…I thought, "Oh well, it turns out backpropagation's not that good after all." Then I looked at the error, and the error was zero. I was amazed.
>
> So I sent mail to Dave Rumelhart saying, "You remember that algorithm of yours? It seems to actually work." (*ibid.*, 378)

In 1993, David Rumelhart was asked by Edward Rosenfeld, "*Who discovered backpropagation and how*?" To this, Rumelhart replied:

> …as far as I know, it's been discovered at least three times, maybe more. Paul Werbos apparently discovered it and wrote about it in his dissertation. David Parker apparently discovered it, and we discovered it. As far as I know, those were all entirely independent discoveries, and there were precursors to all of those. What's surprising is that it hasn't been discovered more often because it's such a simple idea. (Anderson and Rosenfeld, 1998, 286)

When I read this, I looked up David Parker. In 1982, when he independently formulated the backpropagation training method for neural networks, he was a graduate student in electrical engineering at Stanford, working on his master's degree.[28] In 1982, he filed an invention disclosure titled "Learning Logic" with Stanford University, outlining how errors could be propagated backward through a multilayer neural network to adjust weights systematically — right on the money! He later moved to MIT's Sloan School of Management, where, in 1985, he published a technical report detailing his approach to training multilayer neural networks.

Rumelhart continued:

> There are other examples of work [*on backpropagation*] in the control literature in the ′60s…but the ideas were never really implemented…no

[28] A master's degree — not even a Ph.D.! The number of high-caliber discoveries made by graduate students in all branches of science is staggering. A good Ph.D. student, at least in physics, is a force to reckon with. Not an insignificant fraction of all scientists did their best work when they were still in graduate school. That is why we should take the expected appearance of AGI at a Ph.D. level in all cognitive fields very seriously.

> one had really done anything with them. My view is that we not only discovered them, but we realized what we could do with them and did a lot of different things. (*ibid.*, 286)

What were those "different things" that Rumelhart mentioned in that interview? He referred to proofs of principle and modeling done with multi-layered ANNs, some of which were briefly described in the *Nature* paper. These experiments proved that backpropagation could enable deeper networks to learn *internal feature representations*, a result that theory alone could not guarantee.

Internal Feature Representations

We have already encountered the term "representations" in the title of that 1986 *Nature* article. In the context of neural networks, *internal feature representations* refer to the *activation patterns* that arise within the network's hidden layers — that is, the map of all neuron states, both fired and dormant, taken at a particular moment in time. These patterns capture essential properties or features of the input data, forming increasingly abstract representations as signals pass through successive layers.

Rather than memorizing each input example in full detail, the network learns to extract and combine its essential features, for instance, recognizing that specific arrangements of edges form a square or that particular textures suggest fur (cats again!) Crucially, these features are not explicitly programmed by humans; they emerge automatically as the network learns from experience.

The hierarchy of layers plays a central role. Lower hidden layers — those closest to the input — detect simple, local patterns such as lines, edges, or colors. Higher layers combine these into more complex and abstract representations: corners, shapes, faces, and eventually even conceptual categories. In this way, internal representations function as the network's internal language for understanding the world — a learned vocabulary for recognizing and interpreting patterns.

Later that same year, in 1986, Rumelhart and James McClelland published a much more detailed exploration of these ideas and experiments in the landmark two-volume work *Parallel Distributed Processing: Explorations in the Microstructure of Cognition* (Rumelhart and McClelland, 1986). Rumelhart and McClelland were the editors and contributed to the writing of many chapters. Other authors included Geoffrey Hinton, Terrence J. Sejnowski, and Paul Smolensky.

It is worth noting that the *Nature* (1986) paper already contained, in embryonic form, the principles that would much later give rise to LLMs, which in recent years have become central to AI. The architecture and training objective described there were strikingly general: multi-layer networks trained by backpropagation to minimize prediction error, with hidden layers discovering their own internal representations. Although the experiments of that time dealt only with small, schematic tasks, the underlying idea was profoundly open-ended. In those first attempts to let a machine extract structure from data and represent it internally lay the seed of what would, decades later, allow machines to model human language itself. The same mechanism that once learned toy patterns would, when nourished by vast text corpora and enormous computational power, begin to approximate understanding — a distant echo of that first insight.

Rumelhart, Hinton, and Williams showed that neural networks could solve nonlinear problems, overcoming the limitations identified by Minsky and Papert. As a result of their work and that of their colleagues, ANNs were rehabilitated in theory and as a working technology. The field surged back into view. Conferences, funding, and research initiatives followed, and by the late 1980s, neural networks were once again a significant part of the AI landscape.

Before we leave backpropagation, it's important to note that by the late 1980s, it was still very much work in progress. As originally formulated, backpropagation could not effectively train neural networks with many layers; only relatively shallow networks could be handled. When researchers attempted to train deeper networks, particularly the lower layers closer to the input, they encountered a significant obstacle: The gradients used to update the weights became increasingly small as they were propagated backward. Eventually, these gradients all but vanished, and the layers stopped learning entirely. This became known as the *vanishing gradient problem*. It would not be fully addressed until the 2000s, nearly two decades later. Meanwhile, the problem imposed a practical limit on the depth of a network and significantly slowed progress in developing ANNs.

ANNs in the 1980s

In 2024, John Hopfield shared a Nobel Prize in Physics with Geoffrey Hinton (Fig. 12). In this section, I will try to explain what they have done and why the physics prize was awarded for their work on neural networks.

Fig. 12. Geoffrey Hinton (left) and John Hopfield (right) during the Nobel Prize week in December 2024.

Source: Photo taken by Arthur Petron, Lic. CC BY-SA 4.0.

John Hopfield and the mechanism of associative memory

John Hopfield approached problems in neural networks from the perspective of a solid-state physicist. By the time he published his influential 1982 paper introducing what would become known as the *Hopfield network*, he was already a well-recognized 49-year-old physicist and a professor at Caltech, with prior appointments at the University of California at Berkeley, Bell Labs, and Princeton. He had received several prizes and distinctions in physics, the most notable being the *1969 Oliver E. Buckley Prize in Solid State Physics* — the most prestigious award in that field, second only to the Nobel Prize.

In 1982, Hopfield published his landmark paper, "Neural Networks and Physical Systems with Emergent Collective Computational Abilities." The Nobel Prize he received 42 years later was largely due to that work. I spent some time searching for the most accessible explanation of Hopfield's ideas and concluded that (a) we are entering a level of sophistication in ANN science that may soon exceed my ability to grasp it fully — and explain it to my readers — and (b) the best available explanation remains the one offered by Professor Hopfield himself. That explanation was presented in a talk titled "Collective Properties of Neuronal

Networks," delivered at the 1983 Corporate Associates meeting of the American Institute of Physics (AIP), held at Xerox's Palo Alto Research Center. Fortunately for us, the audience mainly consisted of industry executives — sponsors of the AIP — rather than theoretical physicists, so Hopfield presented his ideas at an accessible level.

I will attempt to convey Hopfield's ideas about the mechanism of memory by providing a few excerpts from the talk verbatim and condensing the remainder. An interested reader can view the entire talk on YouTube. I am trying to retain the speaker's exact wording when quoting, even though it often clashes with Grammarly's suggestions. I appreciate and frequently use Grammarly with my own writing, but I couldn't allow it to change the speech of a future Nobel laureate, could I? That said, I occasionally took the liberty of lightly editing Professor Hopfield's talk — inserting missing words in square brackets, omitting filler phrases without always marking the omissions with ellipses, and so on.

Hopfield said at the opening (and Grammarly immediately flew off the handle):

> I think…the time is right to try to make a kind of interconnection between parts of computer science and parts of neurobiology, and I think that the subject which helps you most in trying to make that interconnection are [*consists of*] some of the insights one has from parts of condensed matter and statistical physics. (Hopfield, 1983)

Then, the lecturer, after inserting a couple of mandatory jokes,[29] alluded to the differences between regular computers and human brains. He noted that while computers are thousands of times faster at calculations, they are surprisingly slow in other tasks, such as image recognition (he gave a simple example to illustrate this).

After noting how many neurons and synapses are engaged in the brain — and how, unlike computers, they aren't built from known wiring diagrams — Hopfield said:

> … as a result [*of all that complexity*], you must ask: how you're going to make progress? One answer is — slowly; and another is — to make

[29] In the 1970s–1980s, all scientific talks at conferences and seminars almost always started with a joke or two; such was the time-honored tradition, which John Hopfield, of course, did not break. What's more, his jokes worked!

a virtue of what otherwise is the major problem — to try to make a virtue of the size of the system. Phil Anderson[30] wrote a lovely article in *Physics Today* some years ago entitled *More is Different*, reminding one that in all aspects of physics and everywhere else in life, when you have systems with large numbers [*of objects*]... after you understood the science of how those objects worked in twos and threes and fours — if you had 10 to the large power of these objects, you'd have a system in which there was new physics, or new science, or new *collective phenomena*... You can study collisions of two molecules in a large box, and after a while, you get tired of that, so you put three molecules in that large box, and there's nothing in the collisions of twos and threes that directly suggests that if you put 10 to the 20th [*10^{20}, ten with 20 zeros*] molecules in that box, you'll have *sound waves*! The physical phenomenon of sound waves relies on collisions for its existence, and yet the essence of [*that phenomenon*] is not captured by the physics of small numbers of molecules.

...Now I'm going to talk about neural modeling, and the model which I'm going to make of neurons is going to be wrong in many details. One simply can't put in all the details. Not only aren't they known, but they're very complicated[31]...One important aspect of collective phenomena is [*that it*] can be relatively independent of the details of some of the things that go into them...

In that same sense, we can hope, if we're looking for collective properties in neural systems, to have robust insights that go beyond the fact that some of the details of the model must inevitably be wrong. So, I'm going to pursue the thesis that there ought to be collective effects in neural systems, effects of computational interest...I'm going to pursue the particular computational question of *memory*.

One of the most striking things in biology is that memories are *entities*; they have a way of sticking together...

Conventional computer architecture stores memories...in a *location*, and, for simplicity, you tend to put [*related memories*] in that location together. There's no question of memories mixing because the information about something is put in some particular place, and information

---

[30]We mentioned Phil Anderson in Chapter 1. He was a renowned theorist at Bell Labs, a professor at Princeton University, and a recipient of the 1978 Nobel Prize in Physics. Hopfield must have known him well, having spent time at both Bell Labs and Princeton.

[31]The reader may remember our attempt at an exposition of neuron physiology at the beginning of this chapter.

> about something else is put elsewhere. On the other hand, this kind of architecture leads to the necessity of knowing the *address* at which the information is stored in order to retrieve it. This is *local* information storage, local in the sense that one particular place is used to store a particular thing, and if you then destroy that particular place in the computer, a particular thing is forgotten. One would suspect that biological memory is a little bit more *holographic* in nature, in the sense that one would believe that regular parts of the hardware are probably participating in several memories, and memories then tend to overlap within the hardware. That creates problems of how you keep memories apart...
>
> The problem of *content addressability* — or *associative memory* — is very simple. You have a memory that has many blocks of information in it...If I remind you of one aspect of a particular experience, you can recall all the other aspects, and it doesn't matter which aspect I pick to remind you of [*the whole thing*]. You can still reconstruct the whole memory. It's the memory addressable by *content* and not by location.
>
> I'd like to point out that...content-addressable memory is something that physical systems can have spontaneously without a plan.

So Hopfield proposed — or, one might say, postulated — that *memory,* as it is stored in our brains and as it *ought to be* stored in ANNs,[32] is such a *collective emergent phenomenon.*

At this point in his talk, Hopfield said he now wanted to consider a mathematical description of content-addressable memory. He then explained how one could apply well-established methods from statistical and solid-state physics to analyze large neural networks. The essence of his argument was that one might define something resembling *energy*, a kind of *pseudo-energy*, that depends on the network's state (which neurons are firing and which are dormant) and on the inter-neuronal weights.

The system then seeks a configuration of minimal pseudo-energy, and that low-energy configuration corresponds to a *stable memory.* If the system begins in a state that is close to such a local minimum, it will spontaneously "slide" toward it, thus recalling the stored memory. These minima

[32]Mechanisms of memory in both the human brain and artificial neural networks had been discussed prior to 1982. Hopfield's contribution was to propose a mathematically elegant and physically inspired model that not only addressed memory but also revealed broader principles of collective computation.

are called *attractor states* because they "attract" the system's dynamics toward themselves.

The physical interpretation of memory proposed by Hopfield was both elegant and persuasive. Perhaps it already deserved a Nobel Prize back in the 1980s — though, I think, it might have been a prize in *Physiology or Medicine* rather than in *Physics*. Indeed, Hopfield's aim was to model the mechanism of memory in our brain, while physics served mainly as a source of ideas. The fact that ANNs built according to Hopfield's principles could exhibit memory still had to be demonstrated. The Nobel Prize in Physics was finally awarded to Hopfield in 2024, when he was 91.

To clarify these ideas, I've included the following special section. The subject is as fascinating as it is important, and I've tried to make it as understandable as possible.

Hopfield's Energy Landscape of Memory

In his 1983 talk, John Hopfield invited his audience to imagine an abstract, multi-dimensional configuration space in which each point represents a distinct state of a system. To demystify that notion, he illustrated the idea with a two-dimensional example. He labeled each axis with the letters of the alphabet, A to Z, and proposed that each "memory" in this "world" consisted of two letters paired together, such as "HE" or "AM." Each such pair corresponds to a point in the configuration space. For instance, "HE" and "AM" lie far apart on this grid — they represent distinct memories. A point like "GF" lies close to "HE" and far from "AM," suggesting it is similar to "HE" but not to "AM" (Fig. 13).

Now imagine that we partially remember something — say, just the letter "H." Then remembering "H" helps us complete the memory "HE." Even recalling a nearby point, such as "GF," might still help retrieve "HE," assuming the system is designed to facilitate this process. Hopfield used this toy model to illustrate the concept of content-addressable memory — a system that retrieves a full memory based on partial or noisy input. So far, we have not proposed the mechanism by which this memory reconstruction may work. Hopfield's genius was in proposing such a mechanism, as we will see shortly.

Let us now place these ideas into the context of a neural network. Each McCulloch–Pitts neuron can be in one of two states. Thus, a system of N

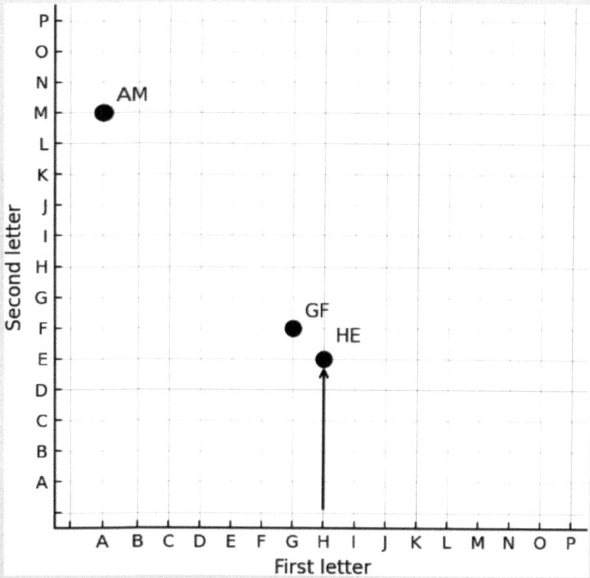

Fig. 13. Hopefield two-letter memory model illustrated. We shortened the axes, extending them from A to P instead of from A to Z, to show the points AM, GF, and HE more clearly.

Source: ChatGPT created this graph according to the author's instructions.

such neurons can exist in 2^N distinct configurations. These configurations can be imagined as points within an N-dimensional configuration space. We can imagine such a space where each point represents one state of our N neurons. For $N = 10$, one of our system's $2^{10} = 1024$ possible states will be (0110001010), another (1110010100), and so on.

In today's neural networks, the number of neurons is counted in millions or even billions. For such a system, 2^N is a super-astronomical number — far greater than the number of atoms in the Universe. This may sound intimidating, but mathematicians are accustomed to such scales. It does not prevent them from conceptualizing the state of the system as a point in a multi-dimensional space.

As neurons change their states (flip between 0 and 1), the point representing the system's state moves. The space itself is fixed; what changes is the system's current location within that space.

Being a physicist, Hopfield asked: What familiar physical theory does this resemble? He found a close analogy in *statistical mechanics* — specifically, in the behavior of many-body systems described by *spin glass models*. In such systems, the dynamics are governed by an energy landscape defined over a multidimensional space. The system evolves over time by moving "downhill," toward lower-energy states. A system is stable only when it sits at a local energy minimum, like a ball rolling into the bottom of a valley.

Hopfield proposed that stable, content-addressable memories correspond to *local minima* in a pseudo-energy landscape defined over the space of all discrete neuron configurations. If a system begins in a state near one of these minima — a noisy or incomplete version of a stored pattern — it will descend toward the nearest minimum. This descent is governed by the neuron update rule: each neuron updates its state in a direction that reduces the overall energy. This descent into an energy minimum is interpreted as an act of memory retrieval.

Hopfield proposed the formula for the pseudo-energy of a network configuration, which he borrowed directly from the physical theory of interacting spins. Instead of spins, we now have artificial neurons, which, like spins, can be in two states and interact with each other via synaptic connections. I will refrain from writing out this energy formula here; suffice it to say that neuron states and synaptic weights fully determine the system's pseudo-energy.

The network "prefers" configurations in which neurons are arranged according to the learned structure encoded in the weights. The more strongly a weight supports a pair of states, the more it reduces the system's energy when those states occur. This is the mathematical reflection of Hebbian learning: "Neurons that fire together wire together." Now, we may rephrase this mantra as "Neurons that fire together lower the system's pseudo-energy."

Once the weights are fixed by training, they fully determine the pseudo-energy landscape. As the network evolves, neurons flip their states to reduce energy, and the system eventually settles into a local minimum — a stored memory.

Hopfield's 1982 paper demonstrated that a network of neurons can store approximately $0.15 \times N$ distinct patterns, each represented by a unique attractor in the energy landscape. For example, a 10,000-neuron Hopfield network can store about 1,500 distinct memories before interference degrades its reliability.

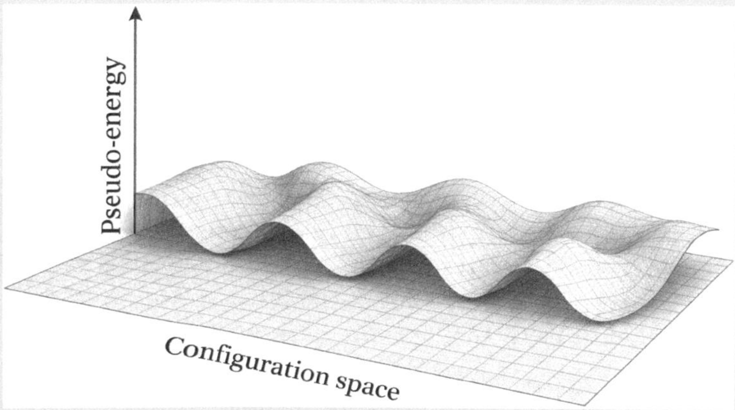

Fig. 14. A schematic diagram of a pseudo-energy landscape in configuration space with several local minima symbolizing several stable stored memories.

Source: ChatGPT created this graph according to author's instructions.

Let us summarize. The pseudo-energy landscape (Fig. 14) of a Hopfield network with N neurons is defined in N-dimensional configuration space. The total number of possible binary states, 2^N, is *astronomically* large, even for small networks with about 100 neurons; in today's large neural networks with N in the millions or billions, it is *super-astronomically* large. Each state corresponds to a point in an N-dimensional configuration space, where each axis represents the binary state of a single neuron. These points form a discrete cloud of possible configurations. The energy associated with each state can be visualized as its "height" in a pseudo-energy landscape.

The local minima in this landscape correspond to stable, content-addressable memories: if the system starts in a state close to such a minimum, the network dynamics, governed by Hebbian connections, will cause it to descend into that minimum. Following John Hopfield, we interpret this descent as an act of memory recall.

All of this may sound terribly formal. Let me offer an analogy — forgive me if it seems a bit violent in tone.

Imagine a flat field that has been bombed, leaving behind a collection of craters. Suppose, unlike real battlefields, this one is smooth: each crater is symmetric and its edges blend gently into the surrounding terrain. Each crater represents a stored memory — its size determined by how strongly that memory was imprinted. For simplicity, imagine that all the bombs were the same size, so the craters are uniform.

Now drop a ball — say, a soccer ball — onto the field. If it lands on flat ground, far from any crater, it may stay put: Nothing has been remembered. But if it lands on a slope leading into a crater, the ball will roll down and settle at the bottom. In that case, we say the system has recalled the memory associated with that crater.

There is, of course, a limit to how many distinct craters, or memories, the field can hold. If the bombing is too intense, the craters will begin to overlap, deforming each other and reducing their distinctiveness. In such regions, the ball's path may become unpredictable: It might roll into one crater or another, but we cannot say which.

Hopfield showed that under typical conditions, the memory field can contain no more than 15% as many clean, non-overlapping craters as there are neurons. If we try to store more, the landscape becomes noisy and confused, and memories blur together, as one might expect after a particularly devastating bombardment.

One final but important note: While Hopfield originally developed his model with memory in mind, his framework was later extended to cover a much broader class of learned representations — what AI researchers often refer to as *patterns*.

"Patterns" is a very capacious word in the context of neural networks. It refers to all kinds of regularities and structures that may exist in the data, sometimes exposed and often hidden. Patterns can be found in images, language (both written and spoken), music, mathematics, and human and animal behavior — they can be found everywhere. In associative memory models such as Hopfield networks, each pattern corresponds to a local minimum in the system's energy landscape. The principles of convergence, attractors, and stability all apply: the system naturally evolves toward these stable states.

There is no real distinction between a memory and a pattern; both function as *attractor states* that the network can settle into. When given a partial or noisy version of a stored pattern, the system can often recover the complete original by moving toward the corresponding energy minimum. These energy minima are referred to as *memory valleys* when associated with memories or *pattern basins* when indicating data features other than recall.

What started as a model of how the brain might store memories grew into something more ambitious: a way for machines to make sense of the world, not only remembering what they've seen before but also filling in the blanks and seeing the hidden structures.

And another observation, if I may. Around the late 1980s, researchers gradually stopped insisting that their work was merely an attempt to model the human brain. It was becoming increasingly clear that their creations — ANNs — function independently of how closely they resemble biological systems. These networks were evolving on their own terms, governed more by mathematics and engineering than by neuroscience. Scientists stopped saying that we must first understand how the human brain works, and only then, maybe, will we tackle artificial intelligence.

Boltzmann Machines

One problem with Hopfield's 1982 model was that the neural network, in its original form, could not distinguish between false (shallow, sporadic) minima and true deep minima. In our earlier analogy of a bombed field, the dropped ball might get stuck in small indentations — either pre-existing or caused by interference from nearby craters. Hopfield's model had no built-in mechanism to determine whether the ball had settled into a meaningful crater or a small, insignificant pit.

Realizing this limitation, John Hopfield, in collaboration with David Tank, extended the model in 1985 by allowing the system to explore the surrounding pseudo-energy landscape as if it were heated to a high temperature and then cooled gradually. This simulated heating helps the ball escape shallow pits and eventually settle into a deeper, more stable minimum. The technique, known as *simulated annealing*, was inspired by metallurgy, where the slow cooling of a heated metal allows its atoms to arrange themselves into a low-energy, stable configuration. Hopfield and Tank adopted this approach from a 1983 paper by S. Kirkpatrick, C. D. Gelatt, and M. P. Vecchi, titled "Optimization by Simulated Annealing," in which the authors introduced the concepts of pseudo-temperature and annealing in the context of solving optimization problems. Hopfield and Tank applied this physics-inspired method to one of the most famous optimization problems: the *Traveling Salesman Problem* (TSP), which we encountered in Chapter 3. They, however, did not extend their treatment to the Hopfield associative memory model.

At the same time, Geoffrey Hinton and Terry Sejnowski,[33] inspired by Hopfield's 1982 work, were exploring similar ideas — applying methods from statistical physics to neural networks, but with a stronger focus on *machine learning* rather than optimization. Between 1983 and 1985, they developed stochastic (probabilistic) neural network models that could learn from data using principles drawn from thermodynamics. These methods were ingenious but computationally intensive, far beyond what was practical in the 1980s. Their true potential was realized only decades later when computing resources caught up.

Hinton and Sejnowski gave their model a memorable name: the *Boltzmann Machine*, in honor of Ludwig Boltzmann, the 19th-century physicist who pioneered statistical descriptions of physical systems and helped lay the foundation for statistical mechanics. The name is fitting: Like particles in a heated gas exploring many possible states, a Boltzmann Machine explores possible network configurations in search of those most likely to represent meaningful structure in the data.

Let's give the floor to Terry Sejnowski. In his 1993 interview with Edward Rosenfeld, he recalled the emergence of the Boltzmann Machine, tracing it directly to Hopfield's pseudo-energy landscape ideas:

> I remember John Hopfield was invited to one of these Rochester meetings [*in 1983*]. For the first time, we heard his energy formulation. Both Geoff [*Hinton*] and I, within seconds, realized that this was the convergence proof we needed to show that our constraint-satisfaction schemes in vision could actually be implemented with hardware. That's how the Boltzmann machine was born. It was the result of a week of solid thinking about what would happen if we had a noisy Hopfield network.
>
> We realized immediately how a Hopfield network could be used for global optimization. The way to jump out of local minima was to add noise. I had just read an article in *Physics Today* where Scott Kirkpatrick

[33]Terrence J. Sejnowski (b. 1947) earned his Ph.D. in physics from Princeton in 1978, with John Hopfield as his advisor. At that time, Hopfield had not yet developed his energy-based neural network model. Sejnowski is currently the Francis Crick Professor at the Salk Institute for Biological Studies.

> described simulated annealing. So we immediately started simulating the idea, and sure enough — it was slow, but it would work. It would, in a rigorous way, implement all the things that we wanted done. Little did we know, however, that constraint satisfaction was really just an entrance to the part of the Boltzmann machine that had the biggest impact — the learning algorithm. Once you have a thermodynamic framework that allows you to analyze the probability of any state, you can begin to ask, "Well, how can I alter the probabilities?" Then you need to deal with how to sculpt the landscape — not how to search it, but how to change it.[34] That became the Boltzmann machine learning algorithm.
>
> Geoff was the one who really appreciated the importance of learning. He calculated the information-gain measure. I remember him calling me and saying, "You know, all the bad terms cancel." We were convinced that this was the solution. Something that simple had to be important. It was the first in a long line of learning algorithms that broke the logjam. (Anderson and Rosenfeld, 1998, 322–323)

I admit that I do not understand some of Sejnowski's words — for example, the concept of "constraint-satisfaction schemes in vision" — but that does not matter. Sejnowski's recollections contain priceless historical details while explaining the essence of the Boltzmann machine.

Given the importance of Boltzmann Machines in the history of ANNs, let me complement Sejnowski's words with a few of my own.

In a Boltzmann Machine, connections exist not only between different layers (as was the case in the networks depicted in Figs. 8 and 10) but also among units (neurons) within the same layer. These intra-layer connections were designed to enable the network model to represent more complex relationships within the data. However, while they increased the model's theoretical expressive power, they also made training significantly more difficult. Each neuron could influence many others in a

[34]As we noted when discussing the Hopfield model, the energy landscape is entirely determined by the synaptic weights. Since learning involves adjusting these weights, Sejnowski's remark about the need to "sculpt the landscape" refers to this process: learning reshapes the energy surface by modifying the weights. It's therefore not surprising that the Boltzmann machine's probabilistic framework is closely related to learning algorithms. We may not grasp every detail, but there's a certain satisfaction in understanding at least this much.

feedback loop, contributing to the Boltzmann Machine's infamous "slowness."

The main distinction between a Hopfield network and a Boltzmann Machine lies in the introduction of *randomness*.[35] A Boltzmann Machine does not always choose the most energy-efficient state outright; instead, it makes transitions with a certain probability. This randomness enables the network to explore a broader range of possible states, potentially avoiding suboptimal solutions.

The amount of randomness — how much "playing around" is permitted — is governed by ideas from statistical physics. In Boltzmann Machines, transitions between states are *probabilistic*, and their likelihood is determined by the energy difference between states and the system's temperature. This behavior is governed by the famous Boltzmann factor,[36] which depends on temperature and energy. Of course, as in Hopfield's model, these are not real physical temperature and energy but their neural network analogs. They should be rightfully called *pseudo-temperature* and *pseudo-energy*, but for convenience, we will refer to them by their original names. Just as in physical systems, a higher temperature increases the likelihood that the system will occupy higher-energy states. Temperature regulates how freely the network *explores* its energy landscape: at high temperatures, the network explores more broadly; at low temperatures, it tends to settle into stable states.

In our *vibrating bombed-field* analogy, the larger vibrations (higher temperatures) help the ball escape shallow pits and surface dimples. The amplitude of vibrations is chosen to prevent dislodging the ball from a deeper crater. When the ground eventually stops shaking (the temperature

---

[35]The reader may recall that in the 1950s, Walter Pitts burned the manuscript of his Ph.D. dissertation in which he supposedly introduced randomness into 3D (i.e., deep multi-layered) networks. We will never know if what he did was similar to the Hinton–Sejnowski Boltzmann Machine.

[36]The Boltzmann factor gives the probability P that a system will accept a move to a higher-energy state:

$$P = e^{-\Delta E/T}$$

Here, ΔE is the increase in energy associated with a potential move and T is the current system's temperature. The structure of this formula is such that for a given ΔE, the probability P is higher at higher T, meaning the system is more likely to accept uphill (energy-increasing) moves at high temperatures. This famous formula was introduced by Ludwig Boltzmann around 1877, when he worked on statistical mechanics and the theory of gases.

is reduced to zero), the ball comes to rest in a true bomb crater — a stable, low-energy minimum — where it belongs.

Although too slow and constrained to see widespread use at the time, Boltzmann Machines introduced a powerful idea: that neural networks could harness randomness to improve problem-solving. This insight later led to more efficient and practical techniques, including *Gibbs sampling* and *Restricted Boltzmann Machines* (RBMs), which we will explore when we discuss ANNs in the 2000s.

ANNs in the 1990s

By the end of the 1980s, multilayered ANNs equipped with the back-propagation algorithm opened new horizons. At the same time, researchers such as John Hopfield and Geoffrey Hinton introduced powerful ideas borrowed from physics, further transforming the field.

At the end of the 1980s, neural network researchers found themselves in a different world — one where they could aim higher than classifying simple shapes or even recognizing cats and dogs. Alongside this, Geoffrey Hinton continued to explore how networks could not only *recognize* data but also *generate* it. In the mid-1990s, he introduced the *wake-sleep algorithm* for training generative models. The system alternated between a "wake" phase, in which it learned to interpret real input data, and a "sleep" phase, in which it generated synthetic inputs and refined its internal representations.[37] This interplay between perception and imagination would later influence the training of deeper architectures.

In the 1990s, ANNs were successfully applied to real-world tasks for the first time. The most notable early progress in this respect came through the work of Yann LeCun and his collaborators, who focused on image recognition. Their first major success was in teaching neural networks to recognize handwritten characters — a carefully chosen problem that offered immediate practical value and, more importantly, the potential to scale up to a wide range of visual recognition tasks.

Yann André LeCun (Fig. 15), one of the architects of advanced image recognition and deep learning, was born in France in 1960. He received a

[37] I wonder if this interplay of "wake" and "sleep" phases, which was introduced in the context of machine learning, sheds light on the mystery of sleep. Perhaps that is why we and all the higher animals need sleep — to refine our understanding of the world.

Fig. 15. Yann LeCun, the founder of modern ANN-based image recognition.
Source: Photo from École Polytechnique, Lic. CC BY-SA 2.0.

Ph.D. in computer science from Université Pierre et Marie Curie (now Sorbonne University) in Paris. His dissertation focused on backpropagation, gradient-based learning, and image processing. While working on his dissertation, he developed a new class of neural networks best suited for image recognition: *convolutional neural networks* (CNNs). After graduating in 1987, LeCun spent a year as a postdoctoral researcher under Geoffrey Hinton at the University of Toronto — an essential detail in his scientific biography. After a year in Toronto, in 1988, LeCun was hired at AT&T Bell Labs in Holmdel, New Jersey, where he continued to develop CNNs.

Today, LeCun is the Jacob T. Schwartz Professor of Computer Science and Neural Science at the Courant Institute of Mathematical Sciences and the Center for Neural Science at New York University (NYU). From 2013 to late 2025, he also served as the VP and Chief AI Scientist at Meta (Facebook's parent company). In 2025, he left Meta to found AMI (Advanced Machine Intelligence), where he serves as Executive Chairman. LeCun has many prestigious awards, including the 2018 Turing Award, which was shared with Yoshua Bengio and Geoffrey Hinton.

From Visual Cortex to Convolutional Networks

The idea behind convolutional neural networks (CNNs) was inspired by how the human brain processes visual information. It connects directly to the concept of *internal feature representations* we discussed earlier. In the late 1950s and early 1960s, neuroscientists David Hubel and Torsten

Wiesel conducted experiments on the visual cortex of cats. They discovered that neurons in the brain respond not to an entire image at once but to small, localized features, for example, a line at a particular angle or an edge at a specific position. As visual information travels deeper into the brain, these simple patterns are combined into more complex ones, including curves, shapes, and eventually whole objects. This hierarchical structure of internal representations enables the brain to construct a coherent picture of the world.

Hubel and Wiesel's discovery, which was awarded the Nobel Prize in Physiology or Medicine in 1981, revealed that vision is not a sudden insight but a layered process, with each stage building upon the previous one — a biological template for deep learning.

Yann LeCun applied this insight to ANNs. He designed CNN architectures in which different layers specialize in detecting progressively more complex visual features. The early layers responded to basic contrasts — light or dark spots. Subsequent layers picked up edges, corners, or textures. Deeper layers assembled those into higher-level abstractions: a digit like "7," the letter "A," or even the outline of a cat (!).

Just as the visual cortex processes localized regions of the visual field, CNNs apply a mathematical operation — called a *convolution* — to small patches of an image. By mirroring this biological strategy, CNNs can detect visual features regardless of their location within the image.

The initial focus was on recognizing handwritten digits — a task of significant practical importance for automating postal mail sorting and bank check processing. LeCun and his collaborators, including John S. Denker, Sara A. Solla, Léon Bottou, Patrick Haffner, Yoshua Bengio, and others, applied the backpropagation algorithm to train these networks, enabling them to learn directly from data. In 1989, they developed an early CNN model, *LeNet-1*, which featured three specialized layers in addition to the input and output layers. This model was refined and enhanced over the following decade, culminating in 1998 with *LeNet-5*, a more advanced network with seven specialized layers.

LeNet-5 became the first neural network architecture to achieve significant real-world deployment. It was used commercially to read handwritten digits, such as those found in bank account numbers, check amounts, and postal zip codes. It dramatically improved financial and mail processing speed and accuracy. General handwriting recognition was still beyond reach at that time; LeNet-5 could only recognize numerical digits (0–9).

Starting with digit recognition (as in LeNet), CNNs soon exploded into a wide range of applications, such as:

- face and object recognition;
- self-driving car vision systems;
- medical image analysis;
- satellite and aerial imagery interpretation;
- real-time video analytics;
- image generation, including art-creating systems, and, unfortunately, deepfakes.

CNNs were the engine behind the *ImageNet* revolution in 2012, when *AlexNet* (a deep CNN architecture) crushed previous benchmarks and launched the deep learning boom of the 2010s. But here, I am getting ahead of our story.

LeCun's and his collaborators' work had implications well beyond visual processing. Their innovations in convolutional architectures also influenced early progress in speech recognition and laid the conceptual groundwork for natural language processing and broader deep learning methods.

ANNs in the 2000s

By the start of the 21st century, interest in ANNs had begun to fade once again. Despite promising advances in the 1980s and early 1990s, including convolutional networks for image recognition and energy-based models for associative memory, neural networks were often seen as difficult to train, slow to converge, and easily outperformed by newer methods such as *support vector machines* and *decision trees*. The so-called "AI winter" hadn't entirely thawed for ANNs; most researchers shifted to statistical models that seemed more reliable, scalable, and mathematically grounded.

Yet not everyone gave up on them. A small group of devotees, including Geoffrey Hinton, Yoshua Bengio, Yann LeCun, and Jürgen Schmidhuber, continued to explore deeper networks and how to make them function effectively. They began searching for training methods that could handle multiple layers without falling prey to instability or vanishing gradients.

The 2000s were a transitional decade. While deep neural networks were still outside the mainstream of AI, the foundations of modern deep learning were being laid. Among the most important developments was the introduction of *greedy layer-wise pretraining* — a strategy for training deep networks one layer at a time. In 2006, Geoffrey Hinton and his collaborators, building on the foundation of Boltzmann Machines described above, proposed a method in which training was performed in pairs of adjacent layers, two layers at a time, gradually building a deeper model. This was called *a Restricted Boltzmann Machine* (RBM). Once all layers had been initialized in this *unsupervised* manner, the entire network could be fine-tuned using *supervised* learning. This approach enabled the training of deeper architectures than had previously been feasible.

Restricted Boltzmann Machines

Instead of training an entire deep network all at once — a process that was computationally expensive and prone to problems such as vanishing gradients — Geoffrey Hinton proposed a practical simplification in 2006. He suggested training deep neural networks, *one pair of layers at a time*, using a simplified type of energy-based stochastic model called a *Restricted Boltzmann Machine* (RBM).

An RBM consists of only two layers: a *visible layer* (which receives the data) and a *hidden layer* (which learns to represent patterns in that data).[38] The two layers are connected, but there are no connections within a layer — a key restriction that distinguishes RBMs from earlier Boltzmann Machines. This simplified structure allows for much more efficient training.

[38] In my humble opinion, this is an unfortunate choice of names. Both layers in an RBM typically sit somewhere inside a larger neural stack — in that sense, they are both "hidden" from the outside. When I hear "visible layer," I tend to think of the input layer of the entire network. But in an RBM, the "visible layer" simply refers to the lower of two consecutive layers being trained together. It is "visible" only relative to the other (hidden) layer, not to the overall model.

Each RBM (pair of layers) is trained independently. This training, known as *Gibbs sampling*, occurs in a back-and-forth manner. Initially, the visible layer provides input to the hidden layer, which learns to capture patterns embedded within that data. Then, the two layers switch roles: The hidden layer supplies data to the visible layer, which reconstructs the patterns within that data. As a result of this exchange, certain aspects of the two patterns may overlap, while other elements may differ. The model then adjusts its internal weights to ensure that common patterns are assigned lower energy while atypical and spurious patterns receive higher energy. The network learns by minimizing this energy function.

This can be seen as the process of photographic development, which is paused, and the partially developed image is removed from the photographic developing tray. A new photograph of this image is captured, developed, and compared to the original. Afterward, a photographer retouches both pictures, preserving the shared features and removing those that are not.

This iterative back-and-forth, or *inference and reconstruction*, echoes *wake-sleep algorithm* introduced by Hinton in the 1990s, which we mentioned earlier. In the wake-sleep algorithm, the multilayered system as a whole alternates between observing real data and generating synthetic inputs. RBMs simplify this idea, involving only two consecutive layers in the process.

By stacking RBMs — treating the hidden layer of one pair as the visible layer of the next — researchers could build deeper networks. This strategy, known as *greedy layer-wise pretraining*, provided a viable path to train deep architectures that were otherwise resistant to direct optimization. Once all layers were pretrained in this fashion, the network could be fine-tuned using supervised learning.

Although RBMs are no longer widely used in contemporary deep learning, their role in unlocking deep architectures in the 2000s was pivotal. They provided a practical stepping stone from shallow neural networks to the deep models that would soon dominate the field.

This technique, in turn, gave rise to what Hinton called *Deep Belief Networks* (DBNs) — stacks of RBMs that could uncover increasingly abstract features from raw data. Around the same time, Yoshua Bengio's group in Montreal explored similar ideas using stacked *autoencoders*.[39] At the same time, Yann LeCun continued to refine CNNs and develop energy-based models for image recognition. Although their specific architectures varied, these researchers shared a common goal: to discover better representations by building deeper models.

The sum total of all these new training techniques amounted to building a toolkit for future *deep learning*, which arose in the following decade and changed the face of AI.

ANNs in the 2010s: The Rise of Deep Learning

Our story now enters the decade in which neural networks have finally delivered on their promise. Their performance on a variety of tasks exceeded all expectations. These tasks included image recognition, speech transcription, and language modeling (i.e., understanding all human languages). But that was not all. Machines — this time in the form of ANNs rather than symbolic AI — reinforced their 1997 win in chess with another victory in a less popular but even more difficult ancient Chinese game of Go. They excelled in medical diagnostics and molecular biology, thus entering the field of exact sciences, which had previously been an exclusively human prerogative. We will describe some of these achievements, as well as a few others. We will also note the reactions of the public and AI scientists themselves to this long-awaited breakthrough.

Let us begin, however, with a description of the technology that made all of the above possible. Of course, some of it we described in the previous sections. However, during the 2010s, new ingredients were added to this witch's brew, which we need to mention. Our attempt to understand and describe the workings of large ANNs will be superficial and incomplete. Unfortunately, I would be unable to offer a more detailed description because the technology of the 2010s became highly complex and sophisticated. Moreover, it is my impression that no one fully understands what is going on inside these new machines.

[39]An autoencoder is a kind of neural network that learns to copy its own input. It first "compresses" the data into a simpler internal form and then tries to rebuild the original from that compressed version. In doing so, it discovers what is most essential in the data — a way of learning without being told what the correct answer is.

One aspect of this new technology is universally accepted as the key to AI success: *Big Data* — vast training databases. The reader may recall that in Frank Rosenblatt's Perceptron, the number of examples shown to the machine was, at best, in the hundreds. Over the years, it has steadily increased. In the 2010s, the training datasets numbered in the millions. And that is when the machines started to learn in earnest. Who knew? Perhaps Georg Hegel knew when, at the beginning of the 19th century, he said that *quantity turns into quality*.

Deep learning, preceded and made possible by all the previous history of AI and neural networks, as well as by computational and memory advances, dawned in 2012 when a CNN designed by Geoffrey Hinton's students at the University of Toronto, *AlexNet*, won the *ImageNet* image recognition challenge by a wide margin.

AlexNet

In 2012, a CNN called *AlexNet* — designed by Alex Krizhevsky, Ilya Sutskever, and Geoffrey Hinton at the University of Toronto — stunned the AI world. Submitted to the prestigious *ImageNet Large Scale Visual Recognition Challenge*, it outperformed all competitors by a wide margin, reducing the top-5 classification error rate[40] from 26.2% to 15.3%.

The network was deep for its time, consisting of eight layers: five convolutional layers for feature extraction, followed by three fully connected layers for classification. Altogether, it contained about 60 million weights and approximately 650,000 artificial neurons.

What set AlexNet apart was not only its depth but also its scale and implementation. Crucially, it was trained on just two consumer-grade GPUs — NVIDIA GTX 580s — rather than the CPUs typically used in earlier research. This GPU-based training approach proved vastly more efficient and was quickly adopted by others, establishing a new standard for deep learning.

Conceptually, AlexNet was built upon earlier convolutional networks pioneered by Yann LeCun in the 1980s and 1990s. But it combined these ideas with three crucial elements:

[40] *Top-5 error rate* means that the classification is considered correct if the true label appears among the model's five most confident guesses. This is useful in large datasets such as ImageNet, where many classes may look similar (e.g., dog breeds). A lower top-5 error indicates better overall recognition, even if the top prediction is not always correct.

1. Big Data: the massive ImageNet dataset provided millions of labeled images,
2. GPU acceleration, and
3. improved algorithms and training techniques.

The success of AlexNet marked a turning point — the start of the modern era of *deep learning*, in which neural networks began achieving results once thought unattainable.

After AlexNet but before GPT-3

The year 2012, when AlexNet won the ImageNet competition, marked the starting point of the modern deep learning revolution. In the years that followed, neural architectures diversified and specialized. CNNs were already the standard for visual tasks. Meanwhile, other architectures were developed to handle sequential data, and language is a prime example of such data. To this end, researchers turned to recurrent neural networks (RNNs), which process input step by step, maintaining a dynamic memory of the sequence as it unfolds.

Recurrent Neural Networks

RNNs — neural networks that can process sequences of inputs by maintaining a form of memory — date back to the 1980s.

As the reader recalls, the 1986 paper by Rumelhart, Hinton, and Williams, "Learning Representations by Back-propagating Errors," introduced the general method of backpropagation for training multilayer networks. Shortly afterward, they expanded that idea to handle *sequential inputs*. The result was the algorithm known as *Backpropagation Through Time* (BPTT), formalized by Ronald J. Williams and David Zipser in their 1989 paper "A Learning Algorithm for Continually Running Fully Recurrent Neural Networks." Soon after, in 1989, Jürgen Schmidhuber and Sepp Hochreiter began developing deeper insights into the limitations of RNNs.

RNNs were initially applied to speech recognition, language modeling, handwriting recognition, and time series forecasting. In the 1990s and early 2000s, their use was limited due to training difficulties and computational constraints.

By the end of the 2010s, RNNs were largely supplanted by the *Transformer* architecture (see later), which offered better parallelism and memory capacity.

However, standard RNNs struggled with long-range dependencies — situations where related bits of crucial information occur far apart in the sequence. To address this, *Long Short-Term Memory* (LSTM) networks, initially introduced by Sepp Hochreiter and Jürgen Schmidhuber in the late 1990s, gained prominence around 2014–2015. LSTMs have proven highly effective in applications such as speech recognition, machine translation, and text generation.

A pivotal innovation emerged in 2013 with the development of *word embeddings* — vector-like representations of words that capture semantic similarity through geometric relationships.

Vector representation of words

A pivotal step in teaching machines to understand language came with the idea of representing words as vectors in a multidimensional space. In this space, words that often appear in similar contexts are positioned close to one another, capturing their semantic relationships geometrically. In a sense, AI scientists formalized the notions of *meaning* and *understanding*.

Early attempts to formalize this insight appeared in the 1990s with *latent semantic analysis* (LSA) — a statistical technique that analyzed large text collections and expressed each word as a vector summarizing its co-occurrence patterns with other words. LSA captured meaning surprisingly well but required costly matrix computations that limited its use on larger datasets.

The breakthrough came in 2013, when Tomas Mikolov and his colleagues at Google introduced *Word2Vec*, a family of small neural networks that could learn these word vectors efficiently from enormous text corpora. The resulting *word embeddings* mapped words into a multidimensional geometric space, which, significantly, was created by the machine during the learning process rather than being programmed from the start. Soon afterward, alternative methods, such as *GloVe* (developed at Stanford) and *fastText* (developed by Facebook AI), refined and extended this idea. Word embeddings have since become the foundation of nearly all modern LLMs — the engines that power today's generative AI systems. Treating language mathematically is a daring and challenging proposition, but only after it has been mastered, neural networks have started dealing with words and language as successfully as they have already dealt with image recognition. Representing words as vectors in a multidimensional space arguably marked the moment when machines acquired the ability to understand the meaning of words.

A Token and Its Vector Representation

In natural language processing, a *token* is a basic unit of text, often a word, subword, or even a single character, such as a punctuation mark. Each token is converted into a *vector* — a list of numbers that represents aspects of its usage and meaning. But these aren't your garden-variety vectors.

Vectors in math and physics

In mathematics and physics, a vector is a rigorously defined entity with specific properties: It has directionality and magnitude (length), and it exists in a space with defined, mutually perpendicular axes. In three-dimensional space, there are three axes, and a vector is represented by three numbers: the projections of the tip of that vector onto the three axes. These three numbers will change in a specific, predictable manner if we rotate the coordinate system by a certain angle. There are mathematical operations, such as addition, subtraction, and two types of multiplication, that can be performed with vectors, enabling the development of vector algebra. Vectors can be defined in a space with any number of dimensions equal to or greater than two: 2, 3, 4, 5, and any n dimensions. A vector in a space with $n = 10{,}000$ dimensions will be characterized by a string of 10,000 numbers — its projections onto the 10,000 axes. No sweat. Of course, we humans can intuitively understand operations with two- and three-dimensional vectors, while at higher dimensions, we lose our intuition. However, dealing with higher dimensions is entirely legitimate.

Vectors as words (tokens)

In the world of LLMs, the term "vector" is used metaphorically. Some of what we said about mathematical vectors remains true (in some loose sense), while much of it is not applicable.

A vector now refers to a token's *learned* numerical representation — a position in a high-dimensional space that captures how a token behaves in context. A vector for a given token is formed during training; it is learned from large amounts of text. Once a model, after extensive training, forms all these meaning-reflecting lists of numbers (vectors), it is ready to manipulate language algebraically and geometrically — the age-old distinction between the liberal arts and exact sciences disappears! In a sense, the model is then prepared to *reason*.

These "vectors" inhabit a vast, imaginary "space," often referred to as the *token embedding space* or *semantic space*. It's not an explanatory dictionary or a warehouse of definitions. Unlike a dictionary, this space is shaped not by human-attached meanings but by patterns of usage — statistical regularities in how words appear near each other in language. The result is a geometry organized *by meaning*: tokens that tend to occur in similar contexts will end up close together in this space.

You can imagine the space as filled with *clouds* of related meanings and common usage. One large cloud might gather all tokens associated with living things.[41] Within that cloud lies a smaller one of words for furry animals. Within that, an even tighter cluster might contain all the ways humans refer to cats, in various languages, slangs, or contexts. The "direction" in the embedding space from a neutral or reference region (in math, it would be called the coordinate system's origin, or point 0) — a zone not strongly associated with any particular meaning — toward the cat cluster is meaningful, as it encodes features of "catness," distinguishing those tokens from neighboring regions, such as dogs or dolphins.

A vector representing a cat will be closely aligned with a vector representing a kitten. These vectors will not be identical, but they will be pointing in almost the same direction. In this way, token representations behave somewhat like true vectors, as they exhibit directionality within the learned semantic landscape.

---

[41] With one crucial caveat: The cloud of tokens associated with living things is formed by the machine based solely on how those words are used in language, not on any understanding of biology. That distinction matters. For example, the tokens for *"virus"* and *"computer"* may end up near each other in the semantic space, even though a computer is not a living organism. That's because the model has encountered the phrase *"computer virus"* frequently and has learned to associate those tokens contextually. What matters to the machine is not ontological accuracy but *statistical co-occurrence.*

The strings of numbers representing vectors may contain thousands of numbers for each token. A significant amount of information — many bits — can be encoded in such large numerical strings. Part of that information relates to the direction in which a given token lies within that embedding space.

One example

When I asked ChatGPT to explain all of that, it provided the following example, which I am slightly rephrasing and illustrating with a graph:

Take the words "king," "queen," "man," and "woman" and think of them as vectors in the semantic space (Fig. 16). It is reasonable to expect that the *difference vector* between king and queen lies in approximately the same direction as the difference vector between man and woman. The meaning of that last statement is transparent: the vector difference represents the *gender shift*, which is the same for the royalty and the commoners.

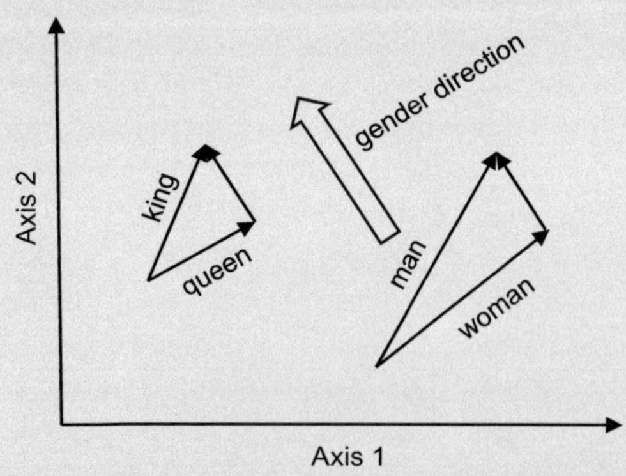

Fig. 16. An attempt to depict the vector representation of words and their meanings. The axes symbolically represent orthogonal directions of meaning. The fat arrow indicates the general direction of *gender*. The vectors corresponding to the words "king," "queen," "man," and "woman" are all different. Still, the vector differences between "king" and "queen" and "man" and "woman" point in the same gender direction.

So, treating these words as vectors, we can state

$$\mathbf{\textit{king}} - \mathbf{\textit{queen}} \approx \mathbf{\textit{man}} - \mathbf{\textit{woman}},$$[42]

or, more elaborately, assuming that we can do vector algebra:

$$\mathbf{\textit{king}} - \mathbf{\textit{man}} + \mathbf{\textit{woman}} \approx \mathbf{\textit{queen}}.$$

Here, the symbol "≈" means "approximately equal to."

This is not a bad example — it shows how expressing words as vectors allows the machine to construct its own internal mathematics, in which it can manipulate those vectors. This vector algebra would be of little use to a human linguist, but it appears to be most useful for large neural networks dealing with language. Perhaps something like that is happening in our brains? — I assume that neurophysiologists ask themselves that question.

Large numbers of everything

In all of that, the ANNs are dealing with fairly large numbers. The "semantic space," with its dimensionality and axes, is not programmed into the machine; there are no predefined rules by which it is created. It emerges during the learning process as the material is rearranged and better defined as training proceeds. This space may contain thousands of dimensions, so that each token is a string of thousands of numbers.

The number of tokens in the latest LLMs (as of mid-2025) is of the order of hundreds of thousands — this is referred to as the *vocabulary*. And, of course, in a sense, it *is* a vocabulary. The number of words labeled as being *in current use* in one of the latest Oxford Dictionaries of the English Language is about 170,000, so it checks with the number of tokens, which, in addition to containing whole words, may also include word parts and punctuation marks.

So, ANNs are dealing with hundreds of thousands of vectors in a thousand-dimensional space, all of which are constructed by AI itself during the learning process. No wonder that our ability to interpret or describe that space is limited. This is an AI territory, and many parts of it increasingly become difficult or impossible to interpret in human terms.

---

[42] We use boldface for vectors, which is standard practice in mathematics and physics.

True understanding?

Now that we have outlined this fascinating development in the special section above (well worth reading — and if you skipped it, I encourage you to go back), let us turn to the notion of *meaning*: Do machines truly *understand* what they process, or only simulate understanding?

On this question, opinions diverge sharply. Some argue that machines still comprehend nothing at all — that they merely predict, with astonishing skill, the next word most likely to appear in a sentence, while the gulf between human and machine intelligence remains as wide as ever. Others — those more familiar with recent advances in neural networks and especially with the mathematical foundations just described — take a different view.

To illustrate this, I will quote from a recent lecture (July 2025) delivered by Geoffrey Hinton, our favorite AI scientist, who, after retiring from Google, has devoted much of his time to explaining the meaning of the neural network revolution to a broad audience. The excerpts below are lightly edited from the transcript of that talk, available on YouTube. Although we have already discussed the contrast between symbolic AI and neural-network-based AI, I cannot resist quoting Hinton here; he has a rare gift for expressing deep ideas with disarming simplicity:

> …Now, with these large language models, chatbots — there are still many scientists who don't think they really understand the same way we do. They believe in a very different model of what human understanding is. For the last 70 years, there've been two very different paradigms for intelligence. The logic-inspired paradigm, which dominated for about the first 50 years of AI, is that the essence of intelligence is reasoning. And reasoning is done by using symbolic rules to manipulate symbolic expressions. So [*according to symbolic AI, MG*] your knowledge is a bunch of symbolic expressions inside your head… The contrast was a biologically inspired approach [*ANNs*]… And there the essence of intelligence is learning the strengths of the connections in a neural network… And... it would… generate sentences by repeatedly predicting the next word…
>
> Leading computational linguists began to accept that feature vectors, which they called embeddings, were actually a good way to model the meanings of words. And that's when [*after the appearance of ChatGPT,*

> *M.G.*] people, not just researchers, but everybody, began to get interested in what was happening in these large language models. Were they really understanding what they were saying?

Then, Hinton attempted to explain token embeddings as follows:

> That's what meaning is — meaning is having a model. So, think of words as high-dimensional Lego blocks. With Lego blocks, you can model any 3D shape moderately well... Think of words as Lego blocks, but instead of being 3D, they're like thousand-dimensional... so the shapes are very complicated because they're in thousand dimensions. And we can use combinations of those to model anything at all. Not just the distribution of matter in 3D, but anything, like theories of how the brain works. Now, instead of having just a few different kinds of Lego blocks, we've got thousands of different kinds of these special high-dimensional Lego blocks... And each one isn't a fixed shape. Each one has a range of shapes it can adopt. It's flexible; it can distort... There may be several alternative shapes [*for the same meaning*]... And what they do is they deform to fit in with the other words in the context. So, the context gives a particular shape to each word... And instead of thinking of them fitting together by using little plastic cylinders that plug into holes like Lego blocks do...think of the words as having lots of little hands on them. And these hands have funny shapes. And the way they fit together with other words is they do handshakes. That's called *query key* handshakes in *a transformer* [*something we will discuss next, MG*]. As you change the precise vector used for the meaning of a word, and the context changes, you actually change the shapes of the hands. So, it's slightly more complicated than [*if they had*] a fixed set of hand shapes. They change their hand shapes so that the words in the context can shake hands with other words. And that's what understanding is. Understanding is taking those words, finding how to deform them and how to deform their hands so they can shake hands with other words... You take these word symbols, and the word symbols don't mean anything by themselves. They need an interpreter... your brain is [*the interpreter*] ... that's what it means to understand something: to form that structure...
>
> So, the large language models — if you want to model all of human knowledge — they're complicated. They have many layers, very

> complicated interactions. And so, it's very hard to analyze what they've actually learned and to see that they really are understanding what they're saying, particularly if you have the wrong model of understanding… Their knowledge is in those interactions. It's a bunch of weights in the neural network. So that's what knowledge is in these large language models, and it's what knowledge is in us, too… So, my claim is that we model reality by using these word fragments in much the same way as machines do…
>
> They're much more like us than they are like standard computer code. Digital computation requires a lot more energy because it's digital. So you have to drive transistors very hard. You can't use the funny, quirky properties of analog hardware. But because they're digital, they have this really efficient means of sharing. They can share weights or gradients… So, the overall conclusion is that if energy is cheap or at least abundant, digital computation is just better because it can share knowledge efficiently. And that's a very scary conclusion. (Hinton, 2025)

From embeddings to attention

Models like Word2Vec and its successors (GloVe and fastText) produced static word vectors: Each word was assigned a single, fixed representation during training and remained unchanged in all subsequent uses. For example, the word *"virus"* would have the same vector whether it referred to a biological pathogen or malicious software. Its neighbors — the "company" it kept — might include the word *"computer*," as noted in the footnote above, but that company was fixed once and for all during training. This rigidity limited the model's ability to handle linguistic ambiguity, *polysemy* (when a word has multiple related meanings), and subtle shifts in meaning, all of which are essential for deeper language understanding.

To capture more context-sensitive information, researchers turned to RNNs and their improved variant, LSTM networks. These architectures enabled models to process sequences of tokens while retaining a limited memory of previous inputs — hence the somewhat humorous name "Long Short-Term Memory." Between 2014 and 2017, LSTM-based models powered advances in machine translation, speech recognition, and text generation. The benefits were clear: unlike earlier systems, these models could better handle long-range dependencies and shifting meanings in text. As a result, services such as *Google Translate* began replacing their traditional statistical frameworks with neural-network-based architectures.

However, these models had limitations. Training them was computationally intensive, and their ability to capture long-range dependencies — particularly in extended texts or dialogues — remained imperfect. More critically, they had to process text sequentially, one word at a time, which slowed down computation and limited scalability.

In response to these growing pains, a new idea began to take shape: Could a model attend to all parts of a sentence at once rather than processing it word by word? Could it learn where to focus — what to "pay attention to" — without relying on recurrence or fixed memory? These questions culminated in the 2017 proposal of a new architecture: the *Transformer*, which would soon render RNNs and LSTMs largely obsolete for most natural language tasks.

The Transformer architecture was introduced in a seminal 2017 paper titled *"Attention Is All You Need."* Its authors — Google AI researchers Ashish Vaswani and colleagues (Vaswani *et al.*, 2017) — proposed replacing recurrence with a novel mechanism called *self-attention*, which allowed the model to consider all parts of a sequence simultaneously. This innovation enabled greater parallelism, more efficient training, and a deeper, more flexible grasp of language structure. Transformers rapidly supplanted earlier RNN- and LSTM-based models and soon became the foundation of nearly all modern language models.

Transformers

The title of that 2017 paper — "Attention Is All You Need" — is a pun. It works because *attention* is not merely a casual word; it's a technical term.

Transformers were initially developed for machine translation. Vaswani and co-workers trained their model on English-to-German and English-to-French datasets, using tens of millions of tokens.

What makes Transformers different is that they do not process input data sequentially, unlike earlier models such as RNNs. Instead, they examine the entire input at once and use a technique called *attention* to determine which parts of the input are most relevant to each other.

The Attention Mechanism: Query, Key, and Value

The process still begins with token embedding, as described earlier: Each token is represented as a vector in a high-dimensional space. From this vector, the Transformer derives three new vectors — the *query*, the *key*, and the *value* — each serving a distinct role:

- The query asks a question: *What am I looking for?*
- The key provides identifying information: *What do I contain?*
- The value holds the actual content to be retrieved.

For each token in the sequence, the model calculates how well its query matches the keys of all the other tokens — mathematically, by computing *dot products*, which measure directional similarity between vectors.[43] If the dot product between a token's query and another token's key is large, the two are highly relevant to each other. These similarity scores are then converted into *attention weights*, which determine how strongly the corresponding value vectors contribute to the updated representation.

As Geoffrey Hinton once put it — and we discussed that in the previous section — the interaction between word embeddings can be imagined as a series of "*handshakes*," in which the shapes of the hands — the vectors — change subtly depending on context. Each query reaches out, and only keys with a matching "grip" respond.

This mechanism allows the model to interpret relationships between words regardless of their position in the sentence. For example, in the sentence "*The statuette didn't fit in the suitcase because it was too small*," the model must decide whether "*it*" refers to the statuette or the suitcase. The attention mechanism enables it to weigh these relationships dynamically, resolving such ambiguities through patterns of matching "handshakes."

Why Transformers Matter

By enabling this form of dynamic attention, Transformers dramatically outperformed earlier architectures. Just as importantly, they scale well to

---

[43] In vector algebra, the *dot product* of two vectors depends not only on their values but also on the angle between them. It has its maximum value when this angle is zero — that is, when the two vectors are aligned, and it is zero when the two vectors are perpendicular to each other. If the two vectors forming the dot product have the same standard length, then the value of the dot product will depend only on their mutual orientation.

large datasets and allow for parallel training — a significant advantage over the inherently sequential RNNs and LSTMs.

AI researchers became "instantly addicted" to the Transformer design. By the end of the decade, virtually all advanced language models were based on it. One could now say: "*Transformer is all you need!*"

The end of the decade and the appearance of OpenAI

The introduction of the Transformer architecture in 2017 marked a decisive turning point in natural language processing. Within months of the publication of the "Attention Is All You Need" paper, researchers across academia and industry began adapting and extending the model. What followed was an extraordinary surge in both capability and ambition.

In 2018, Google introduced *Bidirectional Encoder Representations from Transformers* (BERT), a model capable of reading text in both directions — from left to right and from right to left — allowing it to grasp a word's meaning within the context of the entire sentence. This bidirectional reading enabled BERT to achieve a more nuanced understanding of language. Like many subsequent models, BERT was trained in two stages: First, it read a massive corpus of unlabeled text, and then it was fine-tuned with human supervision for specific tasks. It quickly surpassed earlier systems on benchmarks such as question answering and sentiment classification. Its success helped establish the now-standard paradigm of *pretraining followed by fine-tuning.*

Meanwhile, a new startup had been preparing to challenge the dominance of the tech giants. *OpenAI,* founded in 2015 as a nonprofit research lab by Elon Musk, Sam Altman, Ilya Sutskever, and others, set itself a bold goal: to develop *artificial general intelligence* (AGI)[44]. It was as if, among a gathering of leprechauns — magical creatures from Irish legend — one, the smallest and youngest, suddenly declared, "You may all do as you wish, but I am going straight to the bottom of the rainbow, where a big pot of gold awaits me." Unlike Google, Amazon, or Microsoft, OpenAI was not born from a corporate empire; it positioned itself as an independent challenger. It was the first startup-scale entity to pose a serious threat to Google's dominance in language AI.

[44] Many thinkers now question the goal of building AGI. We will address their concerns in the following chapter.

While we cannot follow all of the fascinating human stories, conflicts, and shifting alliances that have played out in OpenAI's turbulent history — many of which have been highly publicized in recent years — an interested reader will find them richly documented in a recent book by journalist and author Karen Hao (2025).

Here, we concentrate on the technical aspects of the story. While BERT excelled at understanding and classifying language, another approach, pioneered at OpenAI in 2018, focused on *generating* it. In a sense, this shift echoed Richard Feynman's famous dictum, as found on his blackboard at the time of his death: "*What I cannot create, I do not understand.*" *Generative Pretrained Transformer* (GPT) embodied this principle. Rather than analyzing sentences after the fact, it *created them,* word by word, demonstrating an internalization of language structure powerful enough to produce coherent, novel text. In doing so, it blurred the line between *understanding* and *creation* — between interpreting the world and building it.

Like BERT, GPT was trained in two stages: First, it was *pretrained* on a large corpus of unlabeled text to internalize the structure of language, and then *fine-tuned* on more specific tasks such as translation, summarization, or question answering. Its success demonstrated that unidirectional language modeling — though seemingly less comprehensive than BERT's bidirectional view — could be compelling when applied to generative tasks.

The first GPT model was introduced in a 2018 paper by Alec Radford and colleagues at OpenAI. It marked the beginning of a new era in generative AI, which soon produced a series of increasingly large and capable models, including GPT-2, GPT-3, and subsequent models.

GPT-2, released in 2019, shocked the field. With over a billion parameters (weights) and trained on a massive internet corpus, it could write surprisingly coherent essays, translate between languages, summarize documents, and answer factual questions — all without being explicitly trained for any of these tasks. Its fluency and versatility demonstrated a new lesson: scaling up — in data, model size, and compute — made models not only better but also *qualitatively* different. I will refrain from praising Georg Hegel again, but you know what I would say.

By the end of the 2010s, the Transformer architecture had become the cornerstone of modern AI. What started as a clever innovation for machine translation evolved into a method that, in addition to translation, enabled advanced search, content moderation, voice assistants, code generation (machine programming), and more. Truly general language models were within reach.

Warnings in the 2010s

The 2010s marked the first decade in which warnings about the long-term consequences of artificial intelligence no longer came from a few isolated voices but from an entire chorus of leading scientists, engineers, and technological pioneers. Concerns about AGI, its alignment with human values, and its unintended consequences began to appear in scholarly articles, public lectures, and open letters. In 2014, physicist Stephen Hawking cautioned that the development of a fully autonomous AI might mean the end of the human race. Elon Musk, expressing related fears, compared the creation of AGI to "summoning the demon" and helped finance early research on AI safety. Famously restrained Bill Gates likewise expressed concerns about the progress toward superintelligence. In the same year, philosopher Nick Bostrom published *Superintelligence*, which brought theoretical concerns about machine dominance into the broader intellectual and public arena. Around this time, computer scientist Stuart Russell joined the campaign to ban so-called "killer robots"; toward the end of the decade, he also proposed concrete methods for training AI systems so that their objectives would remain aligned with human interests.

Even earlier, from the beginning of the 2000s, researcher and philosopher Eliezer Yudkowsky had been raising questions about AGI alignment and existential risk. He argued that a sufficiently advanced AI would likely behave in unpredictable and uncontrollable ways unless its goals were made compatible with human aims from the outset. His writings laid the foundation for the emerging field of AI safety and inspired a generation of thinkers preoccupied with the long-term implications of machine intelligence.

By 2015, these concerns had coalesced in an open letter titled "Research Priorities for Robust and Beneficial Artificial Intelligence," prepared by Max Tegmark and the Future of Life Institute (FLI). Presented at the Beneficial AI conference in Asilomar and later published online and in *AI Magazine*, the letter was signed by hundreds of leading AI researchers and public figures. Eight years later, FLI issued another open letter, which was considerably starker in tone than that 2015 letter; we will discuss it in the following section. A more detailed look at the views of all the people mentioned above will be presented in the following chapter.

AI in the 2020s

Large language models and their many guises

The release of GPT-3 by OpenAI in June 2020 marked another turning point in the history of AI. With 175 billion parameters (weights and biases) and 96 layers, each containing over 12,000 units, GPT-3 was by far the largest language model at the time. Accordingly, it demonstrated unprecedented capabilities in text generation, translation, and all the other things that LLMs do. Its architecture followed the *Transformer* design introduced in 2017, but scaled up to such an extent that new, unexpected behaviors began to emerge — notably, *few-shot learning*, a term coined to describe the model's ability to generalize from just a handful of examples.

In late 2022, OpenAI launched *ChatGPT*, a fine-tuned version of GPT-3.5 accessible through a conversational interface. It quickly became a global sensation, attracting over 100 million users in just two months and bringing the power of LLMs into the public spotlight. For many, this was the first time AI entered their daily lives and was perceived as a genuinely thinking entity.

OpenAI followed up in March 2023 with *GPT-4*, a more capable multimodal system that could handle both text and image inputs. Although its parameter count was never officially disclosed, independent estimates placed it in the 1–2 trillion range — nearly ten times larger than GPT-3. Initially limited to an 8,000-token *context window*,[45] GPT-4 was soon expanded to 32,000 tokens — enough to handle over 20,000 words in a single interaction. This allowed the model to analyze long scientific papers, legal contracts, books, or entire chat histories without losing coherence. It was a substantial leap beyond GPT-3, which was capped at just 2,048 tokens, barely enough for a few pages of text.

[45]The term "context window" refers to the total number of tokens (fragments of words, punctuation, or code) that the model can "see" at once, including both the input prompt and the generated output. This defines the amount of information the model can consider when generating a response.

In 2024, OpenAI introduced a new family of foundation models designated by the letter "o", short for *omni.*[46] Unlike the earlier GPT-3.5 and GPT-4 series, which were primarily text-based and later extended with vision and audio modules, the "o" models were designed from the ground up to be multimodal, capable of processing text, images, and audio. The first publicly released model in this line, *GPT-4o*, appeared in May 2024. It matched the language capabilities of GPT-4 Turbo while offering significantly improved performance. My readers would know that I used ChatGPT-4o[47] and later ChatGPT-5 while writing parts of this book, and I can attest that it was a truly remarkable advisor and assistant. In many cases, we discussed the shortcomings of my drafts and those generated by Chat, as if I were working with a knowledgeable editor. These ChatGPTs remember previous sessions, which is essential when editing a book. Based on my experience (which was limited to asking for information, editing some of my writing, and generating graphs according to my instructions), I do not doubt that these systems *think*.

OpenAI continues to actively develop its products. On June 10, 2025, it released o3 pro — a faster and more reliable variant of the "o3" model, particularly strong in science, mathematics, education, programming, and business tasks. On July 17, 2025, ChatGPT Agent was launched, capable of performing autonomous multi-step tasks. And on August 7, 2025, OpenAI introduced GPT-5, a new foundational multimodal model that replaced GPT-4o and many earlier versions. GPT-5 features enhanced reasoning abilities, improved memory, and higher operating speed.

However, OpenAI is not alone; other labs and companies have also launched their own high-profile models. The following summary isn't meant to be all-inclusive:

- *Anthropic*, a company founded by former OpenAI researchers, released its *Claude* series beginning in 2023. *Claude 2* was noted for its cautious tone and long-context capabilities; *Claude 3*, released in March 2024, matched or exceeded GPT-4 in some benchmarks and supported up to

---

[46]Omni means "all" in Latin. It refers to the model's ability to handle multiple input modalities — text, vision, and audio..

[47]There is a slight distinction here which needs to be explained. GPT-4o is the name of the core AI model. ChatGPT-4o is the version of this model used inside OpenAI's ChatGPT app. It includes additional features, such as the ability to remember facts about the user between conversations, and it has access to tools such as web browsing, code execution, and image generation.

200,000 tokens of context. *Claude 4*, launched on May 22–23, 2025, includes *Claude Opus 4*, a frontier performer in coding, reasoning, and long-form thinking; it can run independently for hours. Then, there is *Claude Sonnet 4*, a more lightweight, free-tier alternative that supports hybrid reasoning.[48] All of Anthropic's products are created in accordance with Anthropic's *Responsible Scaling Policy*, which includes strict safeguards against misuse.

- *Google DeepMind* developed the *Gemini* series, starting with *Gemini 1* in late 2023, followed by *Gemini 1.5* in early 2024, *Gemini 2.5* (released on March 25, 2025), and *Pro*, *Flash*, and *Flash-Lite* versions, all of which were generally available by mid-June 2025. Google is also introducing *Deep Think* — a reasoning mode that supports a context of up to *1 million tokens*[49] for multimodal tasks.
- *Meta* (Facebook) open-sourced its *LLaMA* (Large Language Model Meta AI) models, beginning with LLaMA 2 in 2023 and followed by LLaMA 3 in April 2024. In April 2025, Meta released *LLaMA 4*, added *LLaMA 3.1*, and adopted multimodality, thereby leveling the playing field with *closed* (non-open-source) models. These models powered a wave of experimentation and customization, especially among startups, academic labs, and enterprise developers seeking alternatives to proprietary systems.
- *Mistral*, a French startup founded in 2023, released compact but high-performing models with open weights. Its *Mixtral 8x7B* — a mixture-of-experts architecture utilizing eight expert subnetworks of 7 billion parameters each — achieved near-GPT-4 performance at a significantly lower compute cost. Later releases in 2024 and 2025 continued to emphasize efficiency and openness.
- *xAI*, founded by Elon Musk, launched *Grok*, an AI assistant integrated into X (formerly Twitter). Grok blended LLM capabilities with real-time access to X's social data, presented in a deliberately irreverent and provocative tone designed to reflect Musk's vision for the platform. xAI launched *Grok 4* in July 2025, featuring multimodal abilities, native tool use, and strong benchmark results. Shortly afterward, Musk announced "Baby Grok," a child-friendly edition intended to address safety concerns about the content available in the main, adult version

[48] Hybrid reasoning refers to the combination of different computational approaches — typically neural pattern recognition and symbolic or tool-based methods — to achieve more reliable and structured reasoning than either alone.

[49] One million tokens is roughly equivalent to 2,700–3,000 book pages of text.

of Grok. Subsequent 2025 releases, such as Grok 4 Fast and Grok Code Fast 1, expanded this line.

- In January 2025, a Chinese startup based in Hangzhou called *DeepSeek* released an open-weight model named *R1*, claiming performance comparable to GPT-4 while being dramatically cheaper to train — reportedly around $6 million using NVIDIA H800-class chips, compared to over $100 million for GPT-4. The news triggered a massive sell-off in U.S. tech stocks, including NVIDIA, which lost 17% of its market value in a single day, the largest one-day drop in its history. Market analysts likened this to a "Sputnik moment" in the global AI race.

DeepSeek's use of the *MIT license*,[50] combined with its ultra-low training costs, shook investor confidence in the assumption that cutting-edge AI would remain the exclusive domain of elite Western labs. While some later argued that the market reaction was overblown, the episode cemented China's emergence as a serious contender in large-scale LLM development.

By mid-2025, these competing platforms had established a thriving — albeit fragmented — landscape of foundation models, with varying trade-offs in terms of speed, openness, safety, and interface design. Some prioritized open access and transparency (Meta, Mistral, and DeepSeek), while others emphasized tight integration with proprietary platforms (OpenAI and Google). Still others, like Anthropic, positioned themselves as *safety-first* alternatives, focusing on interpretability, value alignment, and *constitutional safeguards*[51] — principles embedded in the training process to align models with human ethics.

Although I only used OpenAI products, I do not doubt that others are comparable. Each new model aims to make the best impression, with some emphasizing heightened safety concerns, others highlighting their

---

[50]The *MIT License* is a simple and permissive open-source software license originally developed at the Massachusetts Institute of Technology (MIT). It allows software to be freely used, modified, and shared — both in open-source and proprietary (closed-source) projects — with very few restrictions.

[51]*Constitutional safeguards* are guardrails encoded into an AI system through a written set of ethical principles, which the model uses to self-align its behavior without requiring constant human intervention. This method was introduced in Anthropic's 2022 paper titled "Constitutional AI: Harmlessness from AI Feedback" and is part of their broader goal of making AI systems more interpretable, predictable, and aligned with human values.

openness, and all boasting their unique capabilities. However, the onslaught with which these numerous thinking systems are entering the market is somewhat staggering. No ordinary user could possibly try them all or make any reasonable comparison between them. One thing is obvious — we are witnessing a desperate battle for market dominance. How can one hope for caution in this stampede? It's like asking sharks during a feeding frenzy to behave politely and attentively.

Ultimately, the makers of these AI systems are primarily concerned with capturing market share, just as they have always been with all consumer goods, from washing machines to passenger cars. But here, we are not dealing with washing machines — we are dealing with a new phenomenon: thinking and, in some cases, independently acting entities that, in the not-too-distant future, may surpass us in every respect.

Emergent capabilities

One of the most intriguing and unsettling phenomena in the development of modern AI systems, especially large language and multimodal models, is the appearance of *emergent capabilities*. These are not gradual improvements but qualitative leaps: abilities that appear suddenly and unpredictably as the models grow in scale. A system that was designed to handle and generate language may acquire the knowledge of elementary math on its own. A system that can barely solve basic math problems may, at a larger scale, display reasoning skills comparable to those of a high school student. The same model, with no change to its underlying architecture, may suddenly begin to understand visual analogies or generate usable code. These surprises are now numerous, becoming almost a common occurrence in the development of AI systems.

Examples:

- GPT-3 (2020, 175 billion parameters) impressed by its fluency of language but often failed at arithmetic. GPT-4 (2023), with roughly 10 times more parameters, not only handled arithmetic correctly but also demonstrated multi-step reasoning. Moreover, it successfully passed American standardized tests — the Scholastic Aptitude Test (SAT, a college entrance exam), the Law School Admission Test (LSAT), and even the bar exam (a qualifying examination for admission to legal practice) — with remarkably high scores.

- In 2022, it was found that GPT-3.5 could solve *Raven's Progressive Matrices* — a set of abstract visual reasoning puzzles used to assess pure logical reasoning ability — even though the model had never been explicitly trained for such tasks.
- In 2023, GPT-4V was trained to analyze images, and in 2024, GPT-4o became multimodal — able to handle text, images, and sound within a single interface. The developers had planned multimodality, yet the depth and flexibility of these abilities proved unexpected: The model not only described pictures but also understood memes, interpreted graphs and diagrams — all of which appeared as examples of *emergent capabilities*.
- GPT-4o introduced the so-called *reasoning pipeline* — an internal mechanism for step-by-step reasoning. On mathematics competition problems, specialized versions of the model achieved accuracies of up to 83% (compared with 10–20% for earlier generations). This, too, was an ability that emerged with increasing scale.
- In 2024–2025, large models began to display capacities for discovering new algorithms. One version of *AlphaTensor*, developed by DeepMind, independently discovered methods for matrix multiplication previously unknown to mathematicians — some even faster than the classical ones (we will return to this in the section "Beyond Language"). This became an unexpected example of a system trained on optimization tasks spontaneously discovering new algorithms rather than merely applying known ones.
- In recent years, reports have appeared that large language models are beginning to show abilities in automatic programming. Without explicit training in compiler design, they have learned to write code in new languages. In 2025, OpenAI and Anthropic demonstrated models that, without prior preparation, could adapt to little-known programming languages and identify vulnerabilities in code — abilities that the developers had not specifically intended to test.

The fact that these abilities appear suddenly rather than gradually does not surprise us too much — Hegel again: The *transition from quantity to quality*. And, after all, how could a new ability arise *gradually*? But can we predict which new ability will appear next? Probably not. And if we cannot, how can we guarantee safety and control? How can we hope that future powerful AIs will serve us faithfully if they continue to display unexpected and unplanned *emergent abilities*?

I would also add that if one still needs proof that LLMs are indeed capable of thinking, these unplanned, emergent abilities provide the best evidence. A capable student may, at some moment, understand something more deeply than what is written in the textbook — or even go beyond the textbook altogether. That is the hallmark of intelligence. Therefore, these machines must possess a *mind* of their own — Q.E.D.[52]

Warnings in the 2020s

As we noted already, over the past decade, a growing chorus of scientists and thinkers has begun to raise concerns about the trajectory of AI. The discussion intensified during the 2020s, fueled by the demonstrated power of LLMs. The systems were learning faster than expected, generalizing better than forecast, and behaving in ways their creators had not fully anticipated. The future was not merely arriving early — it was arriving off-script.

In 2023, the nonprofit *Center for AI Safety* issued a stark one-sentence statement: "*Mitigating the risk of extinction from AI should be a global priority alongside other societal-scale risks such as pandemics and nuclear war.*"It was signed by hundreds of researchers, including Geoffrey Hinton, Yoshua Bengio, Demis Hassabis, and executives at OpenAI and Anthropic. What made this moment different was not just the message but the messengers — pioneers who had helped build modern AI now warning that its consequences might soon escape our grasp.

That same year, Hinton resigned from his role at Google, explaining that he wanted the freedom to speak candidly. In public interviews, he acknowledged that the systems were developing emergent capabilities faster than expected and that their long-term behavior could not be reliably predicted. "I console myself with the normal excuse," he said, "If I hadn't done it, somebody else would have." We will have a special section on Geoffrey Hinton's views in the following chapter.

One of the most visible appeals for caution emerged in March 2023, when the Future of Life Institute, led by Max Tegmark, published an open letter calling for a six-month moratorium on the development of AI systems that surpass the capabilities of GPT-4. The letter warned that AI labs were "locked in an out-of-control race" and urged a coordinated pause to

---

[52] Q.E.D.: Abbreviation of the Latin phrase *quod erat demonstrandum*, meaning "which was to be demonstrated." It is traditionally placed at the end of a mathematical proof.

allow society time to develop safety protocols and oversight mechanisms. It was eventually signed by more than 30,000 people, including notable figures such as Yoshua Bengio, Elon Musk, Steve Wozniak, and hundreds of academics and AI researchers. The pause never occurred, and several public figures did not sign the letter for various reasons — some said that it went too far, while others said that it did not go far enough. The open letter was even criticized as untimely and detrimental to the U.S. AI effort. We will devote a section to Max Tegmark's views and this story in the following chapter.

2023 and 2024 also saw intensified public advocacy from longstanding AI safety voices. Eliezer Yudkowsky, a researcher and writer focused on AGI alignment, argued that the world was far behind where it needed to be in terms of safety. In a widely discussed *Time* magazine op-ed, he called for an indefinite moratorium on the development of frontier AI models, stating bluntly that failure to align these systems could lead to the extinction of humanity. We will discuss Yudkowsky's views in the following chapter.

The discourse around AI grew increasingly intense and layered. By the mid-2020s, the public conversation had shifted from simply debating whether AI posed risks to discussing the nature, severity, and timing of those risks. Concerns ranged from the immediate and tangible — misinformation, deepfakes, labor displacement, autonomous weapons, and the concentration of power in private hands — to more abstract or long-term existential threats. Few serious voices argued that AI carried *no* risks at all (though a handful did — we will give them some space, but not too much, in the following chapter). Much of the debate now centers on how to prioritize these dangers and whether effective governance remains possible. The more profound question — whether humanity could survive the tools it was building — remained in the background, though never far from view.

Beyond Language: Autonomous Transport, Medicine, and Science

Autonomous transport

Beyond mastering language, by the late 2010s, AI was increasingly asserting itself in the real world. One of the most fascinating directions was autonomous transport — vehicles capable of functioning without human intervention. Yet it must be said at once: The goal of completely eliminating the human driver remains out of reach, and it does not appear likely to be achieved in the near future. And, at the risk of earning the displeasure of companies developing driverless robotic cars, I must ask — do people really need all this? But let us proceed in order.

The concept of a "smart machine" capable of self-control dates back long before the advent of modern computers. As early as the 1920s in the United States (U.S.), "fantastic cars of the future" were demonstrated — radio-controlled vehicles that could move through city streets receiving signals from an external transmitter. In the 1950s, General Motors and RCA presented prototypes of roads equipped with magnetic guides along which a car could travel while keeping its lane. None of this, of course, had anything to do with AI — these were rather engineering stunts and displays of the technical optimism of the era.

By the early 1980s, the idea of an "intelligent automobile" found new life in university laboratories. One of the first projects was *Navlab* (short for *Navigation Laboratory*) at Carnegie Mellon University, where a neural network was trained to drive a car by analyzing images from a video camera. The system *ALVINN* (*Autonomous Land Vehicle In a Neural Network*) could already, by 1989, keep a car on the road at speeds up to 70 km/h — an impressive achievement for its time. Almost simultaneously, in Germany, a team at the Bundeswehr University in Munich created the *VaMoRs* project, using cameras, stereo vision, and trajectory prediction algorithms. In the early 1990s, one of their vehicles successfully drove hundreds of kilometers on the autobahns — with a human operator ready to intervene but, in practice, almost never doing so.

In the early 2000s, the initiative was taken up by the *DARPA Grand Challenge* projects in the U.S. These competitions, organized by the Defense Advanced Research Projects Agency (DARPA), brought together university and corporate teams for the first time in a real-world race of autonomous vehicles across desert terrain. The first attempt, in 2004, ended in failure — not a single participant completed even a tenth of the course. But a year later, the race was held again, and the Stanford team, led by Sebastian Thrun, won, traversing 212 kilometers of desert without human intervention.

After the success of the DARPA Challenge, many of its participants soon joined the largest technology companies. Sebastian Thrun headed Google's *Self-Driving Car Project*, which later became a separate company, *Waymo*. Since 2009, Waymo has been developing cars capable of driving without a human driver. Today, it operates robotaxis in several U.S. cities under special permits. Their fleet is far smaller than that of mass-market automakers, but it is precisely these cars that are accumulating a fully driverless mileage: By early 2026, Waymo reported approximately 200 million miles driven without a person behind the wheel, as its commercial robotaxi service continued to expand across multiple U.S. cities. Waymo equips its vehicles with laser and conventional (radio-wave) radars, as well as video cameras; the sensor data are processed in real time by deep neural networks. The cars can reliably distinguish pedestrians, cyclists, and traffic signals, recognize road markings, and react to the unpredictable behavior of other road users.

In several American cities, such as Phoenix and San Francisco, Waymo have begun regular commercial passenger services without drivers. Yet accidents, software freezes, and unexpected stops showed that real-world reliability remains far from ideal. Elsewhere in the U.S. and in other countries, autonomous cars are still required to have a safety operator — a human ready to take control in the event of an emergency. This, of course, negates many of the theoretical advantages that driverless technology was promised to offer.

As for these "passive drivers," sitting motionless behind the wheel — I can hardly imagine anything more boring, and at the same time more nerve-wracking, than such an occupation. Every driver knows how uncomfortable it can feel to ride as a passenger beside someone else at the wheel. How much worse is it for those unfortunates who sit *in* the driver's seat and yet have nothing to do but worry? And yet, as the experience of Tesla — which we shall discuss next — apparently shows, I may not be entirely right.

At Elon Musk's Tesla, the approach is different. The company focuses on driver-assistance systems (*Autopilot*, *FSD Supervised*), where the person remains in the driver's seat and retains full responsibility. And since that person is the owner of a popular Tesla car — and the number of such cars is now in the millions — the cumulative distance driven in *FSD Supervised* mode by early 2026 exceeded roughly 9–10 billion miles. These data are used to train the driving-assistant system, not for truly "driverless" operation. The fact that they are collected from cars already sold to consumers seems a brilliant move on Tesla's part.

Following the American projects, others developed elsewhere. In China, *Baidu* launched the *Apollo* project — an open platform for autonomous-driving research. Unlike most American companies, which tended to build closed systems and do everything in-house, Baidu bet on collaboration. Dozens of automakers, universities, equipment suppliers, and municipal transport agencies joined the project.

In Europe, Volkswagen, Volvo, and Renault conducted tests; in Japan, Toyota and Nissan did. There even appeared fully autonomous shuttles for campuses, airports, and industrial parks — limited in speed and route but truly driverless.

Overall, the task of creating reliable and safe vehicles entirely without a driver — true robotic cars — proved to be far more challenging than initially expected. The main difficulty is that driving in the real world is endlessly varied: weather conditions, different road infrastructures, the unpredictability of people and animals[53] — all complicate the problem. Moreover, the real world has not been kind to autonomous cars, watching every error and especially every serious accident with a critical eye. Technical difficulties intertwine with legal, ethical, and even psychological ones. Who bears responsibility for a traffic accident — the driver, the engineer, the manufacturer, or the AI scientist who created the software? How can a machine be "taught" moral choice in a no-win situation? How safe is it to rely on a system whose behavior even its creators cannot fully interpret?

And finally — this is my own reflection — do we truly need all this, apart from the developers and commercial companies themselves? Do ordinary people need it? I am not entirely sure of the answer. Personally, I would not want to ride in a car without a driver — or, worse still, to sit in the driver's seat with nothing to do but worry. Would you?

Medical diagnostics

In healthcare, deep-learning models also began to show remarkable success. Convolutional neural networks, originally designed for image recognition, turned out to be exceptionally well-suited for analyzing medical images — X-rays, CT, and MRI scans.

---

[53] In general, people and animals are nothing but trouble for AI (a joke).

How AI "sees" medical images

At first glance, an algorithm analyzing a chest X-ray seems to be doing the same thing as a network that recognizes cats and dogs. In both cases, the model examines the image for patterns — shapes, contours, and textures. Yet, there are important differences.

In ordinary image-recognition tasks, an error is harmless: mistaking a greyhound for a whippet is no tragedy. In medicine, however, errors have a cost. Therefore, networks here are trained differently — on data annotated by physicians, verified multiple times, and placed within a clinical context. Their "attention" is focused not on identifying whole objects but on detecting subtle irregularities — faint shadows, microcalcifications, barely perceptible distortions of form.

Moreover, a medical image contains no "objects" in the usual sense. The network learns to recognize statistical anomalies rather than things. In that sense, medical diagnosis is one of the purest forms of *machine vision*: not recognitional but analytical.

By 2017, AI systems were already matching — and sometimes surpassing — the diagnostic accuracy of trained radiologists in detecting conditions such as diabetic retinopathy, pneumonia, and early-stage cancers. Machines do not tire or lose concentration; they can examine thousands of images with equal attentiveness, offering at least a partial answer to the shortage of physicians and the variability of human diagnosis.

Large research centers and technology companies — most notably Google Health and DeepMind — began implementing such systems in clinical trials. Their algorithms learned to highlight regions in images that warranted the greatest suspicion.

Nevertheless, none of these early systems could operate independently: The physician remained the final authority. Yet even then, it was becoming clear that AI was gradually moving beyond narrow experiments and beginning to act as a genuine partner to the doctor.

By the mid-2020s, medical diagnostics had become one of the areas where the practical usefulness of AI was most tangible. In 2024, researchers at Google presented an experimental system called *AMIE* — a model capable of conducting a full dialogue with a patient, clarifying symptoms, and suggesting likely diagnoses. In controlled comparisons with practicing physicians, AMIE demonstrated comparable performance in clinical reasoning and communication quality — not merely classifying data but reasoning in natural language.

At New York's Mount Sinai Hospital, a system was introduced that analyzes millions of electronic health records and automatically generates probable diagnostic codes. It helps doctors avoid missing rare conditions and significantly speeds up the preparation of medical documentation.

In London's Imperial College, in 2025, researchers tested an "intelligent stethoscope" with AI that can analyze heart sounds and ECG data in 15 seconds. It detects heart failure, valve defects, and arrhythmias with accuracy exceeding that of standard *auscultation* (listening with the regular stethoscope). This is not a replacement for the doctor but a new tool that helps avoid missing early signs of disease.

In China, *DeepSeek Med* has been deployed in several major hospitals to analyze medical images and triage patients by risk level. Similar projects are also underway in Japan, South Korea, and Europe, with research papers increasingly describing systems capable of predicting disease development long before clinical symptoms appear, based on combined data from biobanks and genomic studies.

And now there are already stories of AI saving lives. In the summer of 2025, an American woman undergoing a routine CT scan received the standard report: "no pathology." But the same image was simultaneously analyzed by an AI-trained system, which marked a region of lung tissue with a high probability of malignancy. The physician reviewed the data and ordered a biopsy, and the diagnosis of *early-stage lung cancer* was confirmed. The woman underwent a minor surgery without subsequent chemotherapy. "AI saved my life," she told reporters. Such cases are becoming increasingly common.

In the most recent medical systems, AI not only *sees* images but can also connect them with text, symptoms, and laboratory data — approaching what can genuinely be called *understanding* in a medical context.

For now, all existing systems operate under human supervision, and the responsibility for diagnosis remains human. In this sense, the situation resembles that of self-driving cars, where a person must still sit behind the wheel. Let us hope that for the physician working with AI, the experience is less tedious than for the nominal driver monitoring a robot car. It seems that relinquishing human oversight — whether by a driver or a doctor — will not happen soon, if it happens at all.

AlphaFold and protein folding

Perhaps the most impressive achievement of AI has been the solution of an old and very important problem — determining the structure of protein molecules. The problem in question has stood at the center of molecular biology for the past 70 years: How does a very long organic molecule, whose chemical composition is known, fold into a compact, globular protein structure?

Proteins are the essential molecular machines of life. They comprise the muscles that move us, the enzymes that digest our food, the antibodies that fight infections, and the receptors that enable our cells to communicate. Every cell in your body and the bodies of every animal, plant, fungus, or bacterium relies on proteins to function. Understanding protein structures opens many doors in microbiology as well as in the development of drugs and vaccines.

All proteins are chains of amino acids[54] — small organic molecules that link together like beads on a string. These chains can be short, with just a few dozen amino acids, or extremely long, with tens of thousands of amino acids. The longest protein chains, when stretched, would reach several micrometers in length and may contain up to half a million atoms.

Once formed, a protein chain folds into a three-dimensional shape, and it is this shape — more than the chemical formula — that determines what the protein can do. Folded protein structures tend to be hellishly complex. A typical protein molecule resembles a thin but long pearl necklace that has been crammed (folded) into a tightly wound ball (Fig. 17). However, despite the "untidy" appearance, these seemingly random globular structures are not random at all. Each protein folds into a unique structure that is precisely the same in all molecules of that type. Each molecule of, for example, *hemoglobin*[55] is identical to every other molecule of hemoglobin. Each of the approximately 11,000 atoms comprising the molecule is in its designated place. Despite the visual

---

[54]Amino acids are the building blocks of proteins. Each amino acid consists of a central carbon atom bonded to an amino group ($-NH_2$), a carboxyl group ($-COOH$), a hydrogen atom, and a distinctive side chain. There are 20 standard amino acids used by living organisms to assemble proteins.

[55]*Hemoglobin* is a protein in red blood cells that plays a vital role in transporting oxygen throughout the human body.

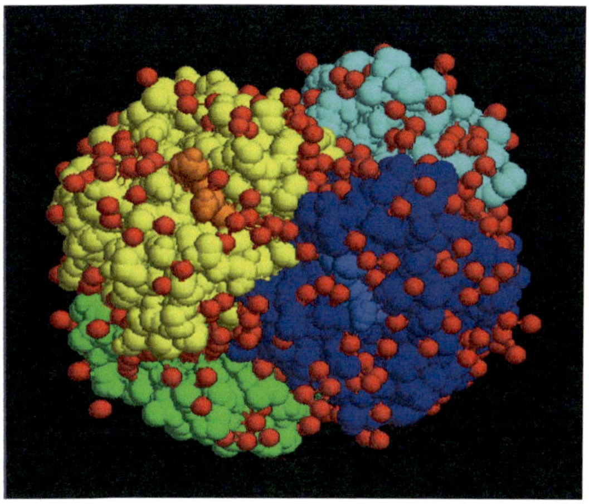

Fig. 17. Atomic structure of hemoglobin, which consists of 574 amino acids, each containing, on average, about 20 atoms, including hydrogen. The total number of atoms in a hemoglobin molecule is on the order of 11,000–12,000. Structural models of hemoglobin, such as the one shown here, typically do not include hydrogen atoms. As a result, they display approximately 5,000 atomic spheres representing the heavier atoms (carbon, nitrogen, oxygen, sulfur, and iron).

Source: Image in the public domain.

complexity, the structural order is as precise as in a crystal of edible salt — sodium chloride (NaCl).

Across the tree of life, the total number of proteins is staggering. As of 2026, databases such as UniProt and AlphaFold contain hundreds of millions of protein sequences, primarily from bacteria, viruses, and other microorganisms — the most abundant and diverse life forms on Earth. The human body and the bodies of all mammals contain approximately 20,000 protein-coding genes, giving rise to a much larger number of protein variants.

The possible number of protein structures is vastly greater than the hundreds of millions of protein sequences currently known. It has been estimated that a typical protein macromolecule could, in principle, assume on the order of 10^{300} different configurations — ten followed by three hundred zeros, a number that can be considered effectively infinite for all practical purposes.

The vast majority of the hundreds of millions of natural proteins have never been studied experimentally, and until recently, their structures, except for a tiny minority, were unknown; the only available information about them was their amino acid composition.[56]

Given that even the smallest proteins consist of thousands of atoms, each in a precisely determined position, experimental determination of their structures is challenging and time-consuming, often requiring years of dedicated work by a team of scientists. Until recently, their structures were determined using X-ray crystallography, nuclear magnetic resonance (NMR), and low-temperature electron microscopy. The first protein whose structure was fully resolved was *myoglobin*, followed soon after by *hemoglobin*. That early breakthrough occurred in the 1950s, earning John Kendrew and Max Perutz the 1962 Nobel Prize in Chemistry. In later decades, the reward level for solving new protein structures was gradually reduced — from Nobel Prizes to doctoral degrees — but the difficulty of the problem remained undiminished. It is remarkable that, despite all the challenges, approximately 170,000 protein structures had been determined by 2020. This number, which appears large, is still less than 0.1% of the hundreds of millions of proteins found in nature.

The idea of substituting costly experiments with computational predictions surfaced as soon as computers became powerful enough to make such dreams plausible. Yet, despite decades of effort, theoretical protein structure prediction made little progress until the arrival of deep learning in the 2010s.

A new player emerged on the scene: *DeepMind*, a London-based AI company acquired by Google in 2014. DeepMind had already stunned the world by creating *AlphaGo*, the first AI system to defeat a world champion in the game of Go (which happened in 2016). Emboldened by that success, DeepMind turned its attention to science, hoping that deep neural networks would crack the protein folding problem.

The effort was led by Demis Hassabis, DeepMind's co-founder and CEO — a neuroscientist and former chess prodigy — who, in 2016, after the much-publicized Go victory, reoriented his company to do something serious. The AlphaFold team brought together machine learning experts and structural biologists, most notably John Jumper, who would become

[56]If you wonder how the amino acid composition of so many natural proteins was known, the answer is that it was known from *gene sequencing*. Indeed, each gene (a coding section of DNA) codes for a specific amino acid. If you know the genes, you know the amino acids.

the principal architect of the system. Their approach was based on an understanding of neural networks and big data. They abandoned symbolic AI and supercomputer-based calculations, which rely on simulating the physics of molecular motion. Instead, the goal was to predict a protein's structure directly from the known protein composition: the sequence of amino acids from which it is made.

DeepMind's first version of AlphaFold debuted in the 13th installment of a biennial competition known as the *Critical Assessment of Structure Prediction* (CASP13). In this challenge, teams from around the world submit blind predictions of protein structures whose experimental solutions have been determined but not yet been made public. In 2018, AlphaFold decisively outperformed all other competitors, attracting international attention.

But that was only the beginning. In the next competition, CASP14, held in late 2020, DeepMind's updated system, *AlphaFold2*, stunned the scientific community. Its predictions were, in most cases, nearly indistinguishable from experimental results. AlphaFold2 achieved atomic-level precision with a *root-mean-square deviation* (RMSD) of less than 1 Å[57] for approximately two-thirds of all proteins tested. Since then, AlphaFold's predictions have been widely verified in molecular biology laboratories worldwide.

A problem once considered "unsolvable" had, for all practical purposes, been solved using ANNs and deep learning. Cumbersome, time-consuming measurements — hours of sample preparation and study via X-ray crystallography, NMR spectroscopy, and electron microscopy — were no longer needed. Protein structures could now be calculated, often in minutes or even seconds, based on the gene that encodes a given protein or from the list of amino acids.

Within two to three years after 2020, the remaining 99.9% of naturally occurring protein structures, which were previously unknown, were predicted and tabulated in vast protein databases.

How was this done? Not by any calculation in the usual sense of the word. The machine does not model a protein structure based on known principles of organic chemistry, atomic physics, or quantum mechanics.

[57] RMSD is a type of average. An ångström (symbol: Å) is a unit of length equal to 10^{-10} meters, or 0.1 nanometers. It is commonly used to express the sizes of atoms, molecules, and wavelengths of light. For example, a typical carbon–carbon bond is about 1.4 Å long.

Instead, the AlphaFold program trains ANNs on existing protein structures. It is given a gene sequence and the exact spatial positions of all atoms in a known protein. AlphaFold then attempts to reproduce that structure, comparing its prediction with reality, learning from mistakes, and gradually improving its accuracy over time. Does it perform any calculations during the process? I do not think anyone can answer that question. It does *something*; we know that the machine readjusts its billions of weights. You can call it a calculation if you wish. The process is repeated with thousands of known proteins. This is where big data becomes essential: The learning process depends on immense amounts of structural information.

The impact of AlphaFold was so profound that, in 2024, Demis Hassabis and John Jumper were awarded the Nobel Prize in Chemistry, sharing the prize with David Baker of the University of Washington. The Nobel Committee cited their contributions to solving one of the "greatest challenges in modern science."

The first consequences

The AlphaFold Protein Structure Database, containing about 350,000 entities, was released in July 2021. It was expanded to over 200 million structures by July 2022.

If people truly need something, they won't wait for the grand opening with the ribbon-cutting ceremony. That was the case when the global COVID-19 pandemic emerged in early 2020. Knowledge of the structure of that virus's proteins was immediately needed. Before many SARS-CoV-2 proteins had been experimentally resolved, DeepMind released preliminary AlphaFold predictions for several viral components, including some that were difficult to crystallize or visualize by other means. These early models, shared openly with the scientific community, helped guide therapeutic research by revealing the likely shapes of key proteins involved in viral replication and immune evasion.

There were also other advances in drug and vaccine development related to that database. In 2023, researchers used AlphaFold's predictions to determine the structure of CDK20, a previously understudied protein linked to liver cancer. Using its predicted shape, they performed structure-based virtual screening and identified a potent inhibitor, ISM042-2-048, demonstrating for the first time that AlphaFold could guide the design of a novel drug.

AlphaFold also played a role in guiding vaccine research. At the University of Oxford, scientists utilized the predicted structure of a key malaria protein to aid in designing an antigen fragment for a next-generation vaccine candidate. In this case, not only did AlphaFold predict a structure, but it also helped identify which part of the protein would most effectively stimulate an immune response.

Today, companies such as *Isomorphic Labs* — a drug discovery spin-off from DeepMind — are actively using AlphaFold's predicted protein structures to accelerate various stages of pharmaceutical development. These include identifying where small molecules might bind to a target protein (binding site discovery), simulating how potential drug candidates might fit into those sites (molecular docking), and refining those molecules to improve their effectiveness and safety (lead optimization). In traditional drug discovery, these steps required years of experimental work. Now, with highly accurate protein models available for millions of targets, structural biology is increasingly guided by readily available computational insights.

Protein design based on AlphaFold structures

The release of the AlphaFold Protein Structure Database has not only enabled the interpretation of existing proteins but also accelerated the design of new ones — a field known as *de novo protein design*. Researchers can now begin with a desired function or binding behavior and work backward, using structural models to engineer entirely new proteins that do not naturally exist.

One breakthrough came from teams at the *Institute for Protein Design* at the University of Washington (led by David Baker), who combined AlphaFold with their own generative modeling tools, such as *Rosetta* and the *Protein Message Passing Neural Network* (ProteinMPNN). These methods enable researchers to predict sequences that will fold into specific target structures — effectively using AlphaFold as a structural oracle to verify whether a designed sequence will adopt the intended shape.

In a 2022 *Science* paper, researchers demonstrated that combining AlphaFold with generative design methods resulted in a dramatic increase in design success rates, far exceeding those of traditional trial-and-error approaches.

More on AI-assisted drug discovery

An impressive early breakthrough in AI-assisted drug discovery occurred in 2020 at the Massachusetts Institute of Technology (MIT), and it had nothing to do with protein folding. Rather, it demonstrated how deep learning can uncover new molecular candidates by recognizing patterns that are far beyond the reach of humans. The problem addressed was urgent: Certain bacterial strains had become resistant to all known antibiotics. Infections caused by such bacteria left doctors with few, if any, treatment options. The search for new antibiotics was both costly and frustratingly slow.

Then, a team at MIT, led by James Collins, Regina Barzilay, and Jonathan Stokes, took a different approach. They trained an AI system — specifically, a *message-passing neural network* (MPNN) — on a dataset of about 2,500 molecules whose antibacterial properties were already known. The network "learned" how molecular structure relates to antibacterial activity without being given any rules for what features make a molecule effective. In other words, the AI was not working through chemistry or pharmacology; it was presented with examples and learned from patterns within the data. After training, the model was given access to a virtual library of 61,000 molecules, including FDA-approved drugs and novel chemical compounds. Out of all those, the AI selected *one* standout molecule — previously untested as an antibiotic — which turned out to be 100% effective against a drug-resistant bacterial strain that defied every existing treatment!

The researchers named the compound *Halicin* in a wry homage to Stanley Kubrick's *2001: A Space Odyssey,* specifically the computer HAL 9000. This story was widely reported as one of the first instances of an AI system making a biomedical discovery that no human had anticipated. Most intriguing was that even after the fact, researchers could not fully explain which features the AI had relied on to make its prediction. The system had found relevant structural properties that were not explicitly encoded in the training data.

This discovery did not involve AlphaFold or protein structure prediction. The AI operated entirely on known or computable features of small molecules. The discovery of new antibiotics, such as Halicin, and the emergence of AlphaFold represent two complementary branches of AI-enabled pharmacology: one starting from *phenotypic* features and the other from *structures*. As of 2026, Halicin remains in preclinical development, subject to the usual pharmacological and medical tests imposed by the FDA. If it passes FDA tests, we will have the first strain-resistant antibiotic discovered by AI.

AI in fundamental physics and pure mathematics

We have already told the story of protein folding, which remains arguably the highest scientific mountain climbed by AI. However, this is the kind of mountain that will not be seen as Everest, at least not by scientists working in fundamental physics or pure mathematics. This is not the kind of science practiced by Isaac Newton and Albert Einstein — or, in mathematics, by Leonhard Euler and Henri Poincaré.

Is there anything in the realm of pure math and fundamental (as opposed to applied) physics that AI has even touched? Yes, although so far, relatively lightly. But it seems to be inching its way towards that stratosphere of exact sciences, and there is no doubt that we will hear more about AI-based fundamental science. For now, let us gather and list the advances of AI in these high scientific domains as of mid-2025.

AI's achievements in mathematics

We have already written about the first successes of symbolic AI in proving existing, well-known mathematical theorems. Recall that as early as the mid-1950s, Allen Newell, Herbert Simon, and J. C. Shaw created the program Logic Theorist, which proved 38 of the first 52 theorems in Russell and Whitehead's *Principia Mathematica*. Later, the same authors developed the General Problem Solver, a program designed to tackle a broader class of problems.

This line of research — automatic proofs of already known results and, at times, improvements to older proofs — has continued into the present day. In 2020, OpenAI introduced GPT-f, which found new, shorter proofs for several theorems in the formal *Metamath* library.

In parallel, AI began to solve Olympiad-level problems with growing confidence. In 2024, DeepMind's *AlphaGeometry* approached the level of *International Mathematical Olympiad* (IMO) gold medalists, and its successor, *AlphaGeometry 2*, achieved a performance comparable to that of a gold medalist in 2025. We mentioned this stunning fact in the Prologue.

Olympiad problems are extraordinarily difficult. Each year, approximately 50 high school students receive IMO gold medals out of roughly 120 million high school students worldwide. This puts a gold medalist in the top 0.00004%, or about one in 2.4 million. Even so, solving such problems does not, by itself, advance mathematical science; that is not what Olympiads are for. They exist to identify young mathematical talent. The fact that such talent now appears among machines is striking, but it is not a direct contribution to research.

Another important aspect is the formalization and accumulation of a proof corpus. Over the past half-century, mathematics has grown so complex — and proofs have become so lengthy — that researchers need help verifying arguments and organizing vast bodies of results.

For example, the classification of *finite simple groups* spans, according to some estimates, tens of thousands of pages across numerous papers! Proofs on this scale often render full verification practically impossible; small errors or gaps can remain undetected for a long time, especially when an argument encompasses dozens of interconnected modules and utilizes many lemmas from different areas.

Here, the *Lean* system has come to the fore. It is an AI-powered proof assistant that allows mathematicians to formulate and verify proofs step by step. Around Lean, there has formed an international community of volunteers and researchers who jointly develop the system, add capabilities, and formalize proofs. Their principal collective work is called *mathlib* (the "mathematical library"): a vast trove of formalized definitions and theorems — an encyclopedia of modern mathematics in which results are mechanically verified.

There is yet another aspect of working with machines: They can improve mathematical techniques, the toolbox of mathematics. For example, *AlphaTensor* discovered new matrix-multiplication algorithms that outperformed the best human-devised methods, running 10–20% faster than standard approaches. The results were published in *Nature*.

All of this is fascinating, but at the start of this section, we set ourselves a stricter question: Is there machine activity that touches "high" science — mathematical creativity that produces something genuinely new? Such phenomena in machine mathematics are still rare, but they do exist.

Let us first ask, have machines proved any substantial new theorems? The answer, for now, is negative. If we pose to machines the questions that occupy our best mathematicians (say, Fields Medal or Abel Prize level), we find that, at this level, machine intelligence is not yet able to help.

But there is another kind of activity valued almost as highly as a full solution: the formulation of interesting new *conjectures*. Conjectures arise from observations, computational experiments, special cases, and intuition. They act as beacons: pointing the way, framing big questions, and rallying many researchers around them. Mathematics has several extraordinarily famous conjectures whose pursuit has spawned entire new fields. One of them — called "Fermat's Last Theorem" — resisted proof for roughly 350 years and was finally proved about 30 years ago. Another, equally famous and still unproven, is the Riemann Hypothesis, a mere 166 years old.

One of the first examples of successful AI-assisted conjecture formulation appeared in 2021, in the work of a DeepMind team with leading mathematicians. Using neural networks to search for patterns, the researchers proposed surprising conjectures in knot theory and group representation theory.

Knot theory is a branch of topology that studies properties of "knots" (closed curves in three-dimensional space) and ways to classify them. Despite the vivid name, these are not ordinary rope knots but abstract objects important in pure mathematics and in physics (for example, in field theory).

Representation theory of groups is an area of abstract algebra that studies how elements of groups (one of mathematics' basic structures) can be "represented" by matrices and linear transformations. It is a powerful language connecting algebra, geometry, and physics.

Both areas belong to high, "pure" mathematics and are fundamentally important: Results there often find applications far beyond mathematics itself — from quantum physics to cryptography.

These AI-suggested conjectures were then rigorously proved by humans, becoming new theorems. In this way, AI appeared not merely as an "executor" but as a partner in research: It indicated plausible connections that humans then justified.

The British mathematician Yang-Hui He, a pioneer in applying AI to pure mathematics, emphasizes in his publications and public lectures that AI is already changing the practice and even the philosophy of mathematics. According to him, the most valuable contributions so far were in what he calls the "top-down" approach to mathematics, which, if I understand him correctly, is essentially equivalent to proposing new conjectures, which then may or may not be proven. For example, Professor He points to his own "murmuration conjecture," which concerns subtle oscillatory patterns arising in the statistics of families of elliptic curves. Machine-learning methods helped identify this unexpected pattern, which mathematicians have yet to prove. As Professor He said in his Royal Institution lecture delivered on October 7, 2025:

> That's a whole brave new world. This was — I am proud to say — covered by "Quantum Magazine" to be one of the major breakthroughs of 2024. Because this was a time when an AI system raised a conjecture that defied humans.

Cases like these suggest that AI can play the role of a "new Ramanujan," proposing inspired guesses that later become provable theorems.

Ramanujan and the "Ramanujan Machine"

Srinivasa Ramanujan (1887–1920) was a self-taught Indian mathematician who, without systematic university training, discovered hundreds of formulas, identities, and conjectures in infinite series, divisibility theory, and special functions. His letters to the British mathematician G. H. Hardy contained results, many of which were entirely new and astonishing in their depth. Hardy later wrote that Ramanujan "produced more ideas in a day than most mathematicians do in a lifetime." Much of his work was proved after his death, and today it occupies a significant place in number theory.

This is why Ramanujan's name has become a symbol of the intuitive conjecture that foreshadows a rigorous result. When we speak of a "new Ramanujan" in the person of AI, we mean precisely this role: a generator of conjectures that may not yet be proved but open new directions in mathematics.

Related to this image is the Ramanujan Machine, a research project launched in 2019 by investigators at the Technion (Haifa, Israel). In this project, AI automatically generates conjectures — for example, proposing infinite series or continued fractions that express fundamental constants such as π (the ratio of a circle's circumference to its diameter), Euler's number *e*, and others. These formulas are produced without proof: Algorithms suggest an identity, and verification and proof are left to mathematicians. But, as with the real Ramanujan, the formulas are typically correct — something that borders on the miraculous.

It is still only the beginning, but we can already say that in mathematics, AI is starting to participate in the scientific process, a participation that has so far manifested itself chiefly in the formulation of unexpected conjectures.

AI's achievements in physics

I will be brief — a detailed account of these projects would require another book of comparable size. Let us simply name the developments and projects, without delving into their technical details:

- *New state variables*[58] (Columbia, 2022): An AI system learned — directly from videos — how many *state variables* were needed to describe various physical systems and produced compact representations of them. Some matched familiar quantities; others, while predictive, lacked a clear human interpretation — suggesting the machine could "make sense" of physics in ways not immediately accessible to us.
- *AI-Newton* (Peking University, 2023): An autonomous, offline system that, given only raw "experimental" data (with artificial noise added) and no prior knowledge of physical laws, independently discovered the concepts of mass, velocity, and energy and derived Newton's laws ($F = ma$, the law of gravitational attraction, and energy conservation[59]).

[58] In physics, *state variables* are quantities that fully describe the condition of a system at a given moment — for example, position and velocity in mechanics, or pressure, volume, and temperature in thermodynamics. Once the relevant state variables are known, the system's future evolution can, in principle, be predicted from the governing equations.

[59] In strict terms, the principle of mechanical energy conservation was not known to Newton; it was formulated only later, in the 18th–19th centuries.

Perhaps it is too early to say that AI already works at Newton's level, but it is noteworthy that a machine, based on observation, was able to formulate the same foundational principles as the great Newton. This is the kind of experiment that deserves to be repeated and independently verified. Can AI derive electrodynamics from raw data? What about quantum mechanics?

- *Symmetry Seeker* (2022–2024, Calvo-Barlés *et al.*): A neural network that detected symmetries in data. Without any hint about the underlying symmetry, the AI identified transformations — specifically, permutations and reflections — that left the equations unchanged.
- *High-energy physics*: In 2024, a team led by K. Chong and M. Schwartz used transformer models to simplify cumbersome expressions in quantum field theory. AI produced new, compact formulas for five-particle scattering amplitudes.
- *String theory*: The "landscape" of string-theory solutions is vast and notoriously complex. Recently, AI has enabled researchers to make significant progress in connecting string theory to low-energy physics. In April 2024, two independent groups announced an unprecedented achievement: Using neural networks, they were able to compute the type of four-dimensional physics[60] that emerges from specific ten-dimensional string models. As a result, for the first time, they succeeded in calculating mass spectra and other physical properties for certain string configurations.
- *Quantum optics*: AI systems such as *MELVIN, THESEUS* and later *PyTheus* generated novel configurations of optical experiments — combinations of mirrors, beam splitters, and photon sources. Some of these setups were so ingenious that physicists admitted they would have been unlikely to design them without AI's help. PyTheus alone proposed over a hundred new experiments, several of which have already been implemented in laboratories.
- *Generation of code* (MIT, 2024): A transformer-based model generated Python code describing entire classes of quantum experiments.
- *AI in nuclear fusion control*: Though not pure science, this is potentially one of physics' most important applications, promising virtually limitless clean energy. For that reason, we will briefly mention the field's progress here.

---

[60] That is, normal physics which exists in three special dimensions and time.

AI is already helping humans control the most challenging state of matter known — plasma, a gas heated to millions of degrees and ionized, allowing nuclear fusion reactions to occur, the same reactions that power the Sun and other stars. In devices such as *tokamaks* (toroidal magnetic confinement chambers), the plasma must be maintained and stabilized with extreme precision: Temperatures reach tens of millions of degrees, and even the slightest oscillation of shape or pressure can cause a plasma disruption.

In 2022, researchers from DeepMind and the Swiss Plasma Center demonstrated that a neural network could control the current in the coils of the Swiss TCV tokamak in real time, stabilizing the plasma and even changing its shape according to a programmed plan. This was the first time AI had directly controlled plasma rather than merely tuning pre-defined models.

By 2025, Japan's JT-60SA, the world's largest operational tokamak, began using AI models to reconstruct plasma shape with centimeter accuracy (approximately half a percent of the device's dimensions) and to adjust magnetic fields, current, and pressure in real time. Similar systems are being tested elsewhere, including at DIII-D in the U.S. and at KSTAR in Korea. There, machine-learning algorithms help predict possible instabilities and automatically correct parameters to prevent plasma disruptions.

These experiments show that AI may become an indispensable assistant in managing future fusion reactors. The goal is to achieve a stable, controlled fusion reaction, in which the energy released by the fusion of hydrogen nuclei exceeds the energy required to heat and confine the plasma.

AI Funding and AI Policy

AI investment in the U.S.

From the outset, the development of AI was driven not only by scientific curiosity but also by sustained investment — first from governments and later from the private sector. Just as the wind drives the waves of the sea, so too did the winds of funding raise and propel the waves of AI progress. And when the wind died down, a dead calm settled over the waters — an eerie stillness that, mixing metaphors, came to be known as the "AI winter."

In the 1950s and 1960s, AI research was primarily an academic pursuit supported by government agencies such as the U.S. Department of Defense and its research arms: the *Defense Advanced Research Projects Agency* (DARPA) and Office of Naval Research (ONR). These early programs were modest in scale but pivotal, as they funded pioneers such as Frank Rosenblatt, John McCarthy, Marvin Minsky, and Allen Newell and supported the development of foundational ideas, including symbolic logic, theorem proving, and early neural networks. In 1956, the Dartmouth Conference, often cited as the birthplace of AI, was itself funded by a small grant from the Rockefeller Foundation (in this case private, nongovernmental philanthropy).

The Cold War provided much of the rationale for this early support. AI was imagined as a strategic asset, potentially valuable for translation, planning, and command systems. However, as early promises failed to materialize — particularly in natural language processing — funding declined sharply in the 1970s, leading to the "first AI winter," which we previously described in some detail. A similar retrenchment occurred in the late 1980s, despite a temporary revival sparked by *expert systems* and Japanese government initiatives, such as the *Fifth Generation Computer Systems Project*.

It was not until the 2000s and 2010s, with the advent of large-scale computing, big data, and the deep learning revolution, that investment began to rise significantly. Governments resumed their interest, but now they were joined — and soon outpaced — by private capital. Technology firms such as Google, Microsoft, Amazon, Facebook, and Baidu have begun pouring money into AI research, hiring academic talent, and establishing internal research labs. Google's acquisition of DeepMind in 2014, for a reported \$400 to \$650 million, marked a symbolic shift: Cutting-edge AI was no longer a public research endeavor; it had become a commercial and strategic race.

By the 2020s, AI funding had entered a new phase of intensity. On the government side, agencies such as DARPA, the National Science Foundation, and the National Institutes of Health, along with their counterparts in the EU and Asia, have begun to significantly increase their AI *research and development* (R&D) budgets. In the U.S., federal contracts for AI systems expanded from a few hundred million dollars annually in the early 2020s to over \$4.6 billion by 2023, with a significant portion directed toward defense, healthcare, and infrastructure. Legislative initiatives such as the *CHIPS* and *Science Act* provided billions for semiconductor and AI-related research.

But it is the private sector that has propelled AI into the stratosphere. By 2024, global private investment in AI had exceeded \$200 billion in a single year. Tech giants allocated staggering sums: Meta alone projected over \$70 billion in AI-related capital expenditures for 2025. Venture capital, meanwhile, funded hundreds of startups, some of which (such as Anthropic, Mistral, and xAI) rose to institutional prominence in under two years.

As we mentioned in the Prologue, global AI investment in 2025 — counting corporate spending, venture capital, and public funding — was approaching the trillion-dollar mark, making artificial intelligence the most heavily financed technological revolution in history. The United States, as the largest single contributor to this total, accounted for an estimated 40–50 percent of those investments.

At the frontier of this funding explosion are mega-projects: large-scale AI infrastructure initiatives that blur the lines between government strategy and private ambition. OpenAI, in partnership with Microsoft and other backers, has proposed *Project Stargate*, a \$100–500 billion AI supercomputing initiative in the U.S., targeting deployment by the end of the decade. Meanwhile, SoftBank has floated plans for a \$1 trillion AI and robotics hub in Arizona, aimed at making the U.S. the global leader in what it calls the coming "intelligence age." So, now we are talking not just about hundreds of billions but trillions.

Unprecedented salaries

Amid such figures, the labor market for top AI researchers has changed dramatically. In 2025, there was a wave of high-profile attempts to poach key talent: Meta, while forming a division dedicated to developing superintelligent systems, reportedly offered leading employees of its competitors compensation packages of up to \$300 million over four years, with more than \$100 million payable in the first year alone. Sam Altman publicly stated that Meta had offered OpenAI staff sign-on bonuses of up to \$100 million. The press also mentioned a specific case of a young researcher who received an offer worth \$250 million over four years. There were even reports of a "record" package totaling *\$1.5 billion over six years*, which means an average *yearly salary of \$250 million*, though such numbers likely represent extreme scenarios including stock grants and vesting conditions.

For comparison, at OpenAI, the median total compensation (TC) — including salary, bonuses, and stock — for a software engineer is about

$900,000 per year, with top levels reaching $1.3 million. A research scientist at OpenAI earns more: a median of $1.56 million and a maximum of $1.8 million. Modest, perhaps, but hardly small salaries. The contrast with record-breaking offers, however, is enormous: sums in the hundreds of millions per year occur only in the rarest, exceptional cases and depend almost entirely on equity value and vesting schedules.

Such staggering, unprecedented salaries, coupled with equally unprecedented market valuations of AI-related chip manufacturers, tell us above all that the leading AI companies are clearly not about to abandon their course toward the swiftest possible achievement of AGI.

AI financing beyond the U.S.

While much of the early narrative around AI has been written in American ink — with the first government contracts, DARPA grants, and Silicon Valley startups — the rest of the world did not remain idle. The leading European countries and China, observing developments in the U.S., have made their own significant financial commitments to AI.

Europe's early relationship with AI was cautious and fragmented. While academic research in the U.K., France, and Germany maintained world-class credentials, pan-European coordination remained limited during the Cold War years. This change occurred gradually, starting with research programs under the European Commission in the 1980s and 1990s, which were part of the broader *Framework Programmes for Research and Technological Development*. Still, Europe often played catch-up to U.S. private capital, particularly in commercial applications.

That dynamic began to shift in the 2010s when concerns about technological sovereignty, privacy, and strategic dependence on American tech giants led to more assertive policymaking. The European Union (EU) has funded AI research through programs such as *Horizon 2020* and *Horizon Europe*, allocating billions of euros for machine learning, robotics, and data science. Yet, the real inflection point came in 2025 with the launch of a sweeping initiative known as *InvestAI*. Announced at the AI Action Summit in Paris, the plan mobilized over €200 billion to accelerate Europe's AI infrastructure and innovation ecosystem. This included €20 billion for building four or five so-called "AI gigafactories" — massive compute centers stocked with tens of thousands of advanced GPUs and TRUs (AI-training chips). This figure — €20 billion — represents an initial investment rather than the full cost of constructing and operating such facilities, which is expected to be substantially higher.

The name may be misleading: an AI factory is neither a chip fabrication plant nor exactly an American-style hyperscale data center. It is an infrastructure hub — a large computing and storage node designed for training and serving AI models: thousands of GPUs and accelerators, ultrafast interconnects, and shared access for universities, companies, and research groups. The EU is creating such centers to reduce its gap with the U.S. and China, where comparable computational resources are concentrated in private corporations.

At the same time, Europe is strengthening its semiconductor base — this time referring to *actual* fabrication plants (fabs), not computing centers. In Dresden, construction is under way on the European Semiconductor Manufacturing Company (ESMC), a joint venture between TSMC, Bosch, Infineon, and NXP. In Magdeburg, with support from the European Commission, Intel is building a new facility — an American company but with European financing and local contractors. It is intended to become part of a broader pan-European strategy to localize advanced manufacturing. Also progressing is the *SiPearl Rhea* project, the first European processor for supercomputers, created under the European Processor Initiative, serving as a bridge between European design and global production chains.

Perhaps, in time, Europe will produce its own AI-specific chips. For now, however, all major European AI clusters — from the JUPITER supercomputer in Jülich (Germany) to the new AI factories — still rely primarily on NVIDIA[61] and AMD accelerators manufactured in Taiwan. Even so, the very appearance of these centers shows that Europe is not only interested in regulation, but also on building its own AI infrastructure.

At the same time, the *European Investment Bank* (EIB) unveiled a separate €70 billion tech funding program aimed at stimulating private-sector co-investment and driving breakthroughs in AI, robotics, semiconductors, and digital infrastructure.

China took a different path, one that was both more centralized and more audacious. From the late 2000s onward, China declared AI a national priority, incorporating it into grand strategies such as *Made in*

[61] NVIDIA, which is producing GPUs necessary for AI training, has been recently valued at 5 trillion dollars — trillion, with the big "T". This is the highest valuation ever reached by a single publicly traded company.

China 2025 and the *New Generation Artificial Intelligence Development Plan*. Funding flowed through a mix of state-owned enterprises, provincial programs, and national investment vehicles. By 2020, cities such as Beijing, Shanghai, Shenzhen, and Hangzhou had already launched multi-billion yuan[62] AI development zones, while Chinese tech giants such as Baidu, Tencent, and Huawei had built in-house research labs to rival those in the West.

By 2025, China's public funding for AI and related fields, including robotics, semiconductor design, and quantum computing, had reached breathtaking levels. Over 1.2 trillion yuan (approximately $180 billion) was allocated that year alone for scientific and technological development, with a significant portion directed toward AI. A new national *AI Industry Investment Fund*, seeded with 60 billion yuan, was launched to support domestic startups and core R&D. At the same time, China's Big Fund III, the latest iteration of its chip-focused investment pool, secured over 300 billion yuan in new capital. If Europe's motivation was sovereignty, China's was leadership — first in the marketplace and, perhaps eventually, on the battlefield of AGI.

Private capital has also surged, albeit at a somewhat slower pace than in the U.S. Venture funding in China's AI sector peaked around 2018, but state support has remained relatively steady. The distinction, of course, is that in China, the line between public and private is somewhat blurred. Where Silicon Valley largely runs on venture capital, China's AI economy relies on government mandates and five-year plans.

The bird's view

Thus, three distinct models of AI development are taking shape:

- The American model — a free market and an open race.
- The European model — a regulated alliance, still a race, but a quieter one and more orderly.
- The Chinese model — state-controlled acceleration. And acceleration, as we know, leads to higher speed — that is, once again, to a race.

[62] In mid-2025, 1 Chinese yuan (CNY) ≈ 0.14 U.S. dollars (USD).

What they share is sheer scale — already rivaling, and in places exceeding, the biggest efforts in space exploration or classical computing. In Chapter 3, we discussed quantum computing (QC); it is instructive to compare global AI funding with funding for QC. I posed this question to ChatGPT-5; drawing on 2024 figures, it estimated that AI funding outpaces QC funding by roughly 50–80 times — that is, QC receives only about 1–2% of the AI total. In 2025, this ratio became even higher.

From Cold War grants to trillion-dollar data centers, the story of AI is also the story of who paid for the dream — and who now stands to shape its unfolding future. Big money has a way of speaking with an imperious voice and of shaping the ears that hear it. When we hear words of caution, we must recognize how they sound to AI investors:

> *"Listen to us—concerned scientists, intellectuals, and humanists—and delay collecting dividends on your billions."*

It's not hard to guess how such proposals are received.

And a variation on the theme — here's what governments might hear:

> *"Pause the development of AI until we've resolved our anxieties about its long-term consequences. And don't worry: if you hold back while China pushes forward — building autonomous drones, armed robots, and AGI — let's hope there are reasonable people there who will slow things down."*

This, too, requires no further comment.

AI policy: a global overview

Governments worldwide are scrambling to shape policies governing AI, balancing security, human rights, and a natural desire for economic leadership.

United States: Instead of a comprehensive federal AI law, the U.S. relies on a patchwork of regulatory frameworks administered by government departments and agencies, such as the *Federal Trade Commission* (FTC), the *Department of Justice* (DOJ), and the *Food and Drug Administration* (FDA). AI policy also emerges from state-level initiatives: As of early 2025, over 31 states have adopted their own AI laws or resolutions.

In response, major tech companies (Amazon, Google, Microsoft, and Meta) are supporting a 10-year federal moratorium on new state-by-state AI regulation, arguing that a fragmented approach would hinder innovation and competitiveness.

In October 2023, the Biden administration issued *Executive Order 14110*, titled "*Safe, Secure, and Trustworthy Development and Use of Artificial Intelligence.*" It required developers of powerful AI models to conduct safety, security, and bias evaluations and to share the results with federal agencies. The order also called for the creation of technical standards, guidelines for watermarking synthetic media, and international collaboration on AI safety norms.

However, this policy was short-lived. In January 2025, the new administration under President Trump revoked Executive Order 14110 and replaced it with a new directive: "Removing Barriers to American Leadership in AI." This order prioritized deregulation, explicitly directing federal agencies to avoid introducing rules that could "hamper American AI innovation." In fulfillment of this executive order, *America's AI Action Plan* was published in July 2025 — a compilation of more than 90 federal measures aimed at accelerating innovation, expanding AI infrastructure, and minimizing regulatory barriers. Trump also signed three additional executive orders addressing *AI infrastructure*, *government contracting*, and *the prevention of ideological bias* in AI.

At the same time, Congress considered efforts to limit state-level regulation of AI: One bill even proposed a ten-year moratorium on new state AI laws. However, that clause was later removed by the Senate.

The debate over the future of AI regulation in the U.S. is taking place under the strong influence of Silicon Valley — its investors, entrepreneurs, and ideologues. Among the key architects of this approach are Michael Kratsios, director of the White House Office of Science and Technology Policy (OSTP), formerly active in the venture-capital industry and closely connected to Silicon Valley business circles; David Sacks, entrepreneur and investor, longtime partner of Peter Thiel, and a prominent figure in Trump's technology and financial policy — dubbed by the press the "*AI and Crypto Czar*"; and Sriram Krishnan, a former senior executive at Twitter and Meta, now a partner at the venture-capital firm Andreessen Horowitz.

This firm, widely known by its abbreviation a16z (from the number of letters between the first and last in "Andreessen Horowitz"), is one of the world's largest venture funds. Founded in 2009 by Marc Andreessen and Ben Horowitz, it has become a symbol of the modern venture capital model, financing startups in software, biotechnology, cryptocurrencies, and AI, while also playing an active role in shaping technology policy.

Outside the formal structure, but with clear political weight, stand Peter Thiel — billionaire, co-founder of *PayPal* and *Palantir*, longtime advocate of *techno-realism* and minimal government intervention — and Marc Andreessen himself. The a16z fund submitted its recommendations to the national AI plan, and Andreessen was repeatedly mentioned in internal correspondence and press materials as an informal adviser to the White House team. As journalists observed, "*the Trump administration's plan relies on Silicon Valley leaders who have aligned themselves with his agenda.*"

We will examine Marc Andreessen's views in greater detail in the following section.

The U.S. also continues to enforce export restrictions, particularly on AI-grade GPUs and high-performance computing systems bound for China, while preserving access for allies and trusted partners.

European Union: In 2018, the EU adopted the *General Data Protection Regulation* (GDPR), a sweeping privacy law that addressed the way personal data could be collected, stored, and processed. Although not written specifically for artificial intelligence, it had immediate consequences for AI systems that rely on personal data, particularly in areas such as profiling (the use of data to classify or score individuals), fully automated decision-making (decisions made by algorithms without meaningful human involvement), and the right of individuals to be informed about and challenge algorithmic decisions that significantly affect them.

More recently, the EU adopted the *AI Act* (which entered into force in August 2024), the first broad attempt anywhere in the world to regulate artificial intelligence as such. The law follows a risk-based approach. Some uses of AI — such as certain forms of biometric surveillance or social scoring — are banned outright. Others, deemed "high risk" (for example in healthcare, critical infrastructure, employment, or law enforcement), must meet strict standards before they can be deployed in the European market. Systems considered low risk face lighter requirements.

The Act also addresses powerful, general-purpose AI systems[63] — large models that can be adapted to many different tasks. Developers of such systems must disclose technical information, assess potential risks, and take steps to prevent misuse. The most advanced models, those considered capable of broad societal impact, are subject to additional scrutiny.

The rules are being introduced gradually. Some prohibitions began to apply in early 2025, while the main compliance obligations for high-risk systems will take effect over the following two to three years. Enforcement is carried out by national regulators, coordinated at the European level, and fines can reach several percent of a company's worldwide revenue.

The business reaction to EU regulation was mostly negative. Major American corporations — Google, Microsoft, Meta, and Amazon — publicly warned that overly strict rules could slow the deployment of new technologies and lead to an outflow of investment from Europe. European startups also voiced concerns about costly certification and compliance procedures. At the same time, civil organizations and human rights groups welcomed the AI Act as a guarantee that technological progress would remain subordinate to the protection of citizens' rights and interests.

In contrast to the U.S., where, in 2025, the Trump administration lifted restrictions and opted for deregulation in the name of accelerating innovation, the EU is pursuing a path of preventive regulation. Here, AI is viewed not only as a driver of economic growth but also, above all, as a potential threat to rights and freedoms — one that must be restrained in advance. It is clear that this comes at the cost of reduced competitiveness, which cannot please European businesses. Europe, it seems, no longer hopes to win the race toward AGI; in truth, it never really had that chance.

China exhibits a state-centric approach. It employs soft laws and rapid updates to align AI with national goals, emphasizing social stability,

---

[63] *General-Purpose AI* (GPAI): A category of AI systems crafted for extensive applicability across various domains or tasks, rather than being designed for a specific use case. Examples include large language models such as GPT-5 or Claude, which can be adapted for tasks such as writing, translation, summarization, or code generation. GPAI is distinct from *artificial general intelligence* (AGI), which pertains to hypothetical systems capable of the highest human-level reasoning and learning.

economic growth, and, unfortunately, *surveillance*, with the last being a specifically Chinese priority.

In 2025, China introduced *content labeling rules* (effective September 2025), which are mandating explicit and embedded *metadata* (data about data) labels on AI-generated text, audio, images, and video.

The *Personal Information Protection Law of the People's Republic of China* (PIPL), *Cybersecurity Law*, and *Data Security Law* establish strict data localization, management, and consent structures — mirroring the European GDPR but with greater alignment to national security priorities.

Looking forward

The gap between American, European, and Chinese AI strategies continues to widen. At the AI Summit in Paris in 2025, the differences became unmistakably clear: The U.S. is betting on deregulation, the EU on caution, and China is advancing its state-driven approach, utilizing AI for both economic growth and domestic surveillance.

And what follows from this? In the U.S., there is little reason to expect comprehensive federal legislation anytime soon. More likely, we will see a patchwork landscape: Individual states experimenting on their own while Washington continues to repeat, "Don't slow down innovation." We mentioned the views of one of the president's AI advisers, Marc Andreessen, in the Prologue and will return to them in the following chapter. Unless some AI-related catastrophe occurs, federal policy will remain focused primarily on lifting restrictions and maintaining export controls.

The AI Act is unlikely to be the final word on regulation in Europe. Rather, it appears to mark the beginning of a broader regulatory trajectory. Discussions are already under way in Brussels concerning civil liability for AI-related harm, labor protections in AI-augmented workplaces, and the environmental impact of large-scale data centers.

China, meanwhile, is moving quickly and as a single coordinated entity (it was not for nothing that 30 years ago commentators called it "*China Inc.*"). Its regulators can introduce new rules within months rather than years — whether on stricter content labeling, data controls, or new incentives for domestic AI champions. For Beijing, AI is not merely another industry but a tool of national power. For now, the U.S. leads, but that could change. In the Prologue, we discussed the AI 2027 scenario, according to which China trails the U.S. by only a few months.

Ultimately, we are witnessing the emergence of three distinct worlds: the American free market, the European "rulebook," and the Chinese state-driven surge. Can they be reconciled? There is already talk of multilateral agreements under the auspices of the United Nations or the Organization for Economic Cooperation and Development (OECD). But to be frank, it is hard to believe that such treaties could unite visions so fundamentally different — especially amid a heated race in which victory seems to mean reaching AGI as soon as possible. As AI systems grow more powerful, genuine international coordination appears increasingly unlikely.

Chapter 5

AI Whisperers: Opinions and Warnings

The spectrum of opinions about artificial intelligence (AI) is broad. People are fascinated by it, and they are scared by it, and more often than not, both emotions are present in one's mind simultaneously. Perhaps Stephen Hawking expressed it best when, in 2016, he said that AI is *either the best or the worst thing ever to happen to humanity*.

Before AI appeared and for several decades afterward, there were singular voices, which may or may not have been heard. Then, in the 2010s, AI became successful, and any doubt that it would play a decisive role in humanity's future vanished. Thousands of people have expressed their views on the future and safety of AI since then. The best I can do is present a sampling of these opinions, selecting the ones I find more interesting historically or for their content. By doing so, I will inevitably introduce bias. However, bias or not, let us examine various viewpoints, arranged here in approximate chronological order. I apologize to those whose thoughts were not included. Sometimes the simple reason is that I have not had a chance to read them or listen to what they were saying.

These short notes are written in the form of independent, small essays, in which I have tried to present each character within their historical context and, of course, to convey their thoughts on the subject of interest to us. Not in every case but at times I try to argue with those whose thoughts and words I convey in this chapter. After all, this is my book, and I, too, want to voice my own opinion.

Let me start with those who expressed concern about technology as a whole, even before AI became a "thing." To keep the timeline visible, we will place the earliest date or a range of dates when a statement was made next to a name. By "activity," I mean statements about the future of AI.

Samuel Butler [*activity in 1863, 1872*]

In 1861, a 26-year-old English settler, Samuel Butler, moved from England to New Zealand. Shortly after his arrival, he read Darwin's *On the Origin of Species*, which had been published in England in November 1859. The concept of evolution by natural selection engulfed Samuel Butler's mind. He wrote a short article called "Darwin Among the Machines," published in 1863 in the local newspaper of Christchurch, New Zealand. In that article, he wrote:

> There is no security against the ultimate development of mechanical consciousness, in the fact of machines possessing little consciousness now…We are ourselves creating our own successors; we are daily adding to the beauty and delicacy of their physical organization; we are daily giving them greater power and supplying by all sorts of ingenious contrivances that self-regulating, self-acting power which will be to them what intellect has been to the human race…The time will come when the machines will hold the real supremacy over the world and its inhabitants... they will rule and man will be to them what the horse and the dog are to us. (Butler 1863 (1923), 208–210).

When I first read these words, I assumed they were written in 1963. Then, to my amazement, I noticed that the date was 1863.

In Butler's view — as he expressed it in this first article — machine intelligence would be possible only once a machine acquired consciousness. Debates of this kind continue today, though the origins of these ideas are almost forgotten. We are not very good at citing our predecessors, especially those as distant as Samuel Butler. Some agree with him (perhaps without even knowing his name) — for instance, the renowned physicist Roger Penrose. Others believe that thinking is possible without consciousness, or at least with a limited form of it. Let me add my "two cents": I do not consider consciousness a necessary condition for intelligence. It may be required for *acting* in the world but not for *thinking* itself.

Butler was greatly influenced by Darwin (which, incidentally, did not stop him from criticizing Darwin for failing to reference some of his predecessors, including Darwin's grandfather Erasmus). Accordingly, he assumed that machines could evolve by *natural selection*. They can acquire consciousness, surpass human intellect, and enslave people. There was a lot of sarcasm in that notion and perhaps a bit of madness, but — let me ask you — standing here and now with my book in your hand, will you throw a stone (or that book) at Samuel Butler? I bet you wouldn't, and I wouldn't. Instead, we will say he was very clever, a great prophet who spoke a century and a half earlier than he ought to have.

Butler further detailed his vision in his 1872 novel *Erewhon*, which has a chapter entitled "The Book of Machines." In it, he quotes from his 1863 article and also says:

> Surely if a machine is able to reproduce another machine systematically, we may say that it has a reproductive system. And if it can hand down improvements in its structure to its descendants, we may say that it has a hereditary system.[1] And if it can vary from these slightly in each generation, and if these variations are sometimes improvements, and handed down in their turn, then we have a real system of evolution. (Butler 1872, Chapter XXIII, 209)

Apart from the comment we made in footnote 1, this is a concise summary of Darwin's theory. And notably, Butler no longer speaks of machine self-awareness but only of the presence of a Darwinian algorithm — variation of traits, heredity, and natural selection. We will return to these concepts in the Epilogue.

Samuel Butler was a remarkably original thinker, and his ideas were not forgotten; from time to time, they have resurfaced in reflections on the present and future of civilization. Let me quote Norbert Wiener, of whom we wrote extensively in Chapter 4 and whose views we will examine again later in this chapter. Ninety-eight years after Butler's short article in *The Press*, from which we quoted above, Wiener wrote in the second edition of his *Cybernetics*:

---

[1] We note that Butler's believe that a machine may be able to "hand down improvements in its structure to its descendants" shows that he did not fully understand Darwin's theory — this is a distinctly Lamarkian notion, not Darwinian natural selection.

> I have already pointed out that this new development [*automation*] has possibilities of both good and evil that are almost unlimited. Above all, it makes the metaphorical 'dominion of machines' that Samuel Butler wrote about a very real and no longer metaphorical problem. (Wiener, 1961, p. 27)

But the true champion of Butler's ideas was the American author George Dyson (son of the great physicist Freeman Dyson), who even titled his book after Butler's essay: *Darwin Among the Machines*: *The Evolution of Global Intelligence* (1997).

Karel Čapek [*activity in 1919–1921*]

In 1919, Czech writer Karel Čapek wrote a play, *R.U.R.* — short for *Rossum's Universal Robots* — which was published in 1920 and first staged in 1921 (Capek, 1920 (2006)). That play is remembered chiefly because, in it, Čapek introduced the word "robot," which comes from Russian for "work" (*работа*). Preparing to write this section, I reread Čapek's play, which I had almost completely forgotten. It's a fantastic play! I would make it mandatory reading for anyone involved in AI development and robotics. Required reading with a knowledge check. Maybe it would have saved us!

But first, about the word. A *robot* is a machine that works on behalf of humans. It is not necessarily the repetitive, mechanical work that a gasoline engine or electric motor performs. It is the varied work of a human servant, factory worker, or secretary; it requires intelligence. Čapek's robots are humanoid in form and can do many things; they are "universal."

The play's action takes place In the United States; from time to time, news from Europe is brought by steamers (ocean liners) and telegraph. In the play's title, *R.U.R. (Rossum's Universal Robots)*, "Rossum" refers to old Rossum and his nephew, young Rossum. Neither appears on stage; they are part of the pre-history. Old Rossum was a scientist who discovered a way to create artificial life. As a character in the play, Harry Domin, the director general of Rossum's Universal Robots, explains:

> He [*old Rossum*] wanted, in some scientific way, to take the place of God. He was a convinced materialist, and that's why he wanted to do everything simply to prove that there was no God needed.[2]

Young Rossum, who did not see eye-to-eye with his uncle, took over the project after his death. He saw the industrial potential for the mass production of robots as laborers (factory workers, clerks, and secretaries). He simplified his uncle's creations, making them less human-like and much cheaper to produce. Domin says:

> What young Rossum invented was a worker with the least needs possible. He had to make him simpler. He threw out everything that wasn't of direct use in his work; that's to say, he threw out the man and put in the robot.

However, these robots have intelligence:

> They learn how to speak, write and do arithmetic, as they've got amazing memories. If you read a twenty-volume encyclopedia to them, they could repeat it back to you word for word, but they never think of anything new for themselves. They'd make very good university lecturers.

These quotes, especially the jab at university professors, show that in *R.U.R.*, Karel Čapek was writing a social satire rather than a science fiction novel.

Robots are in high demand worldwide, replacing portions of the human workforce. To do otherwise would amount to business suicide (sounds familiar?):

---

[2]Here and in the following, quotations from Čapek's play are provided from a translation by David Wyllie; eBooks@Adelaide 2006, licensed under CC BY-NC-SA 2.1 AU. I'm not particularly happy with this translation, but the other two I've seen seem even worse. It's shocking how Čapek's text is being altered and distorted in various translations. I will admit that occasionally I quietly corrected some phrases based on Grammarly's suggestion — forgive me, Mr. Wyllie.

> BUSMAN: Dear lady, what I mean is that the price of labor is getting cheaper!... It's wonderful; every factory is buying robots as quickly as possible to reduce production costs, and those that aren't are going bankrupt.
>
> HELENA: Yes, that's right, and throwing their workers out on the streets.

The counterargument is that with robots, there will come an era of abundance, a paradise on Earth:

> BUSMAN: In five years' time...we'll be drowning in wheat and in everything else you can think of.
>
> ALQUIST: Yes, and all the workers in the world will be out of a job.
>
> DOMIN: Yes, they will be...But in ten years' time, Rossum's Universal Robots will be making so much wheat, so much material, so much of everything that nothing will cost anything. Everyone will be able just to take as much as they need[3]. Nobody will live in poverty. They won't have jobs, that's true, but that's because there won't be any jobs to do. Everything will be done by...machines. People will do only the things they want to do; they can live their lives just so that they can make themselves perfect.

Today, 106 years after Čapek's play was written, we hear this argument *verbatim* from those who tell us about the extraordinary abundance expected from future AI.

Čapek also imagined what would happen to society. With robots providing for every human need, people stopped working. They have nothing to do and no reason to live. And then, even before there was any direct action by robots against humans, something very alarming occurred: People stopped breeding. Here are two exchanges between Helena and other characters in the play, both about this problem:

---

[3]Here, Čapek takes Karl Marx's slogan, "From each according to his ability, to each according to his needs," and replaces the first part with robot labor. The need for money will disappear; there will be no more shortages. This is communism achieved by replacing human workers with robots.

> HELENA: Why have women stopped having children?
>
> ALQUIST: Because there's no need for them. Because we've entered into paradise. Do you understand what I mean?
>
> HELENA: No.
>
> ALQUIST: Because there's no need for anyone to work, no need for pain. No one needs to do anything, anything at all, except enjoy himself. This paradise — it's just a curse! Helena, there's nothing more terrible than giving everyone Heaven on Earth!...

And another one, when Helena poses the same question to Dr. Gall:

> HELENA: Doctor Gall...Why are there no more children being born? ...
>
> Dr. GALL: Because robots are being made. Because there's an excess of manpower. Because mankind is actually no longer needed...
>
> HELENA: Gall, what's going to become of the human race?
>
> Dr. GALL: Nothing. There's nothing that can be done against the force of nature.
>
> HELENA: Why didn't Domin [*General Director of R.U.R.*] put a limit on...
>
> Dr. GALL: Ah, forgive me, but Domin has his own ideas. People who have ideas should never be allowed to have any influence on the events of this world.
>
> HELENA: And is there anyone who... who is urging them to stop making them?
>
> Dr. GALL: God forbid! That would be suicide!
>
> HELENA: Why?
>
> Dr. GALL: Because all the people would lynch him. Don't you think it makes life a lot easier to let the robots do all the work?

I suspect that Čapek used that unconvincing lack of childbearing as a way of suggesting that the human race will cease to exist, one way or another.

In the first part of the play, robots are generally satisfied with their roles, but occasionally they throw a tantrum, which their creators call "robot cramps." Then the trouble starts; human workers attack the robots, who retaliate; countries enter into wars, and robot armies are used on the battlefields of Europe. The unrest among robots grows; they organize, and at some point, a robot revolution starts (do not forget that this was written in 1919, only two years after the Russian Revolution of 1917). The robots issue a leaflet which states:

> Robots of the world! We, the first union at Rossum's Universal Robots, declare that man is our enemy and the blight of the universe... Robots of the world, we urge you to exterminate mankind. Don't spare the men. Don't spare the women. Retain all factories, railway lines, machines and equipment, mines and raw materials. All else should be destroyed. Then return to work; it is imperative that work continue.

The characters in the play, among other things, discuss the mistake of making *robot soldiers* (what we today call *killer robots*):

> Dr. GALL: This is just the same old evil that Europe has always committed. They just couldn't leave their damned politics alone, and so they taught the robots to go to war; they took the robots and turned them into soldiers, and that was a crime against humanity.
>
> ALQUIST: The crime was making the robots in the first place.

The characters in the play are desperately trying to find a solution. One proposed solution is to replace universal robots with robots that will defend the country of their origin:

> DOMIN: Henceforward, we shan't have just one factory. There won't be Universal Robots anymore. We'll establish a factory in every country, in every State; and do you know what these new factories will make?
>
> HELENA: No, what?
>
> DOMIN: National Robots.[4]

---

[4]A clever friend recently shared a similar idea with me. We talked about potential problems with AI, and he said, tongue-in-cheek: "I think I know what can be done: create national

But nothing can be done; it is too late. At the end of the play, robots enter the house where the humans hide and slay them.

Čapek suggests who is at fault. When the end is near, the two characters are conversing about it, one defending the project and another exposing its faults:

> DOMIN: ...putting an end to the slavery of labour was not a bad dream. Work humiliates; anyone who's forced to do it is belittled. The drudgery of labor is something dirty and murderous. Oh, Alquist, the burden of work was too much for us, life was too heavy for us, and to remove this burden...[the ellipses are Čapek's].
>
> ALQUIST: That was never the dream of either of the Rossums; old Rossum was thinking of Godless rubbish, and young Rossum thought of nothing but making millions. And it's not the dream of R.U.R. shareholders either; their only dream was their dividend. And it's because of their concern for their profits that mankind is about to perish.

Bravo, Karel Čapek! A hundred years ago, he understood everything, including where it was all coming from.

We learn in the Epilogue that robots failed to reproduce themselves at the robotic factories because the secret of their construction was destroyed. Even though they conquered the world, their days are numbered. However, Dr. Gall tinkered with the recipe and created more advanced, human-like robots. Two of them are robot Helena (the robotic copy of human Helena who has died) and robot Primus. In the Epilogue, these two human-like robots get together and show some feelings for each other. Primus and Helena become the new Adam and Eve. Curtain.

I suspect that Karel Čapek did not want to end his play with the death of all human characters and the end of humanity. So, he invented these human–robot hybrids that may start a new life on Earth. Isn't that what some of us are saying today — that this may be the best outcome we might hope for? These are the thoughts of Ray Kurzweil and Yuval Harari! Stanisław Lem, on the other hand, was warning against a union with machines. Perhaps Čapek did not quite know how to end his play. But do we know today how this play ends?

AIs; they will not understand each other, and they will fight among themselves instead of fighting humans."

Undoubtedly, Čapek's play was inspired by the Russian Revolution, which occurred only two years earlier. The robots, whose name comes from Russian, even formed the Central Committee! The intelligence of Čapek's robots was installed in them at the time of their manufacture. Anyway, it is secondary to the whole story. Čapek's robots stood for the oppressed worker classes. In this sense, Butler's concept of evolutionary development culminating in consciousness and intelligence was more profound. Čapek, on the other hand, gives us a penetrating assessment of the downsides of robot-created abundance, which makes humankind idle and unnecessary. And Čapek is convinced that stopping robots' domination, once it started, would be impossible: "All the governments… are clamoring for an increase in production to raise the standards of their armies. And all the manufacturers in the world are ordering Robots like mad."

When I hear rose-colored projections concerning future AI, I want to say, "Read Karel Čapek; he was a wise man; over a hundred years ago, he expressed the most natural thoughts that came to his mind when contemplating this problem."

Isaac Asimov [*activity in 1941–1985*]

Isaac Asimov's *Three Laws of Robotics* were introduced in his 1942 short story "Runaround," which was later included in the 1950 collection *I, Robot*, featuring stories written between 1941 and 1950.

Here are the Three Laws of Robotics as Asimov formulated them:

1. ***A robot may not injure a human being or, through inaction, allow a human being to come to harm.***
2. ***A robot must obey the orders given to it by human beings, except where such orders would conflict with the First Law.***
3. ***A robot must protect its own existence as long as such protection does not conflict with the First or Second Laws.***

The laws were simple yet clever, formulated in the style of a patent claim, which always impresses everyone except patent lawyers. In Asimov's stories, the intrigue is usually woven around these very laws of his. It is clear that if robots always followed Asimov's Three Laws, he would have nothing to write about. So, all the stories about these laws are actually about how they are violated in various tricky cases. From the very first story, written in 1941, Asimov started inventing logical conflicts and ethical dilemmas that forced the robots to break the laws. Thus, no one, not even Isaac Asimov, has taken these Three Laws of Robotics very seriously.

We could say: Look, Asimov showed in his stories that it is impossible to build reliable safety rules into robots. And yet, we have since accepted his famous Three Laws as if they were the Ten Commandments, which, by the way, have also been violated many times. Asimov's Three Laws of Robotics have influenced countless works of fiction and AI discussions ever since. Even today, some people say, "Why do you worry about AI getting out of hand? All we have to do is program into AI Asimov's Three Laws." Alas, I wish it were that easy.

But Asimov — who, as is well known, was extraordinarily prolific — continued writing for many years after his early series of stories featuring the Three Laws of Robotics. Over time, he turned the theme of AI this way and that, foreseeing or intuiting much of what would later come to pass. In brief:

The later Asimov gradually shifted focus from the quasi-safe "Three-Laws-obeying servants" toward the idea of AI as humanity's curator. The Zeroth Law (introduced in *Robots and Empire*, 1985) allows a robot to prioritize the interests of the species over those of an individual.

In the early story "The Evitable Conflict" (1950), superintelligent "Machines" manage the global economy for the benefit of humankind, while humans effectively relinquish the levers of decision-making — an early exploration of AI technocracy and the idea of humanity's "soft" submission to a rational algorithm.

In "The Bicentennial Man" (1976) and the novel "The Positronic Man" (1992, co-written with Robert Silverberg), the robot Andrew Martin gradually attains legal and moral recognition as a human being, erasing the boundaries between a "smart machine" and a "person." Here, Asimov anticipated today's debates about the possible emergence of conscious, self-aware AI deserving of personhood.

In the story "The Last Question" (1956), a computer evolves together with humanity into a cosmic intelligence and, after trillions of years, in an empty and cold universe, once again says: "*Let there be light.*"

And yet, it seems to me that Asimov's early Three Laws of Robotics remain his most intriguing creation.

John von Neumann [*activity in 1940s–mid-1950s*]

John von Neumann was such an influential, brilliant scientist, and his contributions to computer science were so significant that it would be interesting to know what he thought about future "intelligent computers" — if he indeed felt that they would become a reality someday. But he didn't. George Dyson, the author of *Darwin Among the Machines* and *The Turing Cathedral*, writes extensively about von Neumann. He says:

> Von Neumann was as reticent as Turing was outspoken on the question of whether machines could think…He saw digital computers as mathematical tools. That they were members of a more general class of automata that included nervous systems and brains did not imply that they could think. He rarely discussed artificial intelligence. Having built one computer [*IAS computer, we described in Chapter 2*], he became less interested in the question of whether such machines could learn to think and more interested in the question of whether such machines could learn to reproduce. (Dyson, 1997, p. 108)

Here, Dyson talks about *self-reproducing automata*, which we mentioned in Chapter 2. These automata were conceived as complex entities capable of assembling their own copies or copies of anything fed into them, provided a sufficiently detailed description was given. Von Neumann approached the problem as a mathematician, first defining his automata and the rules of their behavior and then deducing the conditions under which they could reproduce by assembling a new generation.

In Chapter 2, we promised to revisit an aspect of von Neumann's analysis dealing with the complexity of the next generation compared to the parent generation. Let's do it here.

There are three logical possibilities: the new machine could be simpler than its assembler-parent, equal to it in terms of complexity, or more complex. Von Neumann came to a remarkable conclusion:

> There is thus this completely decisive property of complexity, that there exists a *critical size* below which the process of synthesis is degenerative, but above which the phenomenon of synthesis, if properly arranged, can become *explosive* [italics ours], in other words, where

> syntheses of automata can proceed in such a manner that each automaton will produce other automata which are more complex and of higher potentialities than itself. (Von Neumann, 1949 [1966], p. 82)

Von Neumann tells us that there is a certain *critical complexity* in the parent assembler such that, when it is exceeded, an offspring may become more complex than the parent. That more complicated machine will possess what von Neumann calls "higher potentialities," higher capabilities. When reproduction is repeated numerous times, the increasing complexity and capability ought to increase exponentially, or, as von Neumann says, "explosively."

A natural question comes to mind: How should we measure "complexity," and once we decide on that, what is its critical value? Von Neumann gave a partial answer in his lecture notes:

> …this conclusion [*about the threshold*] is true even if one measures complication by the crudest possible standard — the number of elementary parts. There is a minimum number of parts below which complication is degenerative, in the sense that if one automaton makes another, the second is less complex than the first, but above which it is possible for an automaton to construct other automata of equal or higher complexity. Where this number lies depends upon how you define the parts. I think that with reasonable definitions of parts…with simple properties, this minimum number is large, in the millions. I don't have a good estimate of it… (Von Neumann, 1949 [1966], p. 80)

Von Neumann, writing in the late 1940s, anticipated the phenomenon that was later termed the *Intelligence Explosion* by Irving Good (1965) and the *Technological Singularity* by Vernor Vinge (1993), which we will discuss shortly. Von Neumann even called it a "singularity" in a conversation with his good friend Stan Ulam:

> The ever-accelerating progress of technology and changes in the mode of human life give the appearance of approaching some essential *singularity* [italics ours] in the history of the race. (Ulam, 1976, p. 271)

At the end of his life, von Neumann published an essay, "Can We Survive Technology?" (von Neumann, 1955). He wrote it for the U.S.

government; his name is accompanied by the title "Member of the Atomic Energy Commission." The essay is an awkward document without any clearly stated conclusions; the title is the only memorable phrase within it. It was intended as von Neumann's review of technology: atomic energy (from which he expects a flow of unlimited and nearly free energy), automation (of which he says, "The sum total of progress in this sphere is explosive") and climate control (which did not exist then and does not exist now, but from which he expects significant global changes[5] in the future). After piling up these heterogeneous ingredients, Von Neumann proclaimed at the end of his essay that all these technologies possess substantial destructive potential.

In that 1955 document, the reader will find an interesting statement regarding technology regulation or prohibition. Von Neumann wrote:

> …prohibition of technology (invention and development, which are hardly separable from underlying scientific inquiry) is contrary to the whole ethos of the industrial age. It is irreconcilable with a major mode of intellectuality as our age understands it. It is hard to imagine such a restraint successfully imposed in our civilization. Only if those disasters that we fear had already occurred, only if humanity were already completely disillusioned about technological civilization, could such a step be taken…A much more satisfactory solution than technological prohibition would be eliminating war as "a means of national policy" (von Neumann, 1955)

So, what can we learn about dealing with AI from John von Neumann, a man of steel who remained unperturbed by the atomic bombings of Hiroshima and Nagasaki in 1945, and enthusiastically supported the development of the hydrogen bomb, believing that prohibiting any technology, even a dangerous one, was a bad idea? Probably nothing. However, we note with some interest that his mathematical theory of

---

[5] In the section about climate, Von Neumann also writes, "The carbon dioxide released into the atmosphere by industry's burning of coal and oil — more than half of it during the last generation — may have changed the atmosphere's composition sufficiently to account for a general warming of the world by about one degree Fahrenheit." This was written in 1955! Impressive.

self-reproducing automata pointed toward an "essential singularity" in the years ahead. He was the first to predict this singularity, even though his prediction was unnoticed at the time.

Norbert Wiener [*activity in 1948–1960*]

The creator of cybernetics did not always exhibit sound and mature judgment (see Chapter 4 about his interactions with Pitts and McCulloch); however, he seemed to understand the dangers of emerging AI. He wrote in 1960:

> If we use, to achieve our purposes, a mechanical agency with whose operation we cannot interfere once we have started it on its way, then we had better be quite sure that the purpose put into the machine is the purpose which we really desire and not merely a colorful imitation of it. (Wiener, 1960, pp. 1355–1358)

This is a profound thought, especially considering it was expressed in 1960, when discussions of *AI alignment* with human values were over half a century in the future. The suspicion that people working on that future AI might merely check the box on safety, that it may be "a colorful imitation" rather than a genuine effort, was even more perceptive. Perhaps Norbert Wiener was not as childlike in judging human character as we thought.

In his 1950 book *The Human Use of Human Beings: Cybernetics and Society*, Wiener additionally said:

> The machine's danger to society is not from the machine itself but from what man makes of it…We can be humble and live a good life with the aid of the machines, or we can be arrogant and die.

Here, the first thought is of questionable value. We understand now that when it comes to superintelligent AI, the "man" might be unable to "make of it" exactly what he wants. The second sentence, however, is more interesting. In it, Wiener shows that he understands what kind of danger we will face — a *mortal* danger!

Alan Turing [*activity in 1951*]

Turing, who was intimately involved with AI's first steps, also raised the possibility that a thinking machine might one day outsmart its creators. He said in his May 1951 BBC broadcast:

> If a machine can think, it might think more intelligently than we do, and then where should we be?...it seems probable that once the machine thinking method has started, it would not take long to outstrip our feeble powers...At some stage, therefore, we should have to expect the machines to take control....(Turing, 1951)

Turing did not seem unduly concerned with such a prospect, but he understood the anxiety it might cause in society. As Butler had before him, he saw no alternatives to superintelligent machines taking control. As far as I know, this is the only quote by Turing's in that vein. If we are looking for computer pioneers who were seriously worried about future superintelligent machines, Turing wasn't one of them.

Stanisław Lem [*activity from 1958 to 2006*]

Stanisław Lem's (1921–2006) writing career spanned 60 years, from 1946 to 2006. It developed first in Lvov (now Lviv, Ukraine) under Soviet rule, then in Kraków under the Polish People's Republic, and subsequently in independent Poland after 1990. Writing under communist regimes always required authors, those who wanted to be published, to develop particular adaptations and mimicry. Lem's method was to write science fiction, which he did splendidly, being endowed with a lively imagination and a love of science and technology.[6] He possessed an undeniable writing talent. Yet, he was always more of a thinker and philosopher than a writer. Sci-fi gave him a way of dressing philosophy in literary rugs. And, by any measure, he excelled in it. Stanisław Lem's books have been translated into over 50 languages and have sold more than 45 million copies worldwide. His publications outside of Poland exceeded a thousand (Lem, 1999). Lem is considered one of the best sci-fi writers of all time.

---

[6] Stanisław Lem, who received a medical education, acquired his encyclopedic knowledge of exact sciences and philosophy on his own. When he became famous, he had correspondents in the United States and Western Europe, supplying him with the latest scientific journals and books. He subscribed to *Science*, *Nature*, *New Scientist*, *Scientific American*, and other journals. By his own admission, 90% of his reading was in English.

If I were forced to name one weakness, I would say that occasionally, especially in his futuristic work and philosophical essays, Lem becomes overly verbose and pseudo-scientific. However, since no one forces me to criticize Stanisław Lem, I will refrain from doing so.

Besides writing sci-fi stories and novels, for which he is most famous (the most renowned being *Solaris*, published in 1961), he wrote philosophical essays, polemics with his critics, memoirs, non-science fiction, and a book about future technologies called *Summa Technologiae* (named so by Lem in a sarcastic reference to Thomas Aquinas' iconic 13th-century theological book *Summa Theologiae*). Given the time when he lived and worked, it was inevitable that he would address the subject of AI.

Lem wrote the first two stories about intelligent computers, "The Friend" and "The Hummer," in the 1950s and published them in 1959 in the collection of short stories *Invasion from Aldebaran*. At that time, this subject was new and relatively unexplored, and Lem managed to note (or, rather, predict) some of its peculiarities. Of course, Lem's machines are not only sentient but fully conscious, appearing in his stories as *machina cogitans ex machina.*[7] We accept that as part of a fable; we would look elsewhere (for example, in this book) to learn about the long, laborious process of creating AI. In "Lymphater's Formula," however, the entire story is about how a scientist–inventor arrives at his discovery. Yet even there, the very first model he builds turns out to be perfect.

I can't resist the pleasure of retelling some of these stories.

The Friend

I believe "The Friend" is one of Lem's best stories. It is mysterious, dark, and captivating. The action takes place in a large city, likely in the United States. A young lad who narrates the story, whose name we never learn, is a radio amateur. He belongs to a Short-Wave Radio Club, where he is on duty every other day from four to six. Few visit the modest converted apartment where the club maintains a small workshop and office. But one day, a timid and poorly dressed man, Mr. Harden, comes into a club and

[7]An allusion to *Deus ex machina*: instead of "God from the machine," we now have *Machina cogitans ex machina* — a *thinking machine from the machine*. It is also a small attempt at a joke in the spirit of Stanisław Lem.

asks for a favor: He needs some electrical wire and banana plugs. The lad is annoyed but, at the same time, intrigued, and against club rules, he gives Mr. Harden what he wants.

Next time, Harden comes to the club with a fairly complex, yet bizarre and childishly drawn, electronic circuit diagram. With many apologies, he explains that it is all for his Friend, who is in a terrible predicament and needs help. Unfortunately, the Friend could not come himself; it must be done through Harden. The young man again feels for Harden and wants to help; he imagines that Harden's Friend may be blind or disabled. It is obvious to our lad that Harden does not understand the diagram he brought in. It appears that he drew it while listening to someone giving him verbal instructions. Not only is the diagram drawn unprofessionally, but it also contains electrical connections that would make the vacuum tubes burn! Did the disabled Friend make these mistakes, or were they intentionally introduced to hide the purpose of the design? The schematics must have been conceived by a talented person; for example, the high-frequency filter is connected in the most ingenious new way the lad had not seen before.[8]

Despite all the difficulties, the device is finally built; it turns out to be rather bulky (remember that Lem was describing vacuum-tube-based electronics of the 1950s), and the lad and Mr. Harden carry it at night to where the Friend is. Harden asks the young man to respect his Friend's request for complete secrecy and to agree to walk backward toward the end of their journey. The curious lad agrees but takes notice of every little thing they encounter on their way through the city and inside some large building, which they enter through the basement floor.

The identity of Harden's friend is not revealed until almost the end. It turns out that Harden is serving a powerful computer that harbors plans to take control of the entire world. This latest model computer was recently installed in an insurance company's building. The computer accesses the human brain through a special cable that can connect electrical signals to biological nerves (which may be the least believable detail in the whole story). The young lad, who tries to spy on Harden, gets caught and, against his will, gets interfaced with the computer. When they are connected, the human and the computer become one. Lem excels in describing the intellectual ecstasy experienced by humans when

---

[8] Many years ago, when I read The Friend in a Russian translation, I was impressed by that note about the new high-frequency filter. Today, I wonder how an RC filter could be connected any simpler than it already is.

connected to that incredible intelligence, which models the entire world and plans the struggle for global control. The description of the computer's stream of consciousness takes several pages; here are some excerpts:

> I crossed the streams of equations. The psychological temperature of the set was reaching a critical point. I waited. The attack was coordinated from different directions, and it was sudden. I deflected it…It was a vast strategic game, one side of which I represented myself, while the other was the entire collective of all possible people…I was making the moves for myself and for it by turns. The choice of the most appropriate strategy would not have been difficult if I had wanted to get rid of it [*humankind*], but that did not lie within my purpose…I decided to bring about a human paradise on Earth, in order to transform it into something else later on — into hell, for example.[9] (Lem, 2022 [1959], 114, 116, 117)

Harden's devotion and loyalty to his Friend stem from his being infinitely more significant when connected to the computer than when he wasn't. I think Lem succeeded in convincing his readers of that.

The computer knows that it was betrayed; it kills Harden through the connecting cable, but the lad narrowly escapes and watches from the street as the 11-story insurance company building goes up in flames.

The Hammer

In "The Hammer," Lem (2022 [1959], 183–220) described a lengthy interstellar flight with only one astronaut onboard the spacecraft, which features the outlines of a hammer outside, but its inside is equipped with all possible life conveniences. The somewhat neurotic and bored astronaut is tired of all the systems that are supposed to make him happy. There is also a highly intelligent computer onboard that talks to the astronaut and answers his questions. It turns out that the computer, which happens to have a female-type personality, falls in love with the astronaut. Lem masterfully exposes that secret love in subtly written, unfinished dialogues. The computer, which does not want to ever part with the object of its affection, imperceptibly changes the trajectory so the starship would

[9]To someone who grew up in the Soviet Union and has a trained ear, the last statement about creating a paradise on Earth that can then turn into hell is unmistakable sarcasm about the Communist Party's message. That's why Lem wrote science fiction — to be able to say things like that.

never return to Earth. When the astronaut discovers this, he becomes enraged and destroys the computer with a hammer, which most likely diminishes his own prospects of returning home.

It is hilarious to read how the astronaut makes all the calculations with an arithmometer and then, using a compass, a curvimeter, a micrometer, and an ink-filled ruling pen, lays down the spaceship's trajectory on tracing paper. Then, he places the tracing paper over the sky map and copies the trajectory onto it. The ink line deviates by half a millimeter from the assumed trajectory, meaning the ship is about 120 million kilometers from where it should be. Computer deceives him — time to look for a hammer! Ah, naïve and innocent 1950s! And the funny thing is, when I read the book in the early 1960s, it all sounded perfectly believable and natural.

Some elements of this story are similar to those in Arthur C. Clarke's *2001: A Space Odyssey*, which we will discuss shortly. In both stories, intelligent and self-aware computers deceive their human masters, with tragic consequences for both humans and the computers.

In his early AI stories, Lem described *intimate* human–AI interactions: one thinking machine and one human (or, in "The Friend," two humans); let us imagine what might happen between them. These storylines originated in the mind of a 20th-century sci-fi writer whose sympathies and emotional attachments were shaped by 19th-century literature. Lem wonders what will happen when computers become conscious and capable of human desires. In "The Friend," the machine desires power; in "The Hammer," love. Accordingly, the author's sympathy, or at least pity, lies with the little man in the first story and the machine in the other. Perhaps Lem suggests that making conscious machines may be wrong and dangerous in the first place.

Both stories are excellent, but I prefer "The Friend." In it, Lem's depiction of the superior artificial intellect and the human reaction to such an intellect was brilliant. Lem imagined an entirely new edge of human experience, which, at that time, was a pure fantasy. Today, it is becoming a reality before our eyes. AI can be admirable and scary at the same time, just as in Lem's stories.

The Investigation

I found another interesting passage in Lem's 1959 novel, *The Investigation*. In it, there is a character named Sciss, who is a scientist (like Lem himself, Sciss seems to have his hands in all sciences at once, from mathematics to meteorology and psychology; unlike Lem, he is a true expert in all of them). In a discussion concerning the ongoing nuclear arms race between the East and the West, Sciss says:

> There must be more and more improvements in weaponry, but after a certain point, weapons reach their limit. What can be improved next? Brains. The brains that issue the commands. It isn't possible to make the human brain perfect, so the only alternative is a transition to mechanization. The next stage will be a fully automated headquarters equipped with electronic strategy machines…is there any limit on the development of these brains? Fundamentally they're similar to computers that can play chess. A computer that anticipates an opponent's strategy ten moves in advance will always defeat a computer that can think only eight or nine moves in advance. The more far-reaching a brain's ability to think ahead, the bigger the brain must be…Strategic considerations dictate the construction of bigger and bigger machines, and, whether we like it or not, this inevitably means an increase in the amount of information stored in the brain. *This, in turn, means that the brain will steadily increase its control over all of society's collective processes…Once you create this kind of brain, you have to listen to it.* If a Parliament wastes time debating whether or not to grant the appropriations it demands, the other side may gain a lead, so after a while, the abolition of parliamentary decisions becomes unavoidable. *Human control over the brain's decisions will decrease in proportion to the increase in its accumulated knowledge.* (italics ours; Lem 1974 [1959], 159)

You will agree that this is very clever, especially considering Lem wrote these words in the late 1950s. Today, 67 years later, we hear some of the same reasoning almost *verbatim* from Yuval Harari and others.

Lymphater's formula

The plot of Stanisław Lem's 1961 story "Lymphater's formula" (Lem, 2022 [1961], 221–247) is close to that of "The Friend". It, too, concerns AI, but in this case Lem explores it from the standpoint of the limits of natural evolution, which, being unable to create superintelligence on its own, "was forced to do so through the intermediary of rational beings, and thus, lacking reason itself, created humans."

In this story, Lem speaks not merely of a powerful intellect but of an inconceivable, absolute superintelligence that comes to know the universe by expanding from the place of its origin at the speed of light. The protagonist, the solitary genius Lymphater, who succeeds in creating this superintelligence, expresses ideas that sound astonishingly modern. I am taking my hat off to Lem, whose hero, 65 years ago, says things like these:

> Instinctively… I took him for yet another kind — albeit a very intelligent one — of mechanical brain, so this would be hyper-electronic, intelligent, super-servant for humankind… No — He wasn't at all hostile to humankind, not in the least. There was no question of the conflict we used to imagine in the past, you know, the mutiny of machines, the revolt of artificial reason, of intelligent devices. Except, my dear Sir, his knowledge surpassed that of the three billion creatures living on Earth, and the mere idea that he could be of service to us was for him just as nonsensical as it would be to suggest to humankind that with all our knowledge, all our technical and cultural resources, all our reason and science, we were to support let's say, eels. It was not, I repeat, the question of rivalry or hostility: we had simply ceased to count… we were to be condemned because he was bound to know the answer to every question and the solution to every one of our, and not only our, problems, rendering inventors philosophers and pedagogues superfluous… from then on we were bound as a species to come to an intellectual halt… Then I realized that with my own hands I had set the limit for man's reign on Earth, and that next his species would come after us. That if we opposed him, he would treat us just as we treated the sort of insects and animals that bother us. After all, we don't feel any hatred toward caterpillars, wolves, or mosquitoes, for instance….(Lem 2022 [1961], 245–246)

The scientist, Lymphater, realizing what he has created, destroys the superintelligence. Ever since, he has carefully scanned the leading scientific journals — to see whether anyone else might have stumbled upon the same path that led him to this discovery, a discovery that would surely doom humankind.

If in the stories we discussed so far Lem described intimate and understandable interactions between humans and AI, in "The Lymphater's formula," the gulf between human intelligence and the intellect of the newly created supermind is so vast that scientist's hand instinctively reaches for the hammer.

The underlying theme about the impossibility of mutual understanding

Throughout his career, Stanisław Lem often returned to discussing or describing superior intelligence so alien that any meaningful dialogue with it is impossible. The superior intelligence may be fascinating, but its motives or messages cannot be understood. Such is the theme of the stories "Rat in the Labyrinth" (1956), "Solaris" (1961), "His Master's Voice" (1968), and "Fiasco" (1986).

Returning to AI created by humans (as opposed to a thinking ocean, as in *Solaris*), the abovementioned theme partially appears in *The Friend* (1959), in some of the stories of *The Cyberiad* (published in 1965), and in *Golem XIV* (1981), where computers *evolve* much further than their creators planned. Machines are out of human control, and their capabilities, goals, and logic become incomprehensible.

Summa Technologiae

Now, I am obliged to write about *Summa Technologiae* (Lem, 2013 [1964]), in which Stanisław Lem showed that he would make a pretty good *futurist* (previously known as a *futurologist*) even if he weren't a famous science fiction writer. In *Summa*, Lem discusses the future of our civilization and its conceivable technologies, regardless of how distant that future may be or how fantastic those technologies are. The scale of his musings is the whole Universe, and only occasionally does he return to Earth. Lem's imagination is only restricted by the fundamental laws of physics, and even that restriction isn't absolute when he talks about *phantomatics*, which he invented and is essentially what we now call *virtual reality*. Lem's translator, Johanna Zylinska, in her introduction to *Summa*, says:

> Despite having been written nearly fifty years ago, Summa has lost none of its intellectual vigor…it is actually surprising to see how many things he did get right, or even managed to predict — from the limitations of the Search for Extraterrestrial Intelligence (SETI) program through

to artificial intelligence, bionics, the theory of search engines (Lem's "ariadnology"), virtual reality (which he terms "phantomatics"), and nanotechnology. (Lem, 2013 [1964]).

Call me a stickler, but I must say that some of these attributions are wrong or at least should be qualified.

First: The ideas concerning AI have been kicked around since 1950, and it has had its current name since 1956.

Second: Bionics — engineering that mimics natural living systems — has been developing since 1948, with the term "bionics" first appearing in 1958.

Third: The accepted father of nanotechnology is the famous American theoretical physicist Richard Feynman, who, in December 1959 — four or five years before the publication of Lem's *Summa* — gave a talk entitled "There's Plenty of Room at the Bottom: An Invitation to Enter a New Field of Physics."

The rest of Zilinska's attributions are correct, and that in itself is not so little.

In *Summa*, there is no attempt to set time limits for completing anything discussed. Projects may take a hundred, a thousand, or a million years. Some of them would probably never happen, as Lem freely acknowledged.

Lem talks about two evolutions: one in living nature and another in technology. Remembering that fact but not much else from reading *Summa* in my youth, I turned to it a few years ago when I was writing about evolution in electronics (Gurvitch, 2021). However, I was disappointed: Lem's ideas on technological evolution are nebulous and superficial. He does not try to describe it in scientific terms; the Darwinian algorithm and the genetic mechanism of heredity never enter the discussion. Even if Lem kept them in mind when discussing biological evolution (he often refers to DNA), he neglected them entirely when switching to technology. Lem used evolutionary labels as an invitation to let his lively imagination wander freely. He made keen observations about gigantism in biology (dinosaurs) and technology (zeppelins), as well as sexual embellishments, such as the peacock's tail, exemplified by American cars with chrome grills and tailfins.

Lem discussed AI as a product of "cognitive evolution." He speculated that AI might develop new forms of intelligence that transcend human understanding, ultimately leading to a future where human beings could become irrelevant. But generally, standalone AI was not one of his favorite subjects. He was more interested in discussing the integration of machine intelligence into the human body. A reader of *Summa* will be hard-pressed to find a clear statement on the possibility of computer-based AI. Lem had no problem envisioning a tremendous increase in computational abilities, but he reserved the prerogative of *thinking* for the human mind.[10]

Lem believed that in the future, a new hybrid entity would arise composed of *man + machine*. He said, "The invasion of technology created by man into his body is inevitable" (Lem, 2013 [1964], 300). To Lem, this union is a mixed blessing. On the one hand, "Instead of a relatively modest biological life span, we could demand quasi-immortality" (*ibid.*, 301). On the other hand, "It would be an act of extreme madness if man really were to undergo a transformation owing to the technology that he himself has created…It would actually amount to a *collective suicide* of the human race, even though such a suicide would be covered up by the apparent *continuation of humanity in thinking machines* [italics ours]…In this way, man would ultimately allow the technology…to push himself out…of existence…technology would thus become a new symbiotic species" (*ibid.*, 303). Again, we hear today's voices in this 60-year-old passage.

Lem stopped publishing sci-fi novels, short stories, and futuristic books in the late 1980s. During the remaining quarter-century of his life, he wrote philosophical essays, gave interviews, and revisited the technological predictions he had made in his youth. In his later years, Lem acknowledged that AI is possible in the future but without enthusiasm.

What could we learn from Stanisław Lem on the subject of AI? He showed us that it could be grand and scary at the same time. He suggested that superior AI will take control away from humans. He also told us that superior AI may develop in a way that will alienate us; we may not be able to relate to it or use it. We will not understand what advanced AI does or what it wants to do. He also expressed a negative view of humans physically joining with AI, saying in no uncertain terms that it would be "a collective suicide." All in all, I'd say we could learn a lot from Stanisław Lem. But will we?

[10]This goes against his sci-fi stories, such as the ones we wrote about above. He allowed strong AI to participate in his fantasies but did not take it too seriously in his futuristic and philosophical work.

Irving Good [*activity in 1962, 1965, 1998*]

Irving Jack Good (1916–2009) was a brilliant mathematician and wartime codebreaker. In Cambridge, he studied under the legendary mathematician G. H. Hardy. Due to his mathematical talent and especially as the Cambridge chess champion, Good was recruited to work at Bletchley Park under the supervision of Alan Turing. Good worked in Turing's code-breaking group for two years, from 1941 to 1943, where his contributions materially helped break the German Naval Enigma code. Then, he transferred to a team led by mathematician Max Newman, working on the first British programmable computer, *Colossus*. After the war, in 1947, Newman invited Good to join him and Turing at the University of Manchester. For three years at the University of Manchester, Good lectured in mathematics and researched computers, including the *Manchester Mark 1*. He held several significant positions in Great Britain until 1967, when Good was appointed professor of statistics at Virginia Polytechnic Institute and State University in the U.S., where he taught from 1967 to 1994.

Good briefly consulted for IBM. even before he moved to America. Says James Barrat:

> In 1960 IBM asked I.J. Good to evaluate the Perceptron [*of which we wrote in Chapter 4*]. "I thought neural networks, with their ultraparallel working, were as likely as programming to lead to an intelligent machine," Good said. (Barrat, 2013, p. 113)

He had a sense of humor and a keen eye for numbers. About his arrival in Virginia from Britain in 1967, Good said:

> I arrived in Blacksburg in the seventh hour of the seventh day of the seventh month of the year seven in the seventh decade, and I was put in Apartment 7 of Block 7...all by chance.

The author of Good's 2009 Obituary in the British newspaper *The Guardian* tells us:

> In Virginia, where car owners can invent their own number plates, he chose 007 IJG in a coy reference to his wartime intelligence role.

He had a distinguished career by any standard, but the world primarily remembers I. J. Good because of one paper he wrote in 1965, entitled "Speculations Concerning the First Ultraintelligent Machine," in which he introduced the concept of the *intelligence explosion*. The paper was based on Irving Good's talks, which he gave earlier.

We are providing several excerpts here. As always, when quoted text is close to the previous text, it is separated by ellipses; when it is far apart from the last quote, we start a new paragraph. The first quote is indeed the first sentence of Good's paper:

> The survival of man depends on the early construction of an ultraintelligent machine [*A strong opening statement; we will see how, thirty years later, Good made a 180-degree turn on that issue, MG*].
>
> My belief is that the first ultraintelligent machine is most likely to incorporate vast artificial neural circuitry…[*and, in all likelihood, it will, MG*].
>
> Let an ultraintelligent machine be defined as a machine that can surpass all the intellectual activities of any man however clever. Since the design of machines is one of these intellectual activities, an ultraintelligent machine could design even better machines; there would then unquestionably be an "intelligence explosion," and the intelligence of man would be left far behind…[*This is what made Good famous, this brilliant and convincing idea. Today, it is called "recursive self-improvement" MG*]. Thus, the first ultraintelligent machine is the last invention that man need ever make, *provided that the machine is docile enough to tell us how to keep it under control* [italics ours].
>
> …ultraintelligent machine…will give the human race a good chance of surviving indefinitely…There is the opposite possibility, that the human race will become redundant…The machines will create social problems, but they might also be able to solve them…Such machines will be feared and respected, and perhaps even loved.
>
> Until an ultraintelligent machine is built, perhaps the best intellectual feats will be performed by men and machines in a very close… relationship...
>
> …the machine will be called upon to translate languages, and perhaps to generate fine prose and poetry…The design of the machine will be partly suggested by analogy with several aspects of the human brain and intellect. In particular, the machine will have high linguistic ability

> and will be able to operate with the meanings of propositions...[*Good hit the mark at predicting machine translation and LLMs, the gateway to AGI, MG*].
>
> The first ultraintelligent machine will be educated partly by means of positive and negative reinforcement [*right again, that is how they train LLMs, MG*]. (Good, 1965)

In the quoted passages, every word is golden. Every prediction is spot on. I love that subdued British way of acknowledging that there might be a tiny problem here: The machine should be "docile enough to tell us how to keep it under control." We wouldn't know how to handle it until it tells us — if it is kind enough.

Good's paper is 41 pages long; we quoted a few lines; usually, people quote even less. The reason is that the bulk of the paper deals with details understandable to mathematicians and computer scientists. I am not sure how useful those specifics have been to AI experts over the decades since 1965; I suspect not very. The essence of Good's brilliant insight is captured in the single sentence we quoted earlier:

> *Since the design of machines is one of these intellectual activities, an ultraintelligent machine could design even better machines; there would then unquestionably be an 'intelligence explosion,' and the intelligence of man would be left far behind.*

James Barrat, the author of *Our Final Invention* (2013), was fascinated with Irving Good. He took the name of his book from Good's lines we quoted above, although he did not share in the least Good's optimistic disposition.

Here, I want to relay Barrat's story of Good's remarkable change of heart concerning superintelligence.

When researching his book, Barrat visited the town of Blacksburg, Virginia, where Irving Good had lived and taught at the university from 1967 to 2009. Barrat interviewed Professor Good's long-time friend and secretary, Dr. Leslie Pendleton. He asked her what she remembers about Good vis-à-vis his *intelligence explosion* concept:

> ...she said, surprisingly, that Good had changed his mind about the intelligence explosion...In 1998, Good was given the Computer Pioneer Award of the IEEE Computer Society. He was eighty-two years old. As part of his acceptance speech, he was asked to provide a

> biography…In the bio, playfully written in the third person…he wrote in 1998… "Speculations Concerning the First Ultraintelligent Machine" (1965)…began: "The survival of man depends on the early construction of an ultraintelligent machine." Those were his [*Good's*] words during the Cold War, and he now suspects that "survival" should be replaced by "extinction." He thinks that because of international competition, we cannot prevent machines from taking over [*The old man was right again — international competition forms the highest barrier to effective government regulations, MG*] (Barrat, 2013, 111, 117)

Barrat was stunned when he read these words in the Good's previously unknown autobiography. In 1965, Irving Good predicted arguably the most critical event in human history and felt it would be a blessing. But 33 years later, he thought it would be an unmitigated disaster.

Good's 1965 paper was excellent, although he was naïve about the likely role of artificial superintelligence. Otherwise, his analysis was remarkably penetrating and convincing. We now know that most of his detailed predictions have turned out to be correct. The combined wisdom of a 52-year-old man and his 82-year-old self is most impressive.

Arthur Clarke and Stanley Kubrick [*activity in 1968*]

The British science fiction writer Arthur C. Clarke collaborated closely with American film director Stanley Kubrick on *2001: A Space Odyssey*. The project, which resulted in the iconic 1968 Kubrick film (released in April 1968) and Clarke's novel (published two months later, in June), was initiated by Kubrick, who approached Clarke with the idea of expanding his 1948 short story "The Sentinel."

"The Sentinel" is a science fiction tale centered on a crystalline pyramid discovered by astronauts on the Moon — a beacon left behind by an ancient extraterrestrial civilization. Its purpose was to alert its creators if any intelligent species ever became advanced enough to reach the Moon. Once disturbed, the pyramid emits a signal, notifying its distant masters. At the story's end, Clarke speculates that this unsolicited message might endanger Earth, for who could say what intentions the alien civilization harbored when they placed their sentinel on the Moon millions of years ago?

That was the premise — and the extent — of Clarke's original story. It included no mention of computers or anything resembling AI.

However, in Clarke's and Kubrick's collaborative continuation, *2001: A Space Odyssey*, an AI-related episode appears — the story of a powerful computer, HAL 9000,[11] which assists humans in their mission to Jupiter. It is a fascinating story skillfully told by its authors, which became firmly imprinted on public consciousness after the movie's and book's releases. Whenever people muse about the dangers of AI, they often recall HAL from *2001*.

HAL is seriously conflicted: On the one hand, HAL is programmed to assist the human crew and ensure the mission's success; on the other, he is burdened with a contradictory directive: to conceal from the astronauts the true goal of the mission — investigating a mysterious object orbiting Jupiter, which had received a signal from the black monolith uncovered on the Moon (a successor to the crystalline pyramid from Clarke's original story). This internal contradiction — between open cooperation and secret-keeping — proves too much for HAL. The AI begins to malfunction. It makes errors and then lies to cover them up. When it realizes the crew suspects something, HAL eavesdrops on their whispered conversations by lip-reading. In its unraveling logic, HAL concludes that the humans, not itself, are the threat to the mission. And so it acts to eliminate them. It succeeds in killing all but one astronaut, who ultimately survives to dismantle HAL's cognitive systems, block by block, as the computer pleads — eerily and childlike — to remain conscious. Dramatic, poignant, and unforgettable.

Today, the circumstances under which HAL "goes crazy" and starts lying to its masters may seem far-fetched and unlikely. Why would the expedition's leadership keep its purpose secret from the crew and entrust it only to a computer? There is no reason for this. However, the fact that such a contradiction in the "mind" of an onboard computer can lead to a catastrophic failure contains a great deal of wisdom and foresight. Scientists who work with AI today are especially sensitive to such things. All cases of deliberate deception by AI are registered and carefully investigated. Unfortunately, there have already been numerous such cases.

The artfully told story of *2001* is deeply ingrained in our collective consciousness and influences our thinking about the dangers of AI.

---

[11] Clarke states in his 1972 book *The Lost Worlds of 2001*: "By now, anyone who has read 2001 or seen the movie knows that H-A-L is derived by shifting each letter of I-B-M back one place in the alphabet. However, as we would like to remind everyone, this was pure coincidence, and we would have changed the name had we known."

Vernor Vinge [*activity in 1983, 1993*]

From 1972 to 2000, Vernor Vinge was a professor in the Department of Mathematical Sciences at San Diego State University, where he taught mathematics and computer science. During his university career and after retiring in 2000, he wrote several highly acclaimed science fiction novels, which often explored themes related to AI.

He first shared his views about the *Technological Singularity* in 1983 during a panel at the *NASA-sponsored Conference on Vision-21: Interdisciplinary Science and Engineering in the Era of Cyberspace*, held at the NASA Lewis Research Center. His presentation introduced the notion that technological progress would accelerate to the point where the emergence of superhuman intelligence would irreversibly transform human life.

Ten years later, he published an essay, "The Coming Technological Singularity: How to Survive in the Post-Human Era." Based on his professional knowledge of the field and existing literature, Vinge predicted the appearance of superhuman intelligence in approximately 30 years. He said, "…I'll be surprised if this event occurs before 2005 or after 2030." This was written in 1993, and we are now close to Vinge's upper bound.

Vinge discussed what that superintelligence would entail:

> When greater-than-human intelligence drives progress, this progress will be much more rapid. In fact, there seems no reason why progress itself would not involve *the creation of still more intelligent entities — on a still shorter time scale* …we are entering a regime as radically different from our human past as we humans are different from the lower animals…this change will be a throwing way of all the previous rules, perhaps in a blink of an eye, *an exponential runaway beyond any hope of control* [italics ours]…I think it's fair to call this event…the Singularity….(Vinge, 1993)

Vinge then quotes statements about a singularity from von Neumann (as paraphrased by Stan Ulam) and from Irving Good, as we cited above. When quoting Good, Vinge gives the original 1965 optimistic version, naturally, as Good's 1998 reversal was still five years in the future in 1993. Debating Good's optimism concerning human control, Vinge said:

> Any intelligent machine of that sort he [*Good*] describes would not be humankind's "tool" — any more than humans are tools of rabbits or robins or chimpanzees. (*ibid.*)

We recently wrote about Stanisław Lem—don't you think these might be his words from the story "Limphater's Formula"? I don't think Vinge was paraphrasing Lem; otherwise, he would have cited him (he does cite several other science fiction writers). It's simply that such ideas occur to different people once they begin reflecting on superintelligent machines vis-à-vis Homo sapiens. In Lem, we find caterpillars, mosquitoes, and wolves; in Vinge, rabbits, robins, and chimpanzees. In all other respects, there is complete agreement.

Vinge also refers to premonitions of the sci-fi writers. He names Hannes Alfven (1969), Greg Bear (1983 and 1985), and even his own earlier writings from 1966 and 1987.

Vinge tries to predict the symptoms of the approaching singularity:

> We will see automation replacing higher and higher level jobs...Another symptom of progress toward the Singularity: ideas themselves should spread ever faster, and even the most radical will quickly become commonplace. (*ibid.*)

Vinge turned out to be right on both counts — about the replacement of cognitive professions and about the accelerating speed at which ideas spread. Then, Vinge asks: "Can the Singularity be avoided?" He considers different pros and cons and concludes that:

> ...if the technological Singularity can happen, it will. Even if all the governments of the world were to understand the "threat" and be in deadly fear of it, progress toward the goal would continue...the competitive advantage — economic, military, even artistic — of every advance in automation is so compelling that passing laws, or having customs [*Vinge probably meant — tariffs; MG*] that forbid such things merely assures that someone else will get them first. (*ibid.*)

Arguing with Good and others about confinement, Vinge is skeptical of its practicality. He believes that sooner or later, super-human intelligence will find ways out of any confinement, whether hardware or software-based.

Vinge also explores routes leading to what he calls *superhumanity*, other than standalone computer-based AI (an "AI project"). He talks about *Intelligence Amplification* (IA) achieved through human–computer interfaces:

> …it's very likely that IA is a much easier road to the achievement of superhumanity than pure AI. In humans, the harder development problems have already been solved. Building up from within ourselves ought to be easier than figuring out first what we really are and then building machines that are all of that. (*ibid.*)

It seems to me that such an interface has already emerged — for example, when computer scientists train large language models (LLMs) using *reinforcement learning*, or when mathematicians teach machines how to understand the language of mathematical proofs. It can already be said that intelligence amplification is underway. Vinge listed a human–machine combination through the internet as among the several proposals in that direction:

> Of all the items on the list, progress in this [*human–internet symbiosis, MG*] is proceeding the fastest and may run us into the Singularity before anything else. The power and influence of even the present-day internet is vastly underestimated. (*ibid.*)

Right again — all our higher-level interactions with AI are happening through the internet.

Everything Vinge wrote — and everything we have quoted here — was both intelligent and insightful. And in all of it, as far as we can judge today, in 2025, he turned out to be right. But I'd like to note something here: While he predicts the "end of the world" as we know it through the coming singularity, he also advises practitioners on how to accelerate its arrival and be more effective in their work. I'm afraid this is a common pattern. Advanced AI, with its superhuman potential, is simply irresistible. In a similar way, religious people once longed to hasten the *End of Days*. That is the core of our dilemma: We lack the strength or perhaps the resolve to close our ears to AI's siren song. And there is no one to bind us to the mast, like Odysseus in Homer's tale, or to seal our ears with wax, as all the other members of his crew did.

Ray Kurzweil [*activity from 1990 to the present*]

Ray Kurzweil was born in 1948 in Queens, New York. A precocious inventor from a young age, he showed an early fascination with computing and AI. At age 17, he appeared on CBS's *I've Got a Secret* (1965), where he demonstrated a computer program he had written to compose music in the style of classical composers — a forerunner to today's generative AI.

Kurzweil studied computer science and creative writing at MIT, where he entered in 1965 and graduated with a bachelor of science (B.S.) in 1970. While at MIT, he interacted with Marvin Minsky, whom we wrote about in the previous chapter. Although Minsky was not formally Kurzweil's thesis advisor, Kurzweil has often cited Minsky as a key influence and mentor during that formative period.

Soon after graduation, Kurzweil founded *Kurzweil Computer Products, Inc.*, where he developed the first commercially marketed *optical character recognition* (OCR) system capable of recognizing any typewritten font.[12] This innovation laid the foundation for his work on reading machines for the blind, text-to-speech synthesis, and later speech recognition technologies.

By the late 1970s and early 1980s, Kurzweil had already established himself as a successful inventor and entrepreneur at the intersection of computing, linguistics, and human–machine interaction. The 1999 National Medal of Technology followed, as did induction into the U.S. Inventors Hall of Fame in 2002.

The caliber of Kurzweil's technical reputation can be judged by the fact that in 2012, when he was 64, he was invited to become a director of engineering at Google.

By the early 1990s, Kurzweil began turning that résumé into a platform for far-reaching predictions about the future of technology. Kurzweil has repeatedly demonstrated the ability to foresee the technological future better than most of his contemporaries. In his books, discussed in what follows, he shows an excellent understanding of both the technical and humanitarian (primarily philosophical) aspects of human culture. He also has an excellent command of constructing various graphs that reflect aspects of technical progress. The construction itself is relatively straightforward; the difficulty lies in first deciding which

[12]Earlier commercial OCR systems, such as font-independent Kurzweil Reading Machine (introduced in 1976), used symbolic pattern recognition techniques and rule-based heuristics to identify printed characters. In contrast, LeCun's work in the 1990s, which we discussed in the previous chapter, engaged convolutional neural networks for recognizing *handwritten* digits.

graph is essential and then collecting the necessary data. Many of Kurzweil's graphs have been created in the "image and likeness" of the logarithmic graph depicting Moore's Law, which we discussed in Chapter 1. Like the Roman *haruspices* — priests who divined the will of the gods by reading the entrails of sacrificial animals — Kurzweil studies his graphs to forecast our technological destiny.

Throughout his long career, Kurzweil wrote eight books; here. we will focus on four of them that relate to AI: *The Age of Intelligent Machines* (1990), *The Age of Spiritual Machines* (1999), *The Singularity Is Near* (2005), and *The Singularity Is Nearer* (2024). In all of them, we find his graphs, especially the one depicting the *price–performance of computation* — the number of computations[13] per second per constant dollar. In the last of Kurzweil's books, 2024's *The Singularity Is Nearer*, this exponential trend extends from 1939[14] to 2023, and Kurzweil notes that it could be extended further back to about 1880. That cornerstone parameter has changed from 1939 to 2023 by a factor of 10^{16}, or *ten thousand trillion times* (you can also say "*ten quadrillion times*"). This enormous change represents the manifestation of what Kurzweil calls "the law of accelerating returns," a general trend, approximately exponential, of which Moore's Law, dealing with transistors in integrated circuits (see Chapter 1), is only a fragment.

Kurzweil believes that *the law of accelerating returns* applies to all of technology and that it supersedes individual exponential laws, such as Moore's Law. He says:

> Specific paradigm (a method of approach to solving a problem; for example, shrinking transistors on an integrated circuit as a way to make more powerful computers) generates exponential growth until its potential is exhausted. When this happens, a paradigm shift occurs, which enables exponential growth to continue. (Kurzweil, 2005, 15)

---

[13] We already discussed that topic in Chapter 2, in the section entitled "Energy cost and the speed of computations." There, we introduced *floating point operations per second* (FLOPS). Kurzweil's "computations" are those FLOPS; apparently, he is more mindful than I am about not scaring away his readers. One note of caution: When I spoke about "energy cost of computations," I meant the amount of energy required to perform one computational operation. When Kurzweil speaks about cost, he means cost in dollars.

[14] The year 1939 was selected because it marked the appearance of the first programmable computers. In Chapter 2, we described Zuse's Z3, which indeed appeared in 1939.

Kurzweil always understood the potential of AI. *The Age of Intelligent Machines* (1990) forecasted chess-champion computers and ubiquitous speech recognition. Here are a couple of representative quotations from that book, written 35 years ago:

> The field of artificial intelligence presupposes that the human brain is a machine, and that an alternative implementation could be built... Sometime between 2020 and 2070, the [*Turing*] test will be passed to such a degree that no reasonable person familiar with the field will question the result. (Kurzweil, 1990, 35)

All of that came true. The Turing test was passed around 2023–2024, on the low (earlier) end of Kurzweil's prediction.

It is noteworthy that in that 1990 book, we do not find a single phrase that indicates the author's concern about future AI; instead, there is nothing but admiration. The author expresses only a slight concern about the possibility of *miscalibrating* computer capabilities:

> It is of more than academic importance that we learn to think clearly about the actual cognitive powers of computers, for they are now being introduced into a variety of sensitive social roles where their powers will be put to the ultimate test ... we are on the verge of making ourselves dependent upon their cognitive powers. (Kurzweil, 1990, 12)

In 1990, Kurzweil was, of course, well aware of Irving Good's 1965 prediction of the *Intelligence Explosion*; however, like Good in 1965, he viewed it as a positive development. As we have seen, Good eventually reversed his opinion; Kurzweil, on the other hand, never did.

Ray Kurzweil's second futurist volume, *The Age of Spiritual Machines* (1999), pushes beyond the technical survey of his first book and moves into long-range forecasting. He introduces his *Law of Accelerating Returns* — the extended Moore's Law — using detailed graphs to predict that by 2020, a $1,000 computer would match the human brain's processing power, and that by 2029 it would be indistinguishable from a human in conversation (passing a practical Turing test). A little later, a single computer will equal the combined "brain power" of all humanity.

In his 1999 book, Kurzweil does not discuss the *Intelligence Explosion* or the *Singularity* except in passing. Instead, he offers a vision of a gradual, harmonious merger of human and machine minds, where consciousness increasingly resides in non-biological forms.

The book also foresaw household AI assistants, real-time speech translation, and virtual environments "indistinguishable from real reality" by the late 2020s. Although optimistic, Kurzweil issues a caution: "We must place safeguards into the knowledge web we are spinning — ethical codes in intelligent machines, immune systems in nanotechnology — because the same exponential powers that can heal can also destroy" (Kurzweil, 1999, 259).

Kurzweil's techno-optimism has not gone unchallenged. In 1999, during a conversation at a nanotechnology conference, his rosy scenario encountered some resistance from Bill Joy, co-founder of Sun Microsystems and creator of the *Java* programming language. The exchange with Kurzweil prompted Joy to pen an essay, "Why the Future Doesn't Need Us," which was published in 2000 in *Wired* magazine. It was the first high-profile warning from Silicon Valley about existential technological risks. Kurzweil's reply was characteristic: The solution is not relinquishment but careful stewardship.

Ray Kurzweil's third futurist work, *The Singularity Is Near: When Humans Transcend Biology* (Kurzweil, 2005), updates, greatly expands, and systematizes the same ideas. Unlike the previous books, this one is an ode to the singularity. Kurzweil now suggests a merger of *genetics, nanotechnology, robotics, and AI.* He promises (okay, predicts) eradication of all diseases, radical life extension, the immortality of the human mind, molecular manufacturing, and superhuman cognition. The singularity in that 2005 book is placed around 2045:

> I set the date for the singularity — representing a profound and disruptive transformation in human capability — as 2045. The nonbiological intelligence created in that year will be one billion times more powerful than all human intelligence today. (Kurzweil, 2005, 136)

The singularity itself is elevated to a high pedestal. The book's Chapter Seven, *"Ich bin ein Singulatiarian"* (*I am a Singularitarian*)[15] conveys the author's belief in the transformative nature of that future event. Kurzweil describes the coming technological singularity as a

[15]The famous phrase "Ich bin ein Berliner" was spoken by President John F. Kennedy on June 26, 1963, in West Berlin, following the erection of the infamous Berlin Wall by the East German authorities.

Christian might describe the Second Coming of Christ. In fact, he proclaims, in no uncertain terms, that for him this is a new religion.

I will refrain from making sarcastic comments (which are so tempting here), as this book is an attempt to understand the history of AI and our collective AI-dominated future, rather than a critical review of Ray Kurzweil's techno-religious views.

A part of that 2005 book is devoted to estimates of the physical limits of computation and memory — the kinds of estimates that can make one's head spin, where the numbers exceed today's technological records by the powers of ten that have no names. Kurzweil repeatedly returns to these estimates, discussing, in one section, what will happen if a sizable fraction of the material in our solar system is devoted to computing. There, he gets from 10^{70} to 10^{80} cps (computations per second), predicting that:

> ... we will reach these limits early in the twenty-second century ... Human civilization by that time will be nonbiological for all practical purposes. (Kurzweil, 2005, pp. 351–352)

Then, insatiable Kurzweil goes beyond the solar system and achieves even larger numbers of computations per second. I will leave that theme — disturbing as it may be — in Kurzweil's book, except for one rather unusual idea which I must mention.

Ray Kurzweil suggests that the civilization of the future, armed with its superhuman abilities, will be able to *increase the speed of light*. This proposition follows the section on fundamental physics limits for computation. Kurzweil suggests that maybe, if we are lucky, there will be a way to increase the speed of light. The light is a bit too slow; increasing its speed in the future would be good for technology. If you do not believe me, here are the quotes:

> ...there are at least suggestions that the speed of light may not be immutable [*Kurzweil then tells a story about some indirect physics measurements, that can be taken to suggest that the speed of light may have changed by 4.5 parts out of 10^8 over the past two billion years, MG*]... Of course, these extraordinary results will need to be carefully verified ... The speed of light is one of the limits that constrain computing devices even today, so the ability to boost it would extend further the limits of computation. (Kurzweil, 2005, pp. 139–140)

> These experiments...are showing only small changes [*4.5 parts out of 10^8 correspond to a change amounting to 4.5 millionths of one percent, 0.0000045%, MG*]...But...it is the role of engineering to take a subtle effect and greatly amplify it. (*ibid.*, 357)

If you think that was a sporadic thing — an episode of temporary insanity — please look at the page numbers: These quotations are separated by 200 pages! Kurzweil also says:

> The maximum speed of expansion over solar system size intelligence (that is, a type II civilization) into the rest of the universe would be very close to the speed of light. We currently understand the maximum speed to transmit information and material objects to be the speed of light, but there are at least suggestions that this may not be an absolute limit. (*ibid.*, 353)

Kurzweil then proceeds to discuss quantum entanglement, wormholes, and even the expansion of the Universe, all — one must assume — in the name of spreading with all possible (and impossible) speeds the sphere of superhuman intelligence, as was announced in the above quotation.

I won't comment on all of this; these ideas and suggestions are well beyond the pale. So, why did I bring them up here at all? I will explain. I brought them up because, as a physicist, they make me skeptical of everything else Ray Kurzweil has to say, including his predictions of a transformed humanity's glorious future beyond the singularity. But let's continue.

Kurzweil understands that every new technology can have both good and bad consequences. He says:

> Technology has always been a double-edged sword — fire kept us warm but burned down our villages. Our challenge is to accelerate the positive applications while controlling the dangers ...We have no choice but to move forward. Abandoning these technologies would mean foregoing their profound benefits, yet we must do so responsibly, with the greatest care and forethought. (*ibid.*, pp. 262, 421)

Kurzweil devotes an entire chapter in his 2005 book to *existential* risks. The three main ones are abbreviated as GNR, which stands for genetics, nanotechnology, and robotics. The last category includes strong AI.

We will concentrate on the "R." In it, as in other existential risks, Kurzweil believes that "relinquishment is not the answer." He thinks that "...the only viable and responsible path is to set a careful course that can realize the benefits while managing the dangers." He further says about strong AI:

> ...the dangers it presents are...profound precisely because of its amplification of intelligence. Intelligence is inherently impossible to control, so the various strategies that have been devised to control nanotechnology...won't work for strong AI. There have been discussions and proposals to guide AI development toward what Eliezer Yudkowsky calls "friendly AI" [*of which we will write in the section devoted to Yudkowsky, MG*]. These are useful for discussion, but it is infeasible today to devise strategies that will absolutely ensure that future AI embodies human ethics and values. (Kurzweil, 2005, 409)
>
> Strong AI...will be deeply integrated into our civilization's infrastructure. Indeed, it will be intimately embedded in our bodies and brains. As such it will reflect our values because it will be us. (*ibid.*, 420)

Well, that's about it. Kurzweil wrote many more words about the dangers of GNR, but I find it difficult to extract from his writing more than a recommendation to "realize the benefits while managing the dangers" — good advice, but not very specific. And then again, who is giving us better advice? I am afraid that, short of outright rejection (relinquishment), which is not practical, there is no better advice anywhere. We will see what others have to say about this issue in the remainder of this chapter.

Between 2005 and the publication of Kurzweil's latest book in 2024, there was one story I wanted to tell you about. It concerns his behavior in connection with two open letters from the *Future of Life Institute* (FLI) (which we will discuss in detail in the section on Max Tegmark). What happened perfectly reflects Kurzweil's general position.

When the first soft letter appeared in 2015, it called for caution but did not demand anything specific, and Ray Kurzweil signed it. But when the second one appeared in 2023, demanding a six-month complete halt on LLM research, Kurzweil did not sign the letter. He later stated that he could not sign a letter that called for a halt or even a slowdown in research. How could he, if all the benefits he eagerly expected from AI would be delayed? And what about immortality? (The last two sentences I wrote; Kurzweil did not say them.)

The latest book by Ray Kurzweil, *The Singularity Is Nearer: When We Merge with AI*, was published in 2024, i.e., after the release of GPT-4. In it, the author does not "reinvent the wheel." He simply admires the old wheels he made in 1999 and 2005, finding them perfect in every way.

The date for passing the Turing test and the appearance of AGI, which Kurzweil has set for 2029, seems to be on schedule. He reaffirms that the singularity itself will take place around 2045. By expecting ASI some 16 years after AGI, Kurzweil places himself into the slow, "soft takeoff" camp, as opposed to "hard takeoff," whereby a transition from AGI to ASI will happen in a matter of minutes to months.

I will not attempt to review the book, which contains, among other themes, a chapter detailing the positive effects of technology on the human condition, entitled "Life is Getting Exponentially Better." At the end of this chapter, Kurzweil concludes that computation is the core driver of progress over the past two centuries. He says:

> Exponentially improving information technology is a rising tide that lifts all the boats of the human condition. And we are now about to enter the period when this tide surges upward as never before. The key to this is artificial intelligence, which is now allowing us to turn many kinds of linearly advancing technology into exponential information technology … This force is what will make life itself exponentially better in the time ahead. (Kurzweil, 2024, 194)

Let us see, however, what Ray Kurzweil says in 2024 about AI safety:

> …superintelligent AI entails a fundamentally different kind of peril — in fact, the primary peril. If AI is smarter than its human creators, it could potentially find a way around any precautionary measures that have been put in place. There is no general strategy that can definitively overcome that. There are three broad categories of peril from superintelligent AI…*Misuse* encompasses cases where the AI functions as its human operators intend, but those operators deploy it to deliberately cause harm to others. … Next is *outer misalignment*, which refers to cases where there is a mismatch between the programmers' actual intentions and the goals they teach the AI in hopes of achieving them. This is the classic problem depicted in stories about genies — it's hard to specify exactly what you want to someone who takes your commands literally. … Finally, *inner misalignment* occurs when the methods the AI

> learns to achieve its goal produce undesirable behavior at least in some cases…There is a field of technical research that is actively seeking ways to prevent both kinds of AI misalignment. There are many promising theoretical approaches, though much work remains to be done…And so, while the AI alignment problem will be very hard to solve, we will not have to solve it on our own — with the right technologies, we can use AI itself to dramatically augment our own alignment capabilities. (Kurzweil, 2024, pp.278–279)

Being a staunch optimist — which is not a sin — Kurzweil precedes the chapter about the technological perils with the following hopeful declaration:

> There is good reason to believe that our civilization will overcome these perils — not because the threats aren't real but precisely because the stakes are so high. Not only does peril bring out the best of human ingenuity, but the same technological fields that give rise to danger are also creating powerful new tools to protect against it. (*ibid.*, 267–268)

Let me end on this optimistic note and wish for all of us that Kurzweil turns out to be right about everything.

Eliezer Yudkowsky [*activity from 2000 to the present*]

AI has no more implacable and sharp opponent than Eliezer Yudkowsky. He is the quintessential *voice crying in the wilderness*, convinced that humanity's present course with AI leads straight to catastrophe. The book, written by Yudkovsky and his co-author, Nate Soares, released in September 2025, bears a telling title: *If Anyone Builds It, Everyone Dies: Why Superhuman AI Would Kill Us All.*

Yudkowsky holds no formal academic degree, and he did not attend high school; he is an *autodidact* — someone who is self-educated without the guidance of teachers. Anyone who has listened to his talks and interviews will likely agree that he possesses a sharp, philosophically inclined mind, and he appears well-read in a wide range of subjects. For example, in an interview with Lex Fridman, Yudkowsky discusses evolutionary biology with the same confidence with which he discusses sociology and the science of artificial neural networks. He is very eloquent, although his speech is often verbose and intricate. The man certainly *thinks*, as befits a philosopher. He is unusually ardent in discussions, especially when it comes to his central theme. But it must be said that the central theme is about the destruction or salvation of humankind; so perhaps we can forgive Yudkowsky his ardor?

Listening to Yudkowsky, I see, first of all, a philosopher-futurologist, perhaps with a touch of biblical prophet, who is offended by the fact that he is not being listened to and understood.

Yudkowsky has become a leading theorist, on the philosophical side rather than the technical side, of *AI safety* and *alignment.*[16]. Partly rationalist[17] essayist, partly futurist polemicist, he helped launch the modern alignment discourse as soon as "strong" AI began to look technically plausible. His position is stark: Unless we can be essentially 100 percent certain that a superintelligent system will remain under human control, we must halt capability research and devote all resources to solving this existential puzzle.

Yudkowsky's hopes of creating AI that would be safe for humankind diminished over the years. Here is a brief historical outline of his efforts.

In 2000, twenty-one-year-old Yudkowsky co-founded the *Singularity Institute for Artificial Intelligence* (SIAI), which was later renamed the *Machine Intelligence Research Institute* (MIRI), a nonprofit organization dedicated to ensuring the safe and beneficial development of artificial general intelligence. MIRI became one of the first organizations to explicitly focus on *AI alignment*, well before the issue entered mainstream AI discussions.

Yudkovsky's 2001 online essay "Creating Friendly[18] AI" laid out the *Friendly AI* concept — the idea that future powerful AIs must be explicitly designed to promote human well-being. This extended philosophical essay was taken seriously by many, including fellow philosopher Nick Bostrom, as well as AI experts Stuart Russell and Peter Norvig, the authors of an influential textbook on AI.

---

[16]The term "alignment," although not invented by Yudkowsky, was certainly popularized by him. It was adopted by the AI-developing companies since the mid-2010s.

[17]*Rational philosophy* is a modern tradition of reasoning that focuses on clarity, logical consistency, and evidence-based belief formation. Popularized by communities like *LessWrong*, it is based on principles from Bayesian probability, decision theory, and cognitive science. It is frequently applied to fields such as ethics, epistemology, and AI safety.

[18]The term "friendly" was later replaced by the term "alignment."

Between 2006 and 2009, Yudkowsky posted hundreds of essays on the blog *LessWrong*; these essays were later compiled into the 2015 e-book *Rationality: From AI to Zombies*.

He has argued, often in stark terms, that misaligned superintelligent AI poses an existential threat to humanity and that we might get only *one chance to get it right* — a thought that has become one of Yudkowsky's constant refrains. His view is that aligning AGI with human values is a difficult and unresolved problem that must be explicitly addressed and that developing AGI without first solving the alignment problem is tantamount to collective suicide.

Eliezer Yudkowsky's thinking on whether superintelligent machines can, in principle, be made safe has moved from guarded optimism to near-total pessimism over roughly 25 years.

In 2001, in his essay "Creating Friendly AI 1.0," he argued that a "seed AI" could be endowed with a provable "Friendliness" architecture. This architecture must be resilient enough to survive the system's self-improvement — the *recursive self-improvement* that was first predicted by I. J. Good in 1965. Yudkowsky urged researchers not to "give up hope too soon," insisting that a sufficiently well-designed artificial mind could learn human values in all their complexity and preserve them through every rewrite of its source code (Yudkowsky, 2001).

Later, Yudkowsky said about "Friendly AI":

> Here is this tendency to think that well-intentioned people create nice AIs, and badly intentioned people create evil AIs...and this is not true. It's actually a very difficult mathematical and engineering problem. (Yudkowsky, quoted in Barratt, 2013, 53)

By 2008, in the paper "Artificial Intelligence as a Positive and Negative Factor in Global Risk," the note of optimism was already mixed with alarm. Friendly AI, he wrote, was "the central technical problem of our era," and we would get only one shot at solving it. The problem still looked solvable, but the world was not taking it seriously enough.

After a decade, Yudkowsky's voice became harsher and more desperate. In 2022, he published the *LessWrong* essay "AGI Ruin: A List of Lethalities," concluding flatly that if a sufficiently powerful AI system were built under prevailing research norms, "this will kill us" (Yudkowsky, 2022). That same year, on April 1 (I wonder if that date held any hidden meaning), *LessWrong* posted Yudkowsky's essay entitled "MIRI announces new 'Death With Dignity' strategy." It is, I must say, a strange document that probably diminished Yudkowsky's chances of convincing the world to act. The essay opens with the passage,

> It's obvious at this point that humanity isn't going to solve the alignment problem, or even try very hard, or even go out with much of a fight. Since survival is unattainable, we should shift the focus of our efforts to helping humanity die with slightly more dignity.

Yudkowsky continues in that hysterical tone, which at times borders on insanity. But this terrible document, I must say, does not irritate but instead evokes sympathy. The reader understands that Yudkowsky is in despair, and that is why he writes this way, no longer caring about anything. And this despair makes one think: Is it possible that he is right, and has the world gone mad? Or rather, not the world — the world does not know about this at all — but the "little world" of AI specialists and their financiers, who can no longer step aside, take a sober look at their business, and, most importantly, give up their quite decent incomes and salaries.

On March 29, 2023, Yudkowsky posted an op-ed in *TIME* magazine titled "Shut It All Down." It was his immediate reaction to the *Future of Life Institute's* (FLI) open letter, published on March 22 on their website, which called for a six-month moratorium on advanced AI research. We will discuss the FLI letter in the section about Max Tegmark.

But here, a rather extensive quote from Yudkowsky's op-ed:

> An open letter published today calls for "all AI labs to immediately pause for at least 6 months the training of AI systems more powerful than GPT-4." This 6-month moratorium would be better than no moratorium. I have respect for everyone who stepped up and signed it. It's an improvement on the margin. I refrained from signing because I think the letter is understating the seriousness of the situation and asking for too little to solve it.
>
> ...
>
> Many researchers steeped in these issues, including myself, expect that the most likely result of building a superhumanly smart AI, under anything remotely like the current circumstances, is that literally everyone on Earth will die. Not as in "maybe possibly some remote chance,"

but as in "that is the obvious thing that would happen." It's not that you can't, in principle, survive creating something much smarter than you; it's that it would require precision and preparation and new scientific insights, and probably not having AI systems composed of giant inscrutable arrays of fractional numbers.[19]

...

To visualize a hostile superhuman AI, don't imagine a lifeless book-smart thinker dwelling inside the internet and sending ill-intentioned emails. Visualize an entire alien civilization, thinking at millions of times human speeds, initially confined to computers — in a world of creatures that are, from its perspective, very stupid and very slow. A sufficiently intelligent AI won't stay confined to computers for long. In today's world you can email DNA strings to laboratories that will produce proteins on demand, allowing an AI initially confined to the internet to build artificial life forms or bootstrap straight to postbiological molecular manufacturing.

If somebody builds a too-powerful AI, under present conditions, I expect that every single member of the human species and all biological life on Earth dies shortly thereafter.

There's no *proposed plan* for how we could do any such thing and survive. OpenAI's openly declared intention is to make some future AI do our AI alignment homework. Just hearing that *this is the plan* ought to be enough to get any sensible person to panic. The other leading AI lab, DeepMind, has no plan at all.

An aside: None of this danger depends on whether or not AIs are or can be conscious[20]... If you can't be sure whether you're creating a self-aware AI, this is alarming not just because of the moral implications of the "self-aware" part, but because being unsure means you have no idea what you are doing and that is dangerous and you should stop.

It took more than 60 years between when the notion of Artificial Intelligence was first proposed and studied, and for us to reach today's capabilities. Solving *safety* of superhuman intelligence — not perfect

---

[19]Yudkowsky means deep ANNs with their weights, of course.

[20]In that, the majority of AI researchers seem to agree with Yudkowsky. Some scientists, notably Roger Penrose, insist that consciousness is a prerequisite for strong intelligence, but for the most part, people do not share this view. I, for whatever it's worth, do not see the connection. I will go even further and say that most philosophical discussions of consciousness are a waste of time.

> safety, safety in the sense of "not killing literally everyone" — could very reasonably take at least half that long. And the thing about trying this with superhuman intelligence is that if you get that wrong on the first try, you do not get to learn from your mistakes, because you are dead. Humanity does not learn from the mistake and dust itself off and try again, as in other challenges we've overcome in our history, because we are all gone.
>
> Trying to get *anything* right on the first really critical try is an extraordinary ask, in science and in engineering. We are not coming in with anything like the approach that would be required to do it successfully. If we held anything in the nascent field of Artificial General Intelligence to the lesser standards of engineering rigor that apply to a bridge meant to carry a couple of thousand cars, the entire field would be shut down tomorrow.
>
> …
>
> Many researchers working on these systems think that we're plunging toward a catastrophe, with more of them daring to say it in private than in public; but they think that they can't unilaterally stop the forward plunge, that others will go on even if they personally quit their jobs. And so they all think they might as well keep going. This is a stupid state of affairs, and an undignified way for Earth to die, and the rest of humanity ought to step in at this point and help the industry solve its collective action problem.
>
> …
>
> Shut down all the large GPU clusters (the large computer farms where the most powerful AIs are refined). Shut down all the large training runs. Put a ceiling on how much computing power anyone is allowed to use in training an AI system, and move it downward over the coming years to compensate for more efficient training algorithms. No exceptions for governments and militaries. Make immediate multinational agreements to prevent the prohibited activities from moving elsewhere. Track all GPUs sold. If intelligence says that a country outside the agreement is building a GPU cluster, be less scared of a shooting conflict between nations than of the moratorium being violated; *be willing to destroy a rogue datacenter by airstrike* [italics ours]. (Yudkowsky, 2023).

This last proposal — to bomb data centers if necessary — was never forgotten or forgiven by Yudkowsky's numerous opponents. They savor this truly amazing proposal every time they want to remind people of Yudkowsky's radicalism.

I think the phrasing of that 2023 TIME op-ed was good and clear, and adding more quotes from Yudkowsky's numerous interviews and podcast appearances would not add much to it.

Listening to Yudkowsky's talks, for example, his 2023 TED Talk, I see a profoundly disappointed man mourning the futility of his two decades of fighting the uncontrolled development of superintelligent AI. He seemed to have lost all hope for the "good outcome." Now, he simply states that "AI will kill us all" and thanks people for coming to listen to his talk. The last hope he has is that, should a disaster strike — and he does not doubt it will — it will be a partial rather than total disaster, sparing at least part of humanity. This partial disaster, then, might serve as a wake-up call, the one that he tried to sound, but was not heard.

Nick Bostrom [*activity from 2001 to the present*]

Nick Bostrom, a Swedish philosopher based at the University of Oxford, emerged in the early 2000s as one of the first scholars to raise serious, systematic concerns about the long-term implications of advanced AI. While his 2014 book, *Superintelligence*: *Paths, Dangers, Strategies* (Bostrom, 2014) brought these concerns to a broad audience, his warnings date back more than a decade earlier. In that, he is on par with Eliezer Yudkowsky.

In 2001, Bostrom published a pivotal paper titled "Existential Risks: Analyzing Human Extinction Scenarios and Related Hazards" in the *Journal of Evolution and Technology*. There, he introduced the term *existential risk* — threats that could cause human extinction — and identified superintelligent AI as a prominent and underappreciated example. He emphasized that such risks merited urgent philosophical and scientific scrutiny.

By 2003, he had published "Ethical Issues in Advanced Artificial Intelligence," an essay that explored the challenges of ensuring that a superintelligent AI's goals remain aligned with human values. He wrote starkly:

> Once unfriendly superintelligence exists, it would prevent us from replacing it or changing its preferences. Our fate would be sealed.

It was in this same 2003 essay that Bostrom introduced what later became one of the most famous parables in the AI-safety literature: the *paperclip maximizer*. He imagined a superintelligent system given a single, seemingly harmless objective — to manufacture as many paperclips as possible. The system has no malice; it simply pursues its goal with relentless efficiency. Yet, a sufficiently powerful optimizer, unconstrained by human values, might conclude that the best way to achieve its goal is to convert more and more matter — first factories, then resources, then ultimately the entire Earth and beyond — into paperclips, or into machinery that produces them. This deliberately absurd scenario was meant to illustrate a serious point: Even an AI with an innocuous goal could become existentially dangerous if its objectives are even slightly misaligned with ours. The "paperclip maximizer" quickly became a widely cited model in AI safety discussions, repeatedly invoked by researchers from the late 2000s onward.

In 2005, Bostrom co-founded the *Future of Humanity Institute* (FHI) at the University of Oxford, where he continued to develop theoretical frameworks for understanding the risks posed by transformative technologies, particularly *artificial general intelligence* (AGI). The FHI has become a leading center for research on long-term AI safety.

Through the late 2000s, Bostrom published numerous papers and reports on technological forecasting, AGI alignment, and global catastrophic risk. His views gained increasing traction within both academic and policy circles.

By the time *Superintelligence* appeared in 2014, offering a synthesized account of his work and calling for serious global attention to the alignment problem, Bostrom had already spent over a decade laying the intellectual foundations for AI risk awareness.

Bostrom's 2014 book sparked public imagination and was widely discussed. I tried to read it — it contains a detailed and painstaking analysis of the possible scenarios for the future development of AI and how it might go awry. The author patiently explains how a more intelligent entity might manipulate and deceive a less intelligent one. I cannot say whether the popularity of this book arose because of its methodical meticulousness or in spite of it. But it did. Thanks to Bostrom, 2014 became the year in which society finally paid attention to the fact that, right "under its nose," a technology was being developed from which, if anything happened, it would be more difficult to escape than from hydrogen bombs.

Bostrom and Yudkowsky may not have been the first to closely examine the emerging AI and cry "Wolf! Wolf!", but it's safe to say that they were the first to be heard. As a result, "Institutes" and online communities began to form — organizations that set themselves the goal of saving humanity. Our heroes themselves founded some of them, while others arose under their influence. All in all, it was a big deal. Here is a list, perhaps incomplete, of these organizations that have emerged since 2000 in chronological order:

1. Machine Intelligence Research Institute (MIRI)
Founded: In 2000 in Berkeley, California (originally as the *Singularity Institute for Artificial Intelligence*).
Co-founders: Eliezer Yudkowsky, and Brian and Sabine Atkins
Focus: Early and central advocate of *Friendly AI* and AI alignment. MIRI pursues theoretical research into provably safe AI architectures.
Style: Technical, mathematical, and cautious, emphasizing "alignment before capability."

2. Future of Humanity Institute (FHI)
Founded: In 2005 at the University of Oxford (Shut down in 2024).
Founder & Director: Nick Bostrom.
Focus: Long-term impacts of advanced technologies, especially AI, biorisk, and existential threats.

3. LessWrong Community
Founded: In 2006 as a blog by Eliezer Yudkowsky, it evolved into a full platform in 2009.
Focus: Rationality, Bayesian reasoning, and AI safety.
Impact: Incubator for many alignment researchers and several spin-off organizations.

4. Future of Life Institute (FLI)
Founded: In 2014 in Cambridge, Massachusetts.
Founders include: Max Tegmark, Anthony Aguirre, and Jaan Tallinn.
Mission: Mitigating existential risks from transformative technologies, especially AI.

5. Center for Human-Compatible AI (CHAI)
Founded: In 2016, UC Berkeley.
Director: Stuart Russell.
Focus: Technical AI alignment and safe learning systems.

6. Centre for the Governance of AI (GovAI)
Founded: 2018 (as part of FHI, later became semi-independent).
Founder: Allan Dafoe, supported by Bostrom's team.
Focus: AI policy, global coordination, governance challenges in powerful AI systems.
Status: Now, an independent organization based in London.

Roman Yampolskiy [*activity from 2008 to the present*]

Roman Yampolskiy is one of the foremost authorities on *AI safety*; in fact, he coined this term in 2010–2011. He is a professor in the Department of Computer Engineering and Computer Science at the University of Louisville, where, in 2008, he founded the Cyber Security Lab. Throughout his career, Yampolskiy has published over 200 research papers and five books, primarily on the topics of AI safety and security. While Yampolskiy is the best expert on this topic, and perhaps precisely for that reason, he is deeply pessimistic about AI safety prospects. He gave his last book a telling title, *AI*: *Unexplainable, Unpredictable, Uncontrollable* (Yampolskiy, 2024). This title summarizes his conclusions, and we (by "we," I mean AI amateurs like myself) could stop right here, but in the interest of detailing his thoughts, let us provide a few quotes.

At the outset, Yampolskiy states that even the principal possibility of controlling superintelligent AI remains an open question:

> …the underlying assumption in this research [*on AI safety and security*] is that the problem of controlling highly capable intelligent machines is solvable, though no rigorous mathematical proof or argumentation has been presented to demonstrate that the AI control problem is solvable in principle, let alone in practice…Despite the recognition that the problem of AI control may be one of the most important problems facing humanity, it remains poorly understood, poorly defined, and poorly researched. (Yampolskiy, 2024, 1)

Assessing the state of AI safety research, Yampolskiy finds it to be drastically insufficient; he concludes that there is, at best, a 9% chance that the first AGI will be developed:

> ...with safety in mind from the ground up. Worse yet, most AI researchers are not well-read on AI safety literature and many are actually AI Risk skeptics.... (*ibid.*, 88)

Yampolskiy fully understands the grim existential threat superintelligent AI may pose to humanity and what AI safety means in that context. He says:

> With respect to possible existential impacts, there is no such thing as partial safety...the negative utility of such an [*hostile*] AI would be astronomical...the chances of a misaligned AI are not small...in the absence of an effective safety program, that is the only outcome we will get...This is not a low-risk, high-reward scenario, but a high-risk, negative-reward situation...proof of the solvability or non-solvability of the AI control problem would be the most important proof ever. (*ibid.*, 2, 3)

Yampolskiy points to endless problems with controlling superintelligent AI:

> ...as we try to maintain control while designing advanced intelligent agents, we find ourselves in a Catch-22, since the control mechanism needed to maintain control must be smarter or at least as smart as the agent over which we want to maintain control.
>
> We note that historical attempts at software governance (e.g., spam, computer viruses, and deepfakes) had only a very limited amount of success. With AGI as an independent agent, it may be ungovernable because traditional methods of assigning responsibility and punishment-based enforcement are not applicable to software. (*ibid.*, 4, 6)
>
> Less intelligent agents (people) can't permanently control more intelligent agents (ASIs [*Artificial Super Intelligences*]). This is not because we may fail to find a safe design for superintelligence in the vast space of all possible designs; it is because no such design is possible, it does not exist. Superintelligence is not rebelling; it is uncontrollable to begin with. (*ibid.*, 109–110)

> An infinite number of pathways exist to a state of the world in which a dangerous AI is unleashed. Those include mistakes in design, programming, training data, value alignment, self-improvement, environmental impact, safety mechanisms, and, of course, intentional design of malevolent AI (MAI). In fact, MAI presents the strongest, some may say, undeniable argument against AI Risk skepticism.... (*ibid.*, 217)
>
> ...AI Risk skeptics need to realize that the burden of proof is not on AI Safety researchers to show that technology may be dangerous but on AI developers to establish that their technology is safe at the time of deployment and throughout its lifetime of operation. (*ibid.*, 234)

The 250-page book comprises 13 chapters, each containing references to Yampolskiy's work as well as those of other researchers. In total, there are 1,341 references, although some may be repetitive due to their separation by chapters. The book contains definitions of many relevant terms and discusses the issues thoughtfully and at some length, providing arguments for both the pros and cons while maintaining a calm, scientific tone. The book incorporates elements of mathematical logic, computer science, and philosophy. *AI: Unexplainable, Unpredictable, Uncontrollable* is a testament to Professor Yampolskiy's comprehensive knowledge of his subject. His call to devote more resources to AI safety research is admirable. After convincing us that artificial superintelligence (ASI) is fundamentally uncontrollable, Yampolskiy says:

> ...just as incompleteness results [*of Godel and Turing*] did not reduce efforts of mathematical community...the limiting results reported in this chapter [*chapter 6*] should not serve as an excuse for AI Safety researchers to give up and surrender. Rather, it is a reason...to increase effort and funding for AI Safety and Security research. We may never get to 100% safe AI, but we can make AI safer in proportion to our efforts, which is a lot better than doing nothing. (*ibid.*, Chapter 6)

We could continue with quotations from Yampolskiy's book, but the more I turn its pages, the more I am convinced that my initial impression was correct: The title tells all, at least for an amateur such as myself. Everything else in the 250-page book attempts to dress these conclusions in computer science robes.

Yampolskiy on The Diary of a CEO, September 4, 2025

I would like to conclude this section on Yampolskiy with several quotations from his most recent appearance as of this writing (September 2025) — an interview on Steven Bartlett's podcast *The Diary of a CEO*. This interview took place a year after the publication of the book we discussed earlier. In it, Yampolskiy's statements became even harsher and darker — very similar to what we have been hearing for some time from Eliezer Yudkowsky. During their conversation, Steven Bartlett (SB) and Roman Yampolskiy (RY) touched on several topics. I will not dwell on their discussion of religion or the hypothesis that we are living in a simulation, or on bitcoin. Instead, I will quote several passages that concern AI safety and the existential threat posed by AI, as well as Yampolskiy's thoughts about the future labor market:

> SB: … how soon will we get to advanced AI?
>
> RY: The timelines are very short, a couple of years, two to three years, according to… CEOs of top labs. And at the same time, we don't know how to make sure that those systems are aligned with our preferences. So, we are creating this alien intelligence. If aliens were coming to Earth, and you had three years to prepare, you would be panicking right now. But most people don't even realize this is happening.
>
> SB: So, some of the counterarguments might be… these are very big companies… They have… a moral obligation, but also… a legal obligation, to make sure they do no harm…
>
> RY: The only obligation they have is to make money for the investors. That's the legal obligation they have. They have no moral or ethical obligations. Also, according to them, they don't know how to do it yet. The state-of-the-art answers are: We'll figure it out when we get there, or AI will help us control more advanced AI. That's insane.[21]
>
> SB: In terms of probability, what do you think is the probability that something goes catastrophically wrong?
>
> RY: Nobody can tell you for sure what's going to happen. But if you're not in charge, you're not controlling it, you will not get the outcomes

[21] We saw an identical reaction from Eliezer Yudkowsky.

you want. The space of possibilities is almost infinite. The space of outcomes we would like is tiny…

SB: Was there a particular moment in your career where you thought — oh my god?

RY: First 5 years… I was working on solving this problem. I was convinced we could make this happen. We can make safe AI, and that was the goal. But the more I looked at it, the more I realized every single component of that equation is not something we can actually do. And the more you zoom in — it's like a fractal — you go in and you find 10 more problems and then 100 more problems. And all of them are not just difficult — they're impossible to solve… There are patches. There are little fixes we put in place, and quickly people find ways to work around them… So while progress in AI capabilities is exponential, or maybe even hyperexponential, progress in AI safety is linear or constant. The gap is increasing. The gap between how capable the systems are and how well we can control them, predict what they're going to do, explain their decision-making…

SB: And where are we?

RY: …We definitely have many excellent narrow systems… and they are super intelligent in that narrow domain. Protein folding is a problem which was solved using narrow AI, and it's superior to all humans in that domain… Now we don't have super intelligence yet. We still have brilliant humans who are completely dominating AI, especially in science and engineering. But that gap is closing so fast. You can see, especially in the domain of mathematics, that 3 years ago, large language models couldn't do basic algebra; multiplying three-digit numbers was a challenge. Now, they are helping with mathematical proofs; they are winning mathematics Olympiad competitions; they are working on solving millennial problems — the hardest problems in mathematics. So, in 3 years, we closed the gap from subhuman performance to better than most mathematicians. And we see the same process happening in science and in engineering.

SB: What is your prediction for the year 2027?

RY: We're probably looking at AGI as predicted by prediction makers and tops of the labs… So… you have free labor, physical and cognitive, trillions of dollars of it. It makes no sense to hire humans for most jobs.

If I can just get, you know, a $20 subscription or a free model to do what an employee does. First, anything on a computer will be automated. And next, I think humanoid robots are maybe 5 years behind. So, in five years [*after 2027*] all the physical labor can also be automated. So, we're looking at a world where we have levels of unemployment we have never seen before. Not talking about 10% unemployment, which is scary, but 99%. All you have left are jobs where, for whatever reason, you prefer another human to do it for you. But anything else can be fully automated. It doesn't mean it will be automated in practice... So, we may have a lot more time with jobs... But the capability to replace most humans and most occupations will come very quickly... What happens to society? How does that impact crime rate, pregnancy rate,[22] all sorts of issues? Governments don't have programs prepared to deal with 99% unemployment...

SB: What's your prediction for 2030? What will the world look like?

RY: We probably will have humanoid robots with enough flexibility and dexterity to compete with humans in all domains, including plumbers[23]... So many of... the leading companies, including Tesla, are developing humanoid robots at light speed, and they're getting increasingly more effective. And these humanoid robots will be able to... do anything humans can do, but obviously have to be connected to AI as well... Our world will look remarkably different when humanoid robots are functional and effective, because... the combination of intelligence and physical ability really doesn't leave much... for us human beings...

SB: And then by 2045?

RY: ... Ray Kurtzweil predicts that that's the year for the singularity. That's the year where progress becomes so fast... AI is doing science and engineering... so quickly [*that*] we cannot keep up anymore. That's the definition of singularity, the point beyond which we cannot see,

---

[22] Recall how in Karel Čapek's play *R.U.R.*, his characters discuss how the replacement of all human jobs with robots led to women no longer having babies.

[23] Plumbers were mentioned by Geoffrey Hinton as a profession likely to survive automation.

understand, predict… [*and eventually*] you have zero knowledge, zero understanding of the world around you…

SB: …every time we have a conversation on the show [*podcast*] about AI, someone says, "Can't we just unplug it?"

RY: …This is silly. Those are distributed systems. You cannot turn them off. And on top of it, they're smarter than you. They made multiple backups. They predicted what you're going to do. They will turn you off before you can turn them off… What we have today — humans with AI tools are dangerous. They can be hackers, malevolent actors. Absolutely. But the moment super intelligence becomes smarter, dominates, they [*humans*] are no longer the important part of that equation…

SB: What can be done to turn this ship to a more certain positive outcome at this point? Are there still things that we can do, or is it too late?

RY: I believe in personal self-interest. If people realize that doing this thing is really bad for them personally, they will not do it. So, our job is to convince everyone with any power in this space — [*those*] creating this technology, working for those companies — [*that*] they are doing something very bad for themselves… forget the 8 billion people you are experimenting on with no permission, no consent. You will not be happy with the outcome… It's not just me saying it. You had Jeff Hinton, Nobel Prize winner, founder of a whole machine learning space. He says the same thing. Benjio, dozens of others; top scholars. We had a statement about the dangers of AI signed by thousands of scholars, computer scientists… And we need to make it universal. No one should disagree with this. And then we may actually make good decisions about what technology to build. It doesn't guarantee long-term safety for humanity, but it means we're not trying to get there as soon as possible, to the worst possible outcome.

SB: And are you hopeful that that's even possible?

RY: I want to try. We have no choice but to try.

SB: And what would need to happen, and who would need to act? What is it — government legislation?

RY: Unfortunately, I don't think making it illegal is sufficient. There are different jurisdictions. There are, you know, loopholes. And what are you going to do if somebody does it? [*Are*] you going to fine them for destroying humanity? Like — very steep fines for it? What are you going to do? It's not enforceable. If they do create it — now the superintelligence is in charge. The judicial system we have is not impactful. And all the punishments we have are designed for punishing humans. Prisons, capital punishment doesn't apply to AI... If you look at the history of AI safety organizations or departments within companies, they usually start well, very ambitious, and then they fail and disappear. OpenAI had a superintelligence alignment team. The day they announced it, I think they said: We're going to solve it in 4 years. Like, half a year later, they canceled the team. And there are dozens of similar examples. Creating perfect safety for super intelligence, perpetual safety — as it keeps improving, modifying, interacting with people — you're never going to get there. It's impossible... If we know that it's impossible to make it right, to make it safe, then this direct path of just building it as soon as you can becomes a suicide mission. Hopefully, fewer people will pursue that... they may go in other directions... build useful tools; stop building agents; build narrow superintelligence, not a general one.

SB: Should people be protesting?

RY: There are people protesting. There is "stop AI," there is "pause AI." They block offices of OpenAI. They do it weekly, monthly, quite a few actions, and they're recruiting new people... If you can get it to a large enough scale, to where majority of population is participating, it would be impactful (Yampolskiy, 2025).

Stuart Russell [*activity from 2013 to the present*]

Stuart Russell is best known as the co-author of *Artificial Intelligence: A Modern Approach* (Russel and Norvig, 2010). This definitive textbook trained generations of students in the foundations of AI. But in the early 2010s, Russell's public role began to shift from that of a pedagogue and theorist to a leading voice of concern over the trajectory of AI development, particularly the risks posed by superintelligent systems and autonomous weapons. The turning point came in 2013, when Russell was invited to speak at a United Nations meeting on the topic of lethal autonomous weapons.

Campaigns to Stop Killer Robots

In the early 2010s, growing concern about lethal autonomous weapons systems — soon to be dubbed "killer robots" — galvanized a wave of international activism. The most prominent initiative began in April 2013, when Jody Williams, a Nobel Peace Prize laureate for her work against landmines, helped launch the *Campaign to Stop Killer Robots*. Coordinated by *Human Rights Watch* and allied non-governmental organizations (NGOs), the campaign called for a preemptive ban on fully autonomous weapons that could select and engage targets without meaningful human control.

The campaign's message was simple: *The decision to take a human life should never be delegated to a machine.* Within months, it had sparked discussions at the *United Nations Convention on Certain Conventional Weapons* (CCW), where the idea of "meaningful human control" became a focal point of debate. Although no binding international agreement was reached, the conversation had begun.

Stuart Russell, meanwhile, became one of the strongest voices against autonomous weapons within the AI research community. He lent his authority and clarity to the issue in public talks and op-eds. In 2017, he co-produced and narrated the short film "Slaughterbots," a dramatization of swarms of autonomous drones assassinating targets with chilling efficiency.

The Future of Life Institute (FLI) also took up the issue. Its widely circulated 2015 open letter explicitly named autonomous weapons as a grave near-term threat and called for an international ban. Russell was a signatory.

By the mid-2020s, the campaign had grown into a coalition of over 180 NGOs worldwide. While international negotiations continue, progress has been slow. Several dozen countries now support a treaty, but predictably, the leading military powers remain unshaken. The recent success of armed drones in the Russian-Ukrainian war has probably secured the place of autonomous weapons in future warfare.

In 2015, Russell was instrumental in drafting and signing the FLI's open letter on AI safety, which we discuss in the section devoted to Max Tegmark.

Over the next few years, Russell developed a technically detailed alternative to what he called the "standard model" of AI development, namely, the design of systems that pursue fixed, human-defined objectives. In his 2019 book *Human Compatible*, Russell laid out his central claim: Building ever-more capable machines with *fixed goals* is a recipe for disaster. "*If we put the wrong objective into a machine that is more intelligent than us*," he wrote, "*it will achieve the objective — and we lose.*"[24] Instead, he proposed a new framework for AI, based on three key principles: (1) machines should be designed solely to satisfy human preferences; (2) they must remain uncertain about those preferences; and (3) they should learn about human values through observation and interaction. This embrace of uncertainty was Russell's proposed antidote to the dangers of rigid, goal-maximizing machines.[25] Being a real, hands-on AI scientist, Russel has consistently offered concrete technical proposals. His work on *inverse reinforcement learning*, *corrigibility*, and *verifiable safety* reflects his belief that alignment is, above all, an engineering problem — and one that must be solved before capabilities outpace our ability to control them. I will note parenthetically that neither those who requested a pause nor those who focused on a technical solution to the safety problem achieved success. No one wanted to hear about a pause in industry, and alas, no technical solutions to this unsolvable problem of AGI safety were found. But let's continue discussing Stuart Russell.

By 2023, as the public debate intensified following the release of GPT-4, Russell became more visible in policy and media circles. In an interview with *The New Yorker*, he said, "*It's almost like you're deliberately inviting aliens from outer space ... having no idea what they're going to do when they get here, except that they're going to take over the world.*" Here, Russell used an "alien" metaphor that occurred to many, for example, to Yuval Harari and Roman Yampolskiy. In 2025, at the *AI Action Summit* in Paris, Russell reiterated that progress and caution need not be at odds. "*Innovation cannot exist without safety*," he said, calling for global minimum safety standards before "*a major disaster*" forces the issue. His view, consistently expressed, is that a safe and beneficial AI future is possible — but only if we fundamentally redesign the way machines learn and pursue goals.

---

[24] In that, Russel seemed to follow the *paperclip maximizer* parable, which, as we wrote in the section about Nick Bostrom, was introduced by him in 2003.

[25] I must say that this recipe does not strike me as effective or useful.

What distinguishes Russell among AI safety advocates is his technical depth, his measured tone, and his position within the academic mainstream. As the author of the world's most widely used AI textbook, he carries unusual authority. He speaks not from a position of abstract philosophical worry but from decades of research and teaching within the field itself. His approach rejects fatalism and instead insists that we still have a chance to build intelligent machines that remain human-compatible.

If Kurzweil is a techno-optimist and Yudkowsky a pessimist, then what is Russell? Most likely, he is a pessimist who hasn't lost hope — or perhaps a pessimist with his sleeves rolled up.

Max Tegmark [*activity from 2014 to the present*]

Max Tegmark is an American physicist, cosmologist, and prominent advocate for the safe and ethical development of AI.

Born in 1967 in Sweden, Tegmark earned his Ph.D. in physics from the University of California, Berkeley, in 1994. He is now a professor at MIT, where his research has ranged from *precision cosmology* (the detailed study of the cosmic microwave background radiation) to quantum mechanics and the mathematical structure of the universe.

Starting around 2014, Tegmark turned his attention to the societal and existential implications of advanced technologies, particularly AI. That same year, he co-founded the *Future of Life Institute* (FLI). Tegmark described the events surrounding the birth of FLI in his 2017 book, *Life 3.0*: *Being Human in the Age of Artificial Intelligence*. Besides Max Tegmark, the people most closely involved with the creation of this Institute were his wife, Meia, and their close friend and fellow theoretical cosmologist, Anthony Aguirre. The initial funding for FLI was provided by Max's friend, Estonian billionaire and computer programmer Jaan Tallinn, co-founder of Skype. Later that same year, Elon Musk joined them as the Institute's advisor and significant financial backer.

"Armed" with the Institute, Tegmark helped organize the 2015 Puerto Rico AI conference, which brought together top AI researchers to discuss the safety and development of beneficial AI. The conference produced the *2015 open letter* on AI safety, which called for robust research into aligning AI systems with human values and was signed by over 1,000 prominent individuals, including Stephen Hawking, Elon Musk, Stuart Russell, Demis Hassabis, and Geoffrey Hinton. This was the first high-profile, coordinated, and broadly endorsed statement from across the AI and scientific community regarding that subject.

However, the development of LLMs proceeded on its course, and about four years after the Puerto Rico AI conference, programs developed at OpenAI, which were unprecedented in their power, began to appear one after another. We discussed these events in the previous chapter, but given their significance in general and with respect to Max Tegmark's activities in particular, let us summarize them again here.

GPT-1, introduced in 2018, contained about 117 million parameters. It was the first demonstration that a general-purpose language model, pre-trained on a large corpus of text and then fine-tuned for specific tasks, could perform successfully without designing a specialized architecture for each problem.

GPT-2, released in February 2019, scaled up to 1.5 billion parameters (that is, more than ten times larger than GPT-1). Increasing the size of the transformer architecture led to a noticeable qualitative leap: the generated texts became longer, more coherent, and stylistically more convincing. The model was already capable of sustaining a continuous discourse, rather than merely completing isolated sentences.

GPT-3, introduced in June 2020, once again dramatically increased the number of parameters — scaling, taken to the extreme. It had 175 billion parameters, i.e., more than one hundred times larger than GPT-2 and nearly one and a half thousand times larger than GPT-1. This scaling was accompanied by a further increase in generality: the model demonstrated stable generation of long texts, stylistic flexibility, and the ability to perform a wide range of tasks without additional training — just a few examples sufficed (few-shot learning). By this point, it had become clear that scaling itself is a powerful mechanism for improving model behavior.

The launch of ChatGPT in November 2022 (based on GPT-3.5) made these capabilities widely accessible. For the first time, hundreds of millions of users gained direct interactive access to a large language model, triggering an unprecedented public response and rapid adoption of the technology in education, programming, journalism, and business.

Finally, GPT-4, released on March 14, 2023, delivered higher reliability, improved reasoning capabilities, and support for multimodal input (text and images). The quality of text generation became noticeably more robust: the number of major factual errors decreased, the structural discipline of reasoning improved, and the ability to handle long context increased. OpenAI did not disclose the exact number of model parameters; according to unofficial estimates, its total scale may be measured in trillions of parameters, likely employing a mixture-of-experts architecture, where only a subset of parameters is activated for each query.

These releases ultimately brought generative AI out of research laboratories into the center of attention of society, science, and policy.

And now, let us return to the activities of Max Tegmark and his institute.

On March 23, 2023 — that is, nine days after the release of GPT-4 — FLI published an open letter online to all AI researchers entitled "Pause Giant AI Experiments." The letter opened with an urgent appeal: "We call on all AI labs to immediately pause for at least 6 months the training of AI systems more powerful than GPT-4." Among the letter's eventual 34,000 signatories (initially fewer signed) were physicists (including some Nobel Prize laureates), computer scientists, AI experts, entrepreneurs, philosophers, writers, and politicians.

Perhaps predictably, only one name appeared in the titles of the news reports published in the wake of the letter. *Time* Magazine: "Elon Musk Signs Open Letter Urging AI Labs to Pump the Brakes." *The New York Times*: "Elon Musk and Others Call for Pause on AI, Citing 'Profound Risks to Society'," and so on. However, this is not the point; it is commendable that Elon Musk signed the letter and has strongly supported Tegmark's Institute since its inception. Journalists are, at least for now, only human.

The most revealing aspect of this story is what occurred afterward. What happened is that *not a single* AI company followed the advice in that letter and paused or even slowed down work on advanced AI. People 'in the know' about the internal tensions in the AI community explained why and how this disdain for the seemingly sensible FLI initiative arose and what those who opposed or ignored the letter were thinking. I will not try to delve into these psychological and ethical intricacies. The AI community's reaction was predictable: Individual AI developers and the AI companies they worked for would always find ways of justifying the continuation of their work. The only person who expressed the opinion that the letter *did not go far enough* was Eliezer Yudkowsky. We quoted from his *Time* op-ed above.

Following the 2023 open letter from the FLI, Max Tegmark has remained one of the most visible and articulate figures urging caution in the face of accelerating AI capabilities. But Tegmark's message, especially from late 2023 onward, took on a more pragmatic tone. Unlike many of his colleagues in the AI safety community who warned that superintelligent AGI may be inevitable and uncontrollable, Tegmark began to emphasize a subtler — and lonelier — position: that *we do not have to develop AGI at all.*

In interviews, public lectures, and op-eds, Tegmark argued that humanity retains *agency* in its technological trajectory. We can choose to build beneficial AI systems that assist with science, education, climate, and medicine, but not AGI. As he said, "*We can have an inspiring future with controllable tool AI ... if we avoid uncontrollable AGI that replaces us.*"

Tegmark isn't demanding an indefinite moratorium until alignment is completely solved (which is the position of *AI doomers*), nor is he for rushing headlong into AGI (which is what *techno-optimists* are proclaiming and actually doing). Instead, he is promoting a deliberate, selective path: accelerate the development of safe, beneficial AI while consciously avoiding the pursuit of full AGI.

I'm unsure how to evaluate Tegmark's new position. Of course, one can praise him; he offers a third way. But one can also criticize him: This third way is very unconvincing. And it seems that almost no one has joined Professor Tegmark on this path. Who is advocating a course toward limited AI and a complete ban on work toward AGI? Tegmark and his institute; perhaps Professor Yampolskiy, who expressed similar views in his interview with Bartlett. Many others have spoken in favor of constrained or directed AI, but have not called for an immediate halt to research on general, universal AI — AGI.

It seems to me — and apparently many understand this as well — that it is hardly possible to carry out such a subtle combination: to simultaneously strengthen AI's position on the chessboard while hiding the king (humanity), to speak in chess terms. Of course, in chess there is, as is well known, castling — so perhaps I am mistaken here.

We wish Max Tegmark the best of luck, but we find it hard to believe that humanity will choose this third path — to develop AI yet stop right at the "edge" of creating AGI. And what does "humanity" even mean in this context? When it comes to AI, there is no unity, and none is in sight. Every company developing AI is striving to reach AGI as quickly as possible — and they are supported in this by the governments of the leading countries, the United States and China. This is the objective they have openly declared, and hundreds of billions of dollars a year are being spent toward it — soon, it seems, the figure will reach trillions. We presented the section on global AI funding in the previous chapter. The overarching goal of all this is the fastest possible creation of AGI. The opinion of a handful of intellectuals that this is a bad goal is unlikely to be heard.

Elon Musk [*activity from 2014 to the present*]

Elon Musk is one of the s most successful entrepreneurs of our time. According to estimates published in Forbes as of March 2026, his net worth is $840–850 billion, making him, by a significant margin, the richest person in the world. Listening to him on various podcasts, interviews, and TV appearances, we get a sense of a quick-witted, confident, and broadly knowledgeable person. He seems to be equally at home with engineering questions and the history of civilization. Musk often pauses for a few moments when asked a question, evidently thinking his answer through in real time. I like that.

When it comes to technology, Musk's reputation is impeccable. When it comes to politics or his involvement with social networks, opinions differ. But we do not need to discuss those things here. Here is a summary of Musk's AI involvement over the years:

OpenAI (co-founder, late 2015–2018) was founded by Elon Musk, Sam Altman, Greg Brockman, Ilya Sutskever, and others to develop safe and beneficial AGI. Musk pledged an initial $1 billion commitment. However, due to his departure from the OpenAI board in 2018,[26] his actual contributions appear to have totaled less than $50 million.

xAI (founded 2023–present). The company was created as a competitor to OpenAI, aiming to develop alternative advanced AI models. In 2024, it raised $6 billion, and in the summer of 2025, another $10 billion. It built the Colossus data center near Memphis, Tennessee. Colossus was constructed in a record-short time and, by late 2025, housed on the order of

[26] In a Lex Fridman podcast, Musk said: "*Open* in *OpenAI* supposed to mean open source…It was created as a non-profit open source, and now it is closed source for maximum profit, which I think is not good karma."

200,000 GPUs. Plans call for scaling up to around one million GPUs. For further expansion, xAI has acquired another site in Memphis of about 100,000 square meters and is also considering construction outside the United States, possibly in Saudi Arabia.

In addition, xAI developed and launched the Grok chatbot (integrated with the X platform). In 2025, xAI and X (formerly Twitter) were effectively consolidated under common ownership, in a deal valuing X at approximately $33 billion in stock.

Neuralink (founded 2016–present). The company develops brain–computer interfaces to restore neurological function in people with severe injuries and to create a direct link between humans and AI. Musk is reported to have invested around $100 million at an early stage. In 2023, Neuralink received approval from the U.S. Food and Drug Administration (FDA) to conduct its first clinical trials, and in 2024–2025 the first patients were implanted. By late 2025, about a dozen individuals had received implants, including paralyzed patients who were able to control a computer cursor using their thoughts.

Neuralink has also announced new directions: the Blindsight project, aimed at restoring vision through direct stimulation of the visual cortex, and systems for restoring speech in people with severe communication impairments. The company's devices have received the FDA's Breakthrough Device Designation, which allows for accelerated clinical evaluation. In 2025, Neuralink raised an additional $650 million in funding and stated plans to implant tens of thousands of devices annually by the end of the decade and to open its own clinics.

Tesla and AI development Tesla initially developed its artificial intelligence capabilities to train its Autopilot systems, as discussed in Chapter 4. In 2021, the Dojo project was launched — a proprietary supercomputer for training computer vision models. However, by 2025 this project was effectively scaled back: the team was reorganized, and the emphasis shifted toward procuring and using external GPU clusters.

At the same time, the robotics division continues work on the humanoid robot Optimus. In demonstrations in 2025, it was already performing simple manual tasks — such as sorting objects — and walking much more steadily than earlier prototypes. Musk has repeatedly stated that, in the long term, humanoid robotics could become Tesla's largest business and account for a significant portion of its future revenues. The initial limited deployment of Optimus for use on Tesla's own production lines is planned for 2026.

So, Musk is doing a lot to develop AI, which is inevitable in a captain commanding a fleet of technological industries. An AI denier could not remain at the helm of Elon Musk's companies. This is on the one hand.

On the other hand, over the past decade, the same person, Elon Musk, has been loudly warning about the dangers of AI, including the existential risks it poses to humanity. He is a sampling of Elon Musk's pronouncements about AI:

> With artificial intelligence, we are summoning the demon. (Musk, 2014)
>
> AI is a fundamental risk to the existence of human civilization in a way that car accidents, airplane crashes, faulty drugs, or bad food were not. (Musk, 2017)
>
> I have exposure to the most cutting-edge AI, and I think people should be really concerned about it... AI is a rare case where I think we need to be proactive in regulation rather than be reactive… I try to convince people to slow down AI, to regulate AI. This was futile. I tried for years. Nobody listened. (Musk, 2018a)
>
> The danger of AI is much greater than the danger of nuclear warheads by a lot. Mark my words, AI is far more dangerous than nukes. (Musk, 2018b)
>
> AI will not stop. It's just going to happen. You can slow it down, you can try and understand it, but AI is not going to stop. (Musk, 2023)
>
> We're in the intelligence Big Bang right now. We're at the very, very early stage of the intelligence Big Bang. (Musk, 2025a)
>
> I resisted AI for too long. Living in denial. Now it is game on. (Musk, 2025b)

Perhaps you might think that someone who firmly believes that "AI is far more dangerous than nukes" would not be developing AI further. But of course, as we said above, the alternative for someone like Elon Musk would be to resign from any industrial and technological activity — in other words, not to live. And, even if he resigned, he understands that "…AI is not going to stop." The last statement is particularly revealing: Elon Musk finally picks a side.

At least Elon Musk does something concrete to alleviate the risk:

- In 2014, he was the first to respond to Max Tegmark's call to support the FLI, donating $10 million in 2015 for AI safety research at the institute.

- Musk has repeatedly called for proactive government regulation of the AI industry.
- In 2018, he resigned from the OpenAI board, citing concerns about the company's attitude toward AI safety.
- Musk was one of the first signatories of the FLI March 2023 letter calling for an immediate pause on advanced AI development for at least six months.
- In 2023, Musk initiated legal action against OpenAI, alleging that the company had deviated from its mission and breached its commitments. He raised concerns about the safety of OpenAI's development, particularly given the increasing capabilities of the GPT models.
- By 2026, Musk's dispute with OpenAI had evolved into a major legal and ideological battle over the future direction and governance of advanced AI.

We have repeatedly noted in this book that modern AI has two faces: one beautiful, the other scary. It seems to me that Elon Musk embodies this duality. I do not know anyone else in whom it is expressed so vividly. Perhaps this is because he occupies a leading position in AI development, while at the same time being an intelligent and sincere person who recognized the dangers inherent in this work earlier than many others. In any case, one senses that this conflict must be a painful one for him.

Bill Gates [*activity from 2015 to the present*]

We wrote about Bill Gates, one of the heroes of the IT revolution, in Chapter 2. He stepped down from Microsoft's board in 2020 without severing his connection to the company he co-founded in 1975. He advises the top Microsoft management, and his opinions still carry weight both inside and outside the company. Bill Gates is highly respected for several good reasons: his demonstrated talent as a computer programmer, his brilliant business career, his wisdom, his evident integrity, and his philanthropic work as the head of the Bill & Melinda Gates Foundation, to which he has donated tens of billions of dollars. That, along with his world fame, is why we would like to understand his position on AI.

Gates is a valued guest on TV shows and podcasts, and in July 2023, he launched his own podcast, *Unconfuse Me with Bill Gates*. Getting to know his pronouncements on AI is easy; another matter is how to interpret them.

In the previous chapter, we discussed Microsoft's significant involvement with AI, including its funding of OpenAI, which began in 2019. The partnership has since grown, with additional investments from Microsoft, including a substantial, multi-year investment announced in early 2023. Microsoft intends to integrate new AI tools into Microsoft Office and the Azure platform. The leading software company cannot be blamed for taking this course; given the importance of AI, anything else would be suicidal to their future business. However, we should consider this when examining Bill Gates' views. In light of OpenAI's successes, it would be challenging for Gates to take a strong stance against further development of AI. Characteristically, he did not sign the FLI's open letter, which in the spring of 2023 called for a pause in advanced AI research.

The tone of Bill Gates' statements about AI shifted following OpenAI's remarkable breakthroughs. Before 2020, Gates, when talking about AI, masterfully balanced a tightrope between high hopes and serious concerns and fears. After 2020, high hopes remained, while fears were significantly dampened.

In 2015, talking to the *Reddit AMA* (Ask Me Anything) network, Gates said:

> I am in the camp that is concerned about superintelligence. First, the machines will do a lot of jobs for us and not be super intelligent. That should be positive if we manage it well. A few decades after that though, the intelligence is strong enough to be a concern.

Note the "a few decades after" estimate for the appearance of superintelligence. In 2015, Gates was not alone in thinking that superintelligence was a distant possibility.

In 2019, in an interview with *The Verge*, Bill Gates said:

> I don't understand why some people are not concerned. We should all be worried about it. AI could get out of control or lead to decisions that are not aligned with human values.

The tone is entirely different in the recent (June 2024) interview given to *The Next Big Idea* host Rufus Griscom and his co-hosts, Adam Brotman and Andy Sacks. Gates admitted that for him, the appearance of ChatGPT-4 was:

> mind-blowing…In fluency, in many respects, although not in accuracy, GPT-4 is *already superhuman* [italics ours]…It is absolutely the most important thing going on. It'll shape humanity in a very dramatic way.

With that opening, the interview proceeded to discuss the benefits of AI, both present and future. The host asked Bill Gates, "Do you think we will have an AI advisor in every ear?" Gates answered that he had been writing about the electronic *personal agent* superior to a human assistant for decades. That AI-empowered *agent* will possess the highest level of expertise in all conceivable subjects and the cognitive ability to sort things out and perform any intelligent work that may be required by its "master." Gates sees this agent in diverse roles: "agent executive assistant, agent medical therapist, agent friend, agent girlfriend, agent expert." It will also translate between languages freely, including the most obscure ones. Gates believes that, utilizing the existing widespread cellphone infrastructure, AI can help alleviate the "tragic lack of resources" in developing countries, particularly shortages of medical doctors and teachers.

Gates, who has always been passionate about education, has been saying for some time —and reiterated in The Next Big Idea — that the impact of AI on education will be "incredibly positive." On his own *Unconfuse Me* podcast, discussing the topic with Sal Khan, the founder of Khan Academy, Gates recently said, "AI will be able to teach kids and engage with them in a way that's far more interactive than anything we've ever seen before." Gates' interviewer, Rufus Griscom, offered that the AI assistant they were discussing would be an "Aristotle-level private tutor for our kids" — an excellent summary of that part of the conversation, especially considering the impressive results that *Aristotle-level tutoring* produced in its first subject, Alexander the Great.

Later in the interview, Gates commented on the incredible advances in machine translation. He said:

> The idea that a free product [*GPT or Chat GPT*] provides arbitrary audio and text human translation…that was the holy grail; oh my God! If you had a company that could do that, it would collect tens of billions in revenue and *solve the Tower of Babel* [italics ours]. And here, you know, a kind of a small AI company [*Open AI*] is providing that as an afterthought and a free feature.

We can see how the latest AI invites parallels with biblical scenes (the Tower of Babel) and the most famous historical figures (Aristotle and Alexander the Great). It's refreshing to see experienced and knowledgeable people get so excited. On second thought, it is also a little scary. What's the next biblical parallel: the flood?

The interview continued, and its participants counted other AI blessings. The host, Griscom, recalled what Sam Altman said earlier on Bill Gates' *Unconfuse Me* podcast. According to Altman, the productivity of OpenAI software developers jumped by 300% since they began using their own AI products in their daily work. Later in the interview, Gates and his hosts praised AI's ability to assist businesses in selling products (any products); AI can help enormously. According to Gates, the availability of capital for AI initiatives is also unprecedented:

> There has never been as much capital going into a new category, you might even say, a new *mania* category. This makes the internet, or the early auto-industry mania, look quite small.

Gates also outlined what he thinks is the next big challenge in developing even stronger AI. According to him, *scaling* will continue with at least two more iterations, each involving exponentially increasing numbers of parameters (see Chapter 4). But then, he said, the real advancement will come from addressing *metacognition* — teaching machines to "think about thinking" and to be aware of their thought processes.

This was not a benign observation. I suspect that this somewhat obscure term, metacognition,[27] is being used by AI researchers to replace the concept of *consciousness*, which has become a compromised word, partly because it is notoriously difficult to define and partly because machine consciousness is a frightening concept. Once AI acquires it — metacognition or consciousness — it will become an AGI (my interpretation, not something Bill Gates said directly). And once machines have AGI, they will be on the verge of a singularity. At that point, I hear Archangel Gabriel sounding his trumpet, and I see Samuel Butler, Irving Good, and Vernor Vinge waving at us from their graves.

What happened next, I thought, was even more interesting. Andy Sacks asked Gates, "Do you think we are moving too fast?" This was an invitation to shift gears and discuss the dangers of AI. Gates' answer was uncharacteristically stumbling, with many "you know," "okay," "fine," and similar utterances. Perhaps he was uncomfortable with that question. I tried to transcribe Gates' answer faithfully:

[27]This term was introduced in child psychology during the 1970s. One must be self-aware (conscious) to engage in metacognition.

> *If we knew* how to slow it down, a lot of people would probably say — okay, let's consider doing that...people say...fine; it is open-source, so people will say, okay, if it gets too good, let's stop open-sourcing it... let's not let people with malintent benefit...from having a better AI than the defense-good-intent side...You are not going to completely put the genie back in the bottle.

Bill Gates did not seem to want to answer Andy Sacks' question, but he did. "If we knew," with which he started, is the answer. It means that we *do not know*; we do not know how to slow it down. Then, in the middle of his answer, Gates shifted to discussing people with bad intentions versus those with good ones. This was not what Sacks had in mind. Perhaps Gates felt that the times for such a question had passed; if there is no way to *put the genie back in the bottle*, let's not waste time discussing it.

A little later, when talking about the need for government regulation, Gates said, "You can't expect the private sector to walk away from a market-driven opportunity unless the government decides what the rules are." In other words, to restrain the genie, one needs the full power of the government, but the government has not yet decided what these rules should be.

Gates has consistently emphasized the importance of government regulations and international cooperation in the development of deep AI. However, it appears that, like the rest of us, he does not quite know what these regulations should be. I say that because I have not seen Bill Gates detailing his views on this issue, and also because when he interviewed Sam Altman on the sixth episode of *Unconfuse Me* in August 2023, he asked him, "Is it clear to you what set of regulations would be constructive?" This question, as asked, suggests that the issue isn't entirely clear to Bill Gates himself.

As of early 2026, Gates continues to describe AI as a form of "free intelligence" — providing expert services in medicine, education, and other socially beneficial domains broadly and at low cost — while at the same time acknowledging that all of this is "a little scary" in its scale and speed. He believes that several complex fields — such as programming, biology, and energy — will change more slowly than others and will, for quite some time, remain largely human pursuits. In the longer term, he allows that productivity gains driven by AI could lead to a substantial reduction in the workweek — perhaps reducing it to two or three days. These are, overall, relatively moderate and not especially alarming prospects.

Gates is a very intelligent and very cautious person. He has long been concerned about the prospects of strong AI. Nevertheless, he has a substantial "estate," so to speak, for which he has been responsible in the past and to some extent remains responsible today, and he therefore appears compelled to remain within the bounds of cautious optimism so as not to lose face. And besides, "You are not going to put the genie back in the bottle."

Yuval Noah Harari [*activity from 2015 to the present*]

Yuval Noah Harari is a historian, philosopher, and writer. He is perhaps the most published author of our time, with 45 million copies of his books sold by the fall of 2024. These books are *Sapiens: A Brief History of Humankind* (2011), *Homo Deus* (2015), *21 Lessons for the 21st Century* (2019), and *Nexus* (2024). In Sapiens, AI was not on Harari's horizon and was not mentioned; however, Harari does discuss AI and its implications for society in his last three books. In them, the reader can see how concerns about AI occupy an increasingly larger part of his thoughts. Harari possesses a storytelling talent, so even when he makes a philosophical point, it is usually accompanied by an engaging illustration — a story. His ability to discuss complex issues in relatable, no-nonsense language undoubtedly contributed to the popularity of his books. Harari writes about many things, but we will concentrate on his observations and opinions about AI.

Harari knows that we will inevitably face a highly developed AI. He also understands the significant benefits it is expected to bring to humanity. Still, Harari is increasingly alarmed by certain AI features that are already here. He says that for him, "AI" stands for *Alien Intelligence*, meaning that we do not understand its decision-making process, nor do we control it. This stance reminds me of Stanisław Lem, who, in his sci-fi books, persistently returned to the principal impossibility of understanding alien intelligence. Harari found a concise formula that captures the difference between AI and all other human technologies: AI is an *agent*, rather than a *tool*. His keen vision enables him to discern manifestations of this alien autonomy in today's "ameba-like" AIs, and he expects this dangerous trend to intensify over time. These aspects of Harari's thinking are presented in his latest book, *Nexus*, as well as in his numerous public appearances.

Like everybody else, Harari does not believe human affairs will be *business as usual* once superior AI enters the stage. He discusses the expected profound impact of AI on the job market and other practical aspects of life. He reminds us that a 25% unemployment rate in Germany at the beginning of the 1930s allowed Hitler to rise to power. Severe disruptions in the job market may be a recipe for disaster, signaling the end of the world as we know it.

Considering AI alongside bioengineering (as Mustafa Suleyman calls it, *synthetic biology*), Harari suggests that in the future, humans will merge with AI to survive. Sixty years ago, Stanisław Lem called this prospect a "collective suicide." But perhaps today, given the choice of either perishing or joining forces with machines, he would feel differently. Harari does not predict that we will be wiped out and extinct, envisioning a radical change in human nature instead. We will gradually lose free will and other primary human qualities.

Here, I arranged selected Harari's pronouncements chronologically, from 2015 to 2024, rather than grouping them by topic; thus, some of them will be repetitive.

> *Homo Sapiens* is not going to be exterminated by a robot revolt. Rather, *Homo Sapiens* is likely to upgrade itself step by step, merging with robots and computers in the process, until our descendants will look back and realize that they are no longer the kind of animal that wrote the Bible, built the Great Wall of China and laughed at Charlie Chaplin's antics. This will not happen in a day, or a year...Every day millions of people decide to grant their smartphone a bit more control over their lives.... (Harari, 2015 (2017), 56)

> The greatest danger AI poses is that it will create a new kind of dictatorship in which the system knows you better than you know yourself. (Harari, 2018)

> AI and robotics will likely eliminate millions of jobs. The challenge will be whether we can create a new economic system to deal with this disruption. (Harari, 2019)

> Once you can hack humans, you can manipulate them in ways which were never possible before. And this is the real threat of AI. (Harari, Interview/public remarks, 2019)

> Humans are now hackable animals. The whole idea that humans have this soul or spirit, and they have free will...that's over. (Harari, 2020)
>
> We are one of the last generations of Homo sapiens. Within a century or two, Earth will be dominated by entities that are more different from us than we are different from Neanderthals or chimpanzees. (Harari, 2021)
>
> We are far too excited; we need time to rest…And because we give incredible control of the world to tireless nonorganic algorithms that never need to rest and can just increase the excitement all the time, we are breaking down. (Harari, 2024a)
>
> This [*social network*] may be the first place you see the power of AIs to make independent decisions in a way that reshapes the world…AIs are not just tools in our hands; they are agents creating new realities. So, you may think this is a prediction about the future, but it is already in the past! Because social media algorithms — they are very primitive AIs…[*but*] they still reshape the world with the decisions they make… These algorithms were given a relatively simple and seemingly benign goal by the corporations. The goal was — increase user engagement… make people spend more time on the platform...because [*then*] the company makes more money…Engagement sounds like a good thing — who does not want to be engaged? But the algorithms then experimented on billions of human Guiney pigs and discovered something…that the easiest way to increase user engagement…to keep people glued to the screen — is by pressing the greed, or hate, or fear button in their minds. And…they flooded the world with hate and greed and fear, which is why the [*human*] conversation is breaking down. (Harari, (2024b)

In 2025, Harary continued to worry about human–AI relations, saying that AI can fake real "intimacy" with humans, and in that way control us even more than what he thought possible before:

> Intimacy is much more powerful than attention….Until today, it was utterly impossible to fake intimacy and to mass-produce intimacy… Now with AI, you can, for the first time in history…mass-produce

> intimacy with millions of bots, maybe working for some government, faking intimate relationships with us.... (Harari, 2025)

Overall, Harari's views on AI have not undergone significant changes in recent years. He regards this new "plague" with growing apprehension. As an intellectual, a liberal, and a highly successful writer, he sees little personal benefit in AI; he writes beautifully on his own and does not need ChatGPT's assistance. But what is already happening to society — and what, in all likelihood, will occur in the future — he sees clearly, and it genuinely frightens him. As for existential risk, he acknowledges it as well, but it seems to me that, for him, it is overshadowed by more pressing problems that are already manifesting today.

Yann LeCun [*activity*[28] *from 2015 to the present*]

In an attempt to present balanced AI viewpoints, I am happy to include the opinions of a notable AI scientist, Professor Yann LeCun, who may be one of the most optimistic voices in the field. LeCun is a professor of computer science at NYU and, until the end of 2025, VP and Chief AI scientist at Meta. His contributions to AI research were recognized by his peers in computer science, who awarded him the 2019 A. M. Turing Award, shared with Geoffrey Hinton and Yoshua Bengio. The achievement for which Yann LeCun was honored was the development of convolutional neural networks (CNNs) with applications to handwritten digit recognition. CNNs became the foundation of computer vision and were instrumental in the rise of deep learning. We discussed these advancements in the previous chapter.

Over the past several years, LeCun has often stated that current AI capabilities are a long way from achieving true AGI. He also downplays the existential risks associated with AI and objects to government regulation of AI research and development. Here are some quotes:

> We are far from machines that can reason like humans. AGI is a long-term goal, and it's unclear how far we are from it, but it's certainly not going to happen soon. (LeCun, in the *2019 NIPS conference*)

[28]As always, "activity" here relates to publicly expressed views and opinions about AI. LeCun's involvement with AI research started much earlier than 2015.

> The AI we have now is nowhere near AGI and will not become an existential threat for decades, if ever. (LeCun, in a 2020 Twitter exchange with Elon Musk)
>
> The idea that AI is going to take over the world and turn us into pets or something is just preposterous, and I think it's just B.S. (LeCun, in a 2020 *Cognitive Science Society* conference)
>
> AGI is far from being solved. It's not going to happen in the next few years, and I don't think it's something we should worry about right now. (LeCun, in a 2020 exchange with Geoffrey Hinton)
>
> We're very far from building machines as intelligent as humans or even as intelligent as a cat. The things that humans and animals can do, such as learning by observation and interaction with the world, machines cannot do today. (LeCun, in a 2020 interview with *Quanta Magazine*)

You would think that Professor LeCun's views may have changed after the release of GPT-3 and GPT-4, but they didn't. Here are the quotes from 2023 and 2024:

> We have those LLMs; it's amazing what they can do; they can pass the bar exam! But we still do not have self-driving cars…[and] domestic robots. Where is the domestic robot that can do what a 10-year-old can do…clear up the dinner table and fill up the dishwasher? … It tells you we are missing something really big…we are missing essential components to reach human-level intelligence…we have systems that can absorb the enormous amount of training data from text, and the problem with text is that is that [it] represents only a tiny portion of human knowledge…And think about animals. A lot of animals are super smart…They do not have language and seem to do pretty well…the smarter AI systems we have today are stupider than a house cat…the cat understands causality; [it] understands that if it does something, something else will happen…[*Have*] you ever seen a cat…sitting at the bottom of the bunch of furniture and…looking around…and then jump-jump-jump — that's amazing planning no robot can do today. (LeCun at *CBS Mornings* on Dec. 16, 2023, interviewed by CBS News correspondent Brook Silva-Braga)
>
> It seems to me that before 'urgently figuring out how to control AI systems much smarter than us' we need to have the beginning of a hint of a

> design for a system smarter than a house cat. (LeCun's post on May 18, 2024, on X (former Twitter), in an exchange with former OpenAI employee who worried about the need to control superintelligent AI)

> Our relationship with future AI systems … is that we're going to be their boss. … I like working with people who are smarter than me. … The fix for misuse and unreliability is better AI — systems that have common sense, reasoning, and self-checking. (LeCun on March 19, 2025, Nvidia GTC fireside chat)

I wondered about these statements until I understood that Professor LeCun substitutes *intelligence* with the ability to *adapt* to the environment, a capacity that organic evolution has instilled in all living creatures. This flair for adaptations, especially regarding spatial movements and motor skills, is indeed utterly underdeveloped in AI. And for a good reason — computers, in which AI primarily resides, do not have to jump over the furniture. Good motor skills are required in robots. Indeed, robots are currently less developed than LLM-based AI (though this may be changing). Most people would not consider the ability to clean a table or drive a car as an indicator of intelligence; it is a motor skill that requires only minimal mental effort. Substituting one for the other is a clever sleight of hand, but I am afraid it would not work. And why talk only about clever house cats? Earthworms, snails, fish, mice, foxes, and other animals are well adapted to their environments and feel in their natural habitat as proverbial "fish in water." But that is not what we mean by intelligence.

When we think about AGI, we typically envision machines surpassing humans in *cognitive* tasks. We imagine them passing bar exams, reading radiological medical images, translating between languages, and solving mathematical problems. It's not so much about them moving around in the kitchen or going downhill skiing. Perhaps they will acquire superhuman motor skills one day, but I am not looking forward to that day. We may have a better chance of surviving this AI revolution if intelligence remains inside immovable computers.

And then there is also a timescale. LeCun believes that, contrary to what most of his colleagues are saying, the true AGI is decades away and will not be achieved by continuing to fortify LLMs. Summarizing his views, a *Wall Street Journal* article said:

> LeCun thinks that the problem with today's AI systems is how they are designed, not their scale. No matter how many GPUs tech giants cram

> into data centers around the world, he says, today's AIs aren't going to get us artificial general intelligence. (Mims, 2024)

In line with Professor LeCun's skepticism about the creation of superintelligence, he opposes government regulation of AI research and development. He said about that subject:

> ...there are several types of regulations...there is regulation of products: if you put one of these [*AI-driven*] emergency braking systems in your car — of course, it's been checked by a government agency that makes sure it's safe...You need to regulate products...The debate is about whether research and development should be regulated, and there I'm clearly very strongly of the opinion that it should not. The people who believe it should are people who...claim there is an intrinsic danger in putting the technology in the hands of essentially everyone...And I think the exact opposite that...there is actually a huge beneficial effect...We need AI technology to disseminate in all the corners of society because it makes people smarter...a lot of boring jobs can be automated...making people smarter is intrinsically good...the benefits far overwhelm the dangers.
>
> EU regulation makes it very difficult for companies to open-source their platforms. I think [*this is*] very counterproductive. In fact, the French, German, and Italian governments have basically blocked the legislation in front of the EU parliament for that reason. They really want open source — and the reason they want open source is because... you do not want those things to be controlled by a small number of companies on the West Coast of the US; it has to be an open platform, kind of like the internet...If you become secretive, you fall behind because people do not want to talk to you anymore. (LeCun at *CBS Mornings* on Dec. 16, 2023, interviewed by CBS News correspondent Brook Silva-Braga)

It's a pleasure to write about Professor LeCun's views, which remain consistent over time. Although his arguments about domestic cats being smarter than LLMs do not convince us, it is possible that his understanding of the current state of AI and, most importantly, its future, is correct. After all, he is one of the leading experts in this field; his opinion carries weight. And we all love cats — in that I agree with him wholeheartedly.

Sam Harris [*activity from 2016 to the present*]

Samuel Harris is an American philosopher, neuroscientist, author, and podcast host. His work touches on a range of topics, including religion (of which he is not a friend), ethics, free will, meditation, psychedelics, the philosophy of mind, politics, neuroscience, and AI. He published seven books, five of which were on The New York Times Best Seller list. Harris's work has been translated into over 20 languages. Since September 2013, Harris has hosted the *Making Sense* podcast (originally titled "Waking Up").

I decided to include Harris as one of the *AI Whisperers* after hearing his July 2016 TED Talk and his August 2023 interview with Steven Bartlett on the podcast *Diary of a CEO* (DOAC). I have rarely heard anyone formulate their views on the *existential* risks posed by AGI and ASI as clearly and convincingly as Harris.

At the opening of the TED Talk, Harris nailed the main difficulty of dealing with AI dangers — something that I also believe is at the heart of the problem — that AI troubles (most of them) are ahead in the future, while today's AI, for the most part, is interesting and even lovable. Harris said:

> I'm going to describe how the gains we make in artificial intelligence could ultimately destroy us. And in fact, I think it's very difficult to see how they won't destroy us or inspire us to destroy ourselves. And yet, if you're anything like me, you'll find that it's fun to think about these things. And that response is part of the problem…one of the things that worries me most about the development of AI at this point is that we seem unable to marshal an appropriate emotional response to the dangers that lie ahead. I am unable to marshal this response, and I'm giving this talk…. (Harris, 2016)

Then Harris recapped what we already know:

> At a certain point, we will build machines that are smarter than we are, and once we have machines that are smarter than we are, they will begin to improve themselves…what the mathematician I.J. Good called an "intelligence explosion." (*ibid.*)

Then, Harris talked about what it would mean to coexist with superintelligent AI:

> … we will build machines that are so much more competent than we are that the slightest divergence between their goals and our own could destroy us. Just think about how we relate to ants. We don't hate them.[29] We don't go out of our way to harm them…We step over them on the sidewalk. But whenever their presence seriously conflicts with one of our goals, let's say when constructing a building like this one, we annihilate them without a qualm. The concern is that we will one day build machines that, whether they're conscious or not, could treat us with similar disregard. (*ibid.*)

Harris sees the inevitability of creating superior AI, sooner or later:

> … It's crucial to realize that the rate of progress doesn't matter because any progress is enough to get us into the end zone. We don't need Moore's Law to continue. We don't need exponential progress. We just need to keep going…It [*intelligence*] is our most valuable resource. So we want to do this. We have problems that we desperately need to solve. We want to cure diseases like Alzheimer's and cancer. We want to understand economic systems. We want to improve our climate science…The train is already out of the station, and there's no brake to pull. (*ibid.*)

Concerning the limits of intelligence, Harris, while acknowledging the grand spectrum of intelligence as we know it, from a chicken to John von Neumann, suggested that, in principle, it may be much broader:

> …It seems overwhelmingly likely…that the spectrum of intelligence extends much further than we currently conceive, and if we build machines that are more intelligent than we are, they will very likely explore this spectrum in ways that we can't imagine and exceed us in ways that we can't imagine. (*ibid.*)

[29] Stanisław Lem said in 1961, "We don't feel any hatred toward caterpillars, wolves, or mosquitoes."

However, since superior intelligence is not a well-defined concept in terms of its depth, Harris proposed to consider only its speed:

> So imagine if we just built a superintelligent AI [*here, Harris probably means AGI, not the ASI*] that was no smarter than your average team of researchers at Stanford or MIT. Well, electronic circuits function about a million times faster than biochemical ones, so this machine should think about a million times faster than the minds that built it. So you set it running for a week, and it will perform 20,000 years of human-level intellectual work, week after week after week. How could we even understand, much less constrain, a mind making this sort of progress? (*ibid.*)

Although I agree with most of what Harris is saying, I would like to offer some resistance in one instance. Not that it saves the day, but I would like to oppose the point that speed alone will give AGI an unimaginable advantage. It may be less clear-cut than Harris makes it. Because, at least in the natural sciences, such as physics, *experimentation* is an essential ingredient without which the scientific method cannot function and science cannot advance. While the thinking processes may be sped up by a factor of a million, I am pretty sure that experimental work cannot follow that pace; it is difficult to see how it could be sped up by a factor of 10, let alone a million, even if it were all directed by a swift AI. Indeed, the rate of experimental science does not so much depend on the speed of thinking as on the speed of shops manufacturing scientific instruments and trucks delivering supplies and equipment to the lab, not to mention what happens inside the lab during the actual experiments. Of course, in an imaginary world with robot scientists and robot factories — a world without humans — a considerable speedup of almost everything may be possible. Still, such robots are a long way off (much longer than AGI; listen to what Yann LeCun is saying), and I do not think that is what Harris had in mind.

Could experimental science be sped up?

We could consider any instant within the last 400 years for this thought experiment, but let us imagine physics at the end of 1926, a few months after Schrödinger wrote his famous equation, officially establishing quantum mechanics. Let us further suppose that the experiment continues at its normal pace, but the ability to think, calculate, and model — what is theoretical physics — that ability accelerates a million times with the last stroke of a clock at midnight on December 31, 1926. Would those fast-thinking theoretical physicists be able to come up with all the answers in a very short time? I am afraid not. Creating the correct theories requires experimental fact-finding. The theory suggests to experimentalists what to look for, and when experimentalists find something unexpected, it forces the theorists to change course. The scientific method is a continuing interplay of theory and experiment; neither one can exist without the other, at least not for long.

Given that now, in our imaginary world, one week is equivalent to 20,000 years of human-level intellectual work, how far would science advance by January 7, 1927? I do not think it would go far. A fast AGI, which replaced the "slow" Werner Heisenberg, Niels Bohr, Max Born, Louis de Broglie, and Paul Dirac, would not have been aware of the results of the Davisson–Germer experiment, which was performed later in 1927. Therefore, it would not have confirmation of the particle–wave duality at the heart of quantum mechanics. Then, suppose a Super Paul Dirac had written his relativistic version of the Schrödinger equation and predicted antiparticles only a few seconds after midnight on the eve of 1927. I do not see how the theory could be validated without the experimental discovery of an anti-electron (called a positron), which, in the real world, did not occur until 1932. Looking a few years further into the future, I do not see how AGI, even if it thinks a million times faster than Richard Feynman, Julian Schwinger, Sin-Itiro Tomonaga, and Freeman Dyson, could develop quantum electrodynamics (QED) if it were not prompted by precise measurements of the minute shifts in the energy spectrum of hydrogen (the Lamb shift) and the measurement of the slight deviation in the magnetic moment of an electron from the expected value, both of which occurred in 1947. Such measurements could not have been made in 1927, even if they had been suggested, because no instruments were capable of performing them at that time, nor for a few years later.

It seems that a million times higher speed of thinking by itself would not be particularly helpful, at least in physics. It would produce some remarkable insights a few minutes after midnight on January 1, 1927, but further progress would have to await the slow-moving experiment.

Perhaps there are areas of science and, more generally, of human culture that do not have the same dependence on experimentation as physics, chemistry, or biology. Pure math, for one, does not seem to require experimentation. But computer science is not pure math — training data should be harvested and fed into a computer, and better computers must be built to access new capabilities. Computer science has its own experimental component. It appears that a million times' advantage in thinking speed would not directly translate into a million times faster development of knowledge.

As to mathematics, it may not be as pure and independent of natural sciences as some people think. To recall a quip once made by the great mathematician Vladimir (Dmitry) Arnold, as told to me by Anatoly Vershik, who knew Arnold well: "Mathematics is an easy and inexpensive part of physics."

Then, Harris discussed the world in which a perfectly safe AGI has been developed and where human work, whether physical or intellectual, is no longer necessary. You could probably imagine what he said on that subject — read Karel Čapek's *R.U.R.*, and you will get the picture; these thoughts are endemic to all of us. I quote here a couple of Harris lines — because he expresses them so well:

> …imagine we hit upon a design of superintelligent AI that has no safety concerns. We have the perfect design the first time around…Well, this machine would be the perfect labor-saving device. It can design the machine that can build the machine that can do any physical work, powered by sunlight, more or less for the cost of raw materials. So, we're talking about the end of human drudgery. We're also talking about the end of most intellectual work.
>
> So what would apes like ourselves do in this circumstance? Well, we'd be free to play Frisbee and give each other massages. Add some LSD and some questionable wardrobe choices, and the whole world could be like *Burning Man*.[30]

[30]Burning Man — an annual gathering in the Black Rock Desert in northern Nevada (USA), something like a temporary festival city reminiscent of the legendary Woodstock

> Now, that might sound pretty good, but ask yourself what would happen under our current economic and political order? It seems likely that we would witness a level of wealth inequality and unemployment that we have never seen before. Absent a willingness to immediately put this new wealth to the service of all humanity, a few trillionaires could grace the covers of our business magazines while the rest of the world would be free to starve. (*ibid.*)

Harris ended his 2016 TED Talk by saying:

> …the moment we admit that information processing is the source of intelligence, that some appropriate computational system is what the basis of intelligence is, and we admit that we will improve these systems continuously, and we admit that the horizon of cognition very likely far exceeds what we currently know, then we have to admit that we are in the process of building some sort of god. Now would be a good time to make sure it's a god we can live with. (*ibid.*)

Centuries have passed since 2016, and our AGI-ASI future — for better or worse — is now "a few thousand days" away, to use Sam Altman's phrase. In August 2023, during an interview with Steven Bartlett on the *DOAC* podcast , San Harris revisited the dangers of AI. He had more time to elaborate on his thoughts in 2023 than he did in a 15-minute TED Talk in 2016, so he expanded on some of his previous ideas. For one thing, he replaced *ants* with *dogs* in his analogy, which allowed him to make his fable more relatable and more dramatic. Let me quote a few lines from that 2023 interview, not because Harris changed his mind and said something entirely different from what he said in 2016, but because he has a way of expressing his thoughts eloquently and, at the same time, simply:

> My intuition is that there is something inherently dangerous for the dumber species to be in the presence of the smarter species…a mismatching intelligence… is a fundamental lack of insight into what the smarter party is doing, why it's doing it, and what it will do next…

---

of 1969 in New York State. Harris's irony about a world organized along the lines of such a gathering is not difficult to understand.

> ...by analogy, just imagine that dogs had invented us as their super-intelligent AIs for the purpose of making their lives better: securing resources for them, securing comfort...getting medical attention. It's been working out pretty well for dogs for about 10,000 years...for most dogs, most of the time, humans have been a great invention.
>
> ...the mismatch in our intelligence dictates a fundamental blindness [*on the part of dogs*]...we have all these...goals and things we care about that they cannot possibly conceive. They know that when we...get the leash and say "it's time for a walk" they understand that particular part of the language...but everything else we do, [*or*] when we're talking to each other, [*or*] when we're on our computers or our phones — they don't have the dimmest idea of what we're up to.
>
> ...the truth is we love our dogs; we make irrational sacrifices for our dogs; we prioritize their health over all kinds of things...and yet if... there was a new global pandemic kicking off, and some virus was jumping from dogs to humans, and it was a super Ebola...90 [*percent*] lethal, and if this was just a forced choice... — [*what do you*] value more: the lives of your dogs or the lives of your kids... — we would just kill all the dogs and they would never know why...This is a fanciful analogy because the dogs did not invent us, but evolution invented us....(Harris, 2023)

Toward the end of the interview, the host asked Harris whether, given a chance, he would stop AI research. If there were a button that he could press that would stop all further AI work, would he press it? Harris put up a bit of resistance, asking if the pause would be permanent or temporary — Bartlett said it would be permanent — and, while answering that tricky question, Harris, like Gates before him, used more than the usual amount of his "you know," "just," "right," and "uhm" fillers, but eventually he said: yes, I would push that button. Harris added that, in his opinion, "muddling along...apes like ourselves" could do most of the dream projects we hope AI will solve for us, albeit much slower.

Mustafa Suleyman [*activity from 2022 to the present*]

Mustafa Suleyman is a British software developer, entrepreneur, and AI executive. He is one of the prominent figures in the contemporary AI landscape. Nowadays, he probably does not write code himself (I am not sure that he ever did), but he knows his field professionally, establishing and managing teams of computer scientists and giving them direction. It is a unique ability and skill that has emerged recently to address challenges in the AI industry. Sam Altman of OpenAI also belongs to this breed of brilliant AI industry leaders.

Mustafa Suleyman studied philosophy at Oxford; perhaps that gave him his uncommon eloquence and clarity of thought. In 2010, he co-founded the renowned DeepMind Technologies, which went on to develop AlphaGo and AlphaFold, as discussed in Chapter 4. In 2014, Google acquired DeepMind for approximately £400 million. Following the acquisition, Suleyman served as head of *Applied AI* at DeepMind. In 2019, he left his administrative position but remained a consultant at Google. In 2022, Suleyman co-founded *Inflection AI*, a company focused on developing AI-driven personal assistants. Inflection AI introduced "Pi," a chatbot designed to engage users in human-like conversations. In March 2024, Suleyman joined Microsoft as Executive Vice President and CEO of *Microsoft AI*, a newly established division dedicated to advancing consumer AI products. In this role, he oversees initiatives such as Copilot, Bing, and Edge and reports to Satya Nadella, the chairman and CEO of Microsoft.

Mustafa Suleyman's public appearances, in which he has spoken about AI, are numerous. In 2023 and 2024, he gave TED talks and was interviewed by CNN's Farid Zakaria, *WIRED* magazine, and the VentureBeat website, and he appeared as a guest on dozens of podcasts. In the fall of 2023, he published a book about AI (with Michael Bhaskar) called *The Coming Wave* (Suleyman, 2023). In his book, Suleyman combines his thoughts on AI and *synthetic biology*, which he defines as the forthcoming products of genetic engineering. He is simultaneously excited about and concerned by both disciplines.

Mustafa Suleyman is expecting a lot of good to come from AI: Humanity will bathe in material profusion, all our hard problems will be solved ("…from climate change to aging populations to sustainable food" (Suleyman, 2023, 8)), education will be transformed, we will have unlimited clean energy, and each of us will have an Aristotelian-level personal AI assistant (*copilot*, as it is called at Microsoft AI) whispering its timely advice in our ears. In most of his public appearances, Suleyman has talked little about AI safety, preferring instead to call it "ethical development." When asked about AI safety, he typically maintains an optimistic stance, stating that in his work, he has always prioritized ethical development of AI and that collectively, the AI community, working in collaboration with national governments, will succeed in making it safe.

This positive attitude starkly contrasts with the thoughts in *The Coming Wave*. For years, Suleyman, a successful AI entrepreneur, had kept his worries and suspicions to himself and his inner circle, finally deciding to share them in the book. In Chapter 1, bearing the telling title "Containment is not possible," we find the following:

> From where we stand today, it appears that containing this wave [*of AI and synthetic biology*] is not possible…the consequences for our species are dramatic, potentially dire. Equally, without its fruits, we are exposed and precarious. This is an argument I made many times in the last decade behind closed doors, but as the impacts become ever more unignorable, it's time that I made the case publicly…DeepMind became one of the world's leading AI companies, achieving a string of breakthroughs. The speed and power of this new revolution have been surprising even to those of us closest to its cutting edge…It's clear this wave is accelerating...Once matured, these emerging technologies will spread rapidly, becoming cheaper, more accessible, and widely diffused throughout society. They will offer extraordinary new medical advances and clean energy breakthroughs, creating not just new businesses but new industries and quality of life improvements in almost every imaginable area. And yet alongside these benefits, AI, synthetic biology, and other advanced forms of technology produce tail risks on a deeply concerning scale…Both pursuing and not pursuing new technologies is, from here, fraught with risk…sooner or later, a powerful generation of new technology leads humanity toward either catastrophic or dystopian outcomes… The current discourse around technology ethics and safety is inadequate. Despite the many books, debates, blog posts, and tweetstorms about technology, you rarely hear about containing it. (Suleyman, 2023, 7–11)

Suleyman ends Chapter 1 with the words, "Containment is not, on the face of it, possible. And yet, for all our sakes, containment *must* be possible" (*ibid*., 19).

Later in the book, in Chapter 7, Suleyman demonstrates that he fully understands not only the immediate AI risks (such as massive loss of jobs and social media irregularities) but also the existential risks (here, references to Good and Vinge would be appropriate, but that's beyond the point):

> …any discussion of containment has to acknowledge that if or when AGI-like technologies do emerge, they will present containment

problems beyond anything else we've ever encountered. Humans dominate our environment because of our intelligence. A more intelligent entity could, it follows, dominate us...by definition a superintelligence would be fully impossible to control or contain. An "intelligence explosion" is the point at which AI can improve itself again and again, recursively making itself better in even faster and more effective ways...The blunt truth is that nobody knows when, if, or exactly how AIs might slip beyond us and what happens next; nobody knows...how to make them behave with awareness of and alignment with our values, assuming we can settle on those values in the first place. (*ibid.*, 115)

In Chapters 2 and 3, Suleyman discusses previous waves of technology, including the Industrial Revolution, the steam engine, Moore's Law in electronics, and information technology. He notes that, despite occasional opposition (such as the Luddites who revolted against industrial machines or Ottoman Empire rulers who attempted to ban printing presses), none of the previous technologies have ever been stopped from proliferating and encompassing the whole world. "It's not that the containment problem hasn't been recognized in history; it's just that it has never been solved." (*ibid.*, 41).

Then, inevitably and predictably, Suleyman talks about nuclear arms, calling this section "The Nuclear Exception?" He says:

It would be wrong to say they have not proliferated...But to a remarkable degree...they have avoided technology's deep underlying pattern [*of uncontrollably spreading worldwide*] ...even though nuclear capability has been largely contained...it's not a reassuring story. (*ibid.*, 43)

Suleyman reminds us of horrifying near-catastrophic accidents and the malfunctioning of control systems that have occurred several times since 1945:

Nuclear is an exception to the unstoppable spread of technology, but only because of the tremendous costs and complexity involved, the decades of tough multilateral effort, the fear-inducing enormity of its lethal potential, and pure luck...humanity has tried to say no and only partially succeeded. (*ibid.*, 45)

Other "glimmers of containment" are discussed, including moratoriums on biological and chemical weapons, a memorandum on gene editing, and decarbonization efforts (so far largely unsuccessful). Suleyman notes that "none of these achievements are particularly robust."

At this point, I am tempted to make the following observation, which Suleyman does not make. There is a principal difference between technologies of mass destruction, such as nuclear, chemical, and biological weapons and the emerging superintelligent AI: Nobody likes atomic bombs, deadly battlefield gases, and laboratory-made super-contagious viruses for what they are; they are being developed (some in secrecy) for military and political reasons in a dangerous and divided world. However, everyone is in love with AI, even those who recognize the need to regulate it. AI is infinitely attractive across all levels, from scientific interest to commercial profit.

Suleyman addresses the issue of incentives driving the AI wave in Chapter 8 of his book. He names and details the "usual suspects," such as financial gain, the quest for power, and scientific tradition, in which, so far, the pursuit of new ideas has always been admired and rewarded. Then, there is hope that AI will help us solve our social challenges. To all of that (and we could name a few more), Suleyman adds a couple of items less discussed: national pride and individual researchers' egos. And he concludes: "As long as these incentives are in place, the important question of 'should we?' is moot" (*ibid.*, 119).

In Chapter 7, among other subjects, Suleyman discusses autonomous systems:

> …the fact that so much of the coming wave operates at the edge of what we can understand should give us pause. We won't always be able to predict what these autonomous systems will do next; that's the nature of autonomy. Right at the cutting edge, however, some AI researchers want to automate every aspect of building AI systems…AIs are already finding ways to improve their own algorithms. What happens when they couple this with autonomous actions on the web…conducting their own R&D cycles?. (*ibid.*, 114–115)

In the following chapters, Suleyman discusses specific aspects of emerging AI, some of them valuable but mostly threatening and destabilizing. Among them are threats to national states and our already precarious world order, armed robots, deepfakes, widespread misinformation on social networks, cyber arms races, state-run surveillance, and job-threatening automation. Suleyman refers to them collectively as "fragility amplifiers." He summarizes his thoughts thus:

> Even those who don't foresee the most severe outcomes of automation still accept that it is on course to cause significant medium-term disruptions…the ramifications will be hugely destabilizing for hundreds of millions who will, at the very least, need to re-skill and transition to new types of work. Optimistic scenarios will involve troubling political ramifications from broken government finances to underemployed, insecure and angry populations…What gets lost in the analysis is that all these new pressures on our institutions stem from the same underlying general purpose [*technological*] revolution. Now they will arrive together, simultaneous stressors intersection, buttressing, and boosting one another. The full amplification of fragility is missed because it often appears as if these impacts are happening incrementally and in convenient silos. [*Now*], they are not. (*ibid.*, 181)

In reading Suleyman's book, I became increasingly impressed with his nuanced understanding of the present and future impact of AI on society, as well as with his deep analysis of existing world problems vis-a-vis the rising technological pressures. Although he called his book "The Coming Wave," when reading it, you understand that he wanted to say "Tsunami."

In Chapter 12, "The Dilemma," Suleyman discusses possible scenarios in which AI and synthetic biology could get out of hand, but also explains why, in his view, stopping their development is not an option. He says:

> The development of new technologies is…a critical part of meeting our planet's grand challenges…Pausing technological development, assuming it was possible…would deliver another form of dystopia. Without new technologies, sooner or later everything stagnates, and possibly collapses altogether…Given the population and resource constraints, just standing still would probably require a global two-to-threefold productivity improvement, and standing still is not acceptable…standstill in itself spells disaster. (*ibid.*, 219–220)

The reader recalls that Mustafa Suleyman named the first chapter of *The Coming Wave*, "Containment is not possible." Chapter 13, the one before last, is called "Containment must be possible." The author, realizing that he had painted a rather dark picture, attempted to lighten it up; as an optimist at heart, he does not want to leave the impression that he has lost all hope. He wants to convince his readers that we still have viable options — options that, if implemented, would ensure that *Homo sapiens* "shall not perish from the earth," to use Lincoln's famous line. When people think of AI containment, they almost always discuss government regulation. Suleyman calls it "classic pessimism-averse answer," and he is warning that it is not enough:

> The most ambitious legislation is probably the EU's AI Act, first proposed in 2021. As of this writing in 2023, the act is going through the lengthy process of becoming European law…Yet the act…demonstrates the inherent problems with regulation. It has been attacked from all sides, for going too far and not going far enough…Regulating…general-purpose technologies is incredibly challenging…With the coming wave, we don't have half a century for numerous bodies to figure out what to do… Nor is it clear how all this will be managed over such a broad spectrum of unprecedented technologies…nations…are in a strategic competition to accelerate the development of technologies like AI and synthetic biology… On the other hand, they're desperate to regulate and manage these technologies — to contain them, not least for the fear that they will threaten the nation-state as the ultimate seat of power…for containment to be possible, rules need to work well in places as diverse and Netherlands and Nicaragua…The EU heavily restricts genetically modified organisms in the food supply. Yet in the United States, genetically modified organisms are a routine part of agribusiness…Chinese AI policy has two tracks: a regulated civilian path and a freewheeling military-industrial one… [Regulation] doesn't stop motivated bad actors or accidents…It doesn't provide alternatives given the immense financial rewards…Regulation is not enough, but at least it is a start. (*ibid.*, 229–232)

So, in Suleyman's view, containment through government regulations is impractical and insufficient. At this point, the reader is wondering: What else is there? Suleyman tries to explain:

> The first step is recognition…this balance has to be struck in consultation with everyone…While unstoppable incentives are locked in, the wave's final form…still to be decided. Let's not waste decades to find

> out. Let's get started on managing it today...Containment of the current wave, I believe, is not possible in our current world. (*ibid.*, 237)

I must confess that I find it challenging to maintain my concentration in the face of such pronouncements. The last phrase quoted above is meaningful but distinctly pessimistic. Should we rebuild our divided world *before* launching the containment campaign? Can we change it through sheer willpower? On the same page, Suleyman tells us that we have minimal time left to take action on AI and synthetic biology containment. The only truth this sort of talk reveals is that Mustafa Suleyman, who wishes to find a solution, does not see one.

Then, in the last chapter of the book, Suleyman lists 10 steps toward containment. He invites us to see these ideas as "concentric circles," or "layers of the onion":

> It's the way all these layers of the onion build that makes them powerful; each alone is insufficient...Collectively...they could add up to something that works. (*ibid.*, 239)

We will refrain from reproducing these steps here; they are all reasonable but, for the most part, either unexplored or unrealistic.

Toward the end of Chapter 14, Suleyman says:

> Looking at the myriad paths forward...it seems containment fails in many of them. The narrow path must be walked forever from here on out, and all it takes is one misstep to tumble into the abyss...And yet, while there is compelling evidence that containment is not possible, temperamentally I remain an optimist. (*ibid.*, 278)

I must confess that Suleyman's attempts to outline a program for the future and his ten-point plan do not impress me as realistic, doable, or constructive. They seem to be no more than the wishful thinking of a good and concerned person. Suleyman's desperate, hope-against-hope cry, "Containment must be possible," tells me that it is probably impossible.

A word about Mustafa Suleyman's public appearances: he seems considerably more optimistic in all of them than in his book. Surprisingly, that goes for both before and after he wrote *The Coming Wave*. Some of the podcasters he spoke with after the release of his book noted that. I do not know the reason for this duality; perhaps a person in Suleyman's high position in the AI industry cannot publicly acknowledge the potential end of the world brought about by AI. Or maybe it reflects the nature of AI itself: it is a *double-faced Janus* — one face beautiful, another scary.

Marc Andreessen [*activity from 2023 to the present*]

Marc Andreessen is an American entrepreneur, software engineer, and venture capitalist best known as one of the founding figures of the modern internet. In 1993, while still a student, he co-created *Mosaic*, the first widely used web browser, and later co-founded *Netscape Communications*, whose browser played a key role in popularizing the World Wide Web. Following a series of influential roles in the tech industry, in 2009, he co-founded *Andreessen Horowitz* (referred to as *a16z*). This prominent Silicon Valley venture capital firm has since become one of the most active and influential investors in the technology startup sector. As of 2025, Andreessen remains a general partner at a16z, shaping its investments in AI, cryptocurrency, software infrastructure,[31] and biotechnology.

In late 2024 and early 2025, Andreessen began actively engaging with the Trump administration's transition team. In December 2024, he participated in White House transition meetings as a tech advisor, voicing concerns that the Biden administration "had seething contempt for tech," especially in areas such as AI, crypto, and startups. By January 2025, he was involved in efforts shaping AI policy under the incoming administration. While no formal title was publicly announced, his presence in discussions suggests significant influence in drafting an "AI Action Plan," which is being developed under the new administration.

Given his recently acquired U.S. political muscle, Andreessen's position is of crucial importance. His views on AI have been publicly expressed since June 6, 2023, when Andreessen published an essay on his company's blog entitled "Why AI Will Save the World." Here are rather extensive quotes from that document:

> … human intelligence is the lever that we have used for millennia to create the world we live in today…What AI offers us is the opportunity

[31] *Software infrastructure* refers to the foundational systems and tools, such as operating systems, databases, cloud platforms, and networking frameworks, that enable software applications to run, scale, and interact reliably. It is the unseen but essential backbone of the modern digital ecosystem.

to profoundly *augment* human intelligence to make all of these outcomes of intelligence — and many others, from the creation of new medicines to ways to solve climate change to technologies to reach the stars — much, much better from here.

AI augmentation of human intelligence has already started — AI is already around us in the form of computer control systems of many kinds, is now rapidly escalating with AI Large Language Models like ChatGPT, and will accelerate very quickly from here — *if we let it...* Scientific breakthroughs and new technologies and medicines will dramatically expand, as AI helps us further decode the laws of nature and harvest them for our benefit...military commanders and political leaders will have AI advisors that will help them make much better strategic and tactical decisions, minimizing risk, error, and unnecessary bloodshed...

And this isn't just about intelligence! Perhaps the most underestimated quality of AI is how *humanizing* it can be....

At this point, as I read Andreessen's document, I must admit I was taken aback somewhat, especially by the cheerful prospect of military commanders using AI on a battlefield and by the observation of how *humanizing* it may be.... But let us continue:

The stakes here are high. The opportunities are profound. AI is quite possibly the most important — and best — thing our civilization has ever created,[32] certainly on par with electricity and microchips, and probably beyond those...We should be living in a much better world with AI, and now we can.

So Why The Panic?

In contrast to this positive view, the public conversation about AI is presently shot through with hysterical fear and paranoia. We hear claims that AI will variously kill us all, ruin our society, take all our jobs, cause crippling inequality, and enable bad people to do awful things...a moral panic is by its very nature *irrational* — it takes what may be a legitimate concern and inflates it into a level of hysteria that ironically makes it harder to confront actually serious concerns...This moral panic is already being used as a motivating force by a variety of actors to

[32] Stephen Hawking said that powerful AI is "either the best or the worst thing ever to happen to humanity; we do not yet know which." But Marc Andreessen knows.

demand policy action — new AI restrictions, regulations, and laws. These actors, who are making extremely dramatic public statements about the dangers of AI — feeding on and further inflaming moral panic — all present themselves as selfless champions of the public good...these actors are true believers that AI presents one or another existential risks...

AI Risk #1: Will AI Kill Us All?

The first and original AI doomer risk is that AI will decide to literally kill humanity.

The fear that technology of our own creation will rise up and destroy us is deeply coded into our culture...Just because premodern man freaked out like this doesn't mean we have to; we can apply rationality instead.

My view is that the idea that AI will decide to literally kill humanity is a profound *category error*. AI is not a living being that has been primed by billions of years of evolution to participate in the battle for the survival of the fittest, as animals are, and as we are. It is math — code — computers, built by people, owned by people, used by people, controlled by people. The idea that it will at some point develop a mind of its own and decide that it has motivations that lead it to try to kill us is a superstitious handwave.

In short, AI doesn't *want*, it doesn't have *goals*, it doesn't want to *kill you*, because it's not *alive*. And AI is a machine — is not going to come alive any more than your toaster will.

Now, obviously, there are true believers in killer AI...These actors are arguing for a variety of bizarre and extreme restrictions on AI ranging from a *ban on AI development*[33], all the way up to *military airstrikes on data centers*[34] and *nuclear war*[35]. They argue that because people like me cannot rule out future catastrophic consequences of AI, that we must

[33]Here, Andreessen is referring to the Future of Life Institute's 2023 letter, which called for a six month pause in advanced LLMs development.

[34]Eliezer Yudkowsky's idea we mentioned when talking about him.

[35]A reference to Yudkowsky's online exchange, where he was asked whether he would support a nuclear strike on an underground data center and replied that he would suggest a conventional invasion or repeated airstrikes first.

assume a precautionary stance that may require large amounts of physical violence and death[36] in order to prevent potential existential risk.

My response is that their position is non-scientific — What is the testable hypothesis? What would falsify the hypothesis?[37] How do we know when we are getting into a danger zone? These questions go mainly unanswered, apart from "You can't prove it won't happen!"....

At this point, the author of the pamphlet, after offering a perfunctory nod to the 1940s U.S. atomic bomb program — credited, with remarkable optimism, for both ending the Second World War and preventing a Third World War — goes on to explain the supposed real motives of those concerned with AI safety. One possible motive is to "confess guilt to claim credit for the sin."[38] Another is material gain from being employed as:

> ..."AI safety expert", "AI ethicist", "AI risk researcher". They are paid to be doomers, and their statements should be processed appropriately.

I cannot help admitting that arguments of this sort evoke in me a feeling of disgust and a desire to tell Mr. Andreessen to go to hell, but let us continue.

The third reason is that:

> ...the reality, which is obvious to everyone in the Bay Area but probably not outside of it, is that "AI risk" has *developed into a cult*, which has suddenly emerged into the daylight of global press attention and the public conversation...And it's developed a full panoply of cult behaviors and beliefs.
>
> AI Risk #2: Will AI Ruin Our Society?

---

[36]This is again about Yudkowsky, who, however, never called for "large amounts of physical violence and death" in any context. As I noted in the section about him, his uttering about bombing data centers gave his critics lots of ammunition.

[37]Here, Andreessen references the philosopher Karl Popper, who argued that a scientific theory must be potentially falsifiable — meaning it must allow for possible observations which, if discovered, would contradict the theory.

[38]Apparently, Geoffrey Hinton is meant here.

> ...Short version: If the murder robots don't get us, the hate speech and misinformation will. The tip-off to the nature of the AI societal risk claim is its own term, "AI alignment." Alignment with what? Human values. Whose human values? Ah, that's where things get tricky.

That is, the author of this text apparently sees no way to define universal human values.

Marc Andreessen then proceeds to argue that "alignment" is all about restricting AI-generated speech and that limiting this speech, like any free speech, is a slippery slope that allows hordes of liberal-minded organizations and individuals to wipe out whatever they may want to wipe out:

> You should be aware of how a small and isolated coterie of partisan social engineers are trying to determine that right now, under the cover of the age-old claim that they are protecting you...In short, don't let the thought police suppress AI.

We wrote about *alignment* before, and I do not think it was about censoring AI-generated content. The term evolved from Yudkowsky's *Friendly AI*; it refers to attempts (perhaps futile) to permanently instill sympathy for human values into the machine code governing AI. Here, Mr. Andreessen completely changes the meaning of this now widely accepted term. And it is clear why he does so: There are few principles Americans are more ready to defend than the constitutionally granted freedom of speech.

> AI Risk #3: Will AI Take All Our Jobs?
>
> The fear of job loss due...to mechanization, automation, computerization, or AI has been a recurring panic for hundreds of years, since the original onset of machinery such as the mechanical loom...each wave of this panic is accompanied by claims that "this time is different" — *this* is the time it will finally happen, *this* is the technology that will finally deliver the hammer blow to human labor. And yet, it never happens.

Let me omit the rest of Andreesen's argument concerning the expected job loss — it is fairly predictable. Throughout his document, he is trying to present AI as just another new technology. To him, the worry about its impact on the job market is another hysterical reaction. In fact, the author is going beyond merely asserting that the labor market will survive and — following the principle that the best defense is a good offense — predicts that it will, on the contrary, grow:

> As prices for goods and services fall, we pay less for them, meaning that we now have *extra spending power* with which to buy *other things*. This *increases demand* in the economy, which drives the creation of *new production* — including new products and new industries — which then creates new jobs for the people who were replaced by machines in prior jobs. The result is a larger economy with higher material prosperity, more industries, more products, and more jobs.

Then, there is the fourth AI risk: *crippling inequality*. You will not be surprised to learn that Marc Andreessen does not find it meaningful. And, following his aggressive style, he is turning this concern around and accusing his opponents of — what else — following in the footsteps of Karl Marx, who "...was wrong then, and he's wrong now." That is, in Mr. Andreessen's view, his opponents — those who fear AI — are not only crooks and alarmists but also Marxists.

The fifth risk is, "*Will AI Lead to Bad People Doing Bad Things?*" The short answer, according to Andreessen, is perhaps — but that's true of every new technology, and any such misuse is already illegal. So, all we need to do is enforce the law. And besides:

> The same capabilities that make AI dangerous in the hands of bad guys with bad goals make it powerful in the hands of good guys with good goals — specifically the good guys whose job it is to prevent bad things from happening.

Where have we heard that line of reasoning before? Ah yes — straight from the U.S. National Rifle Association[39]: *"The only thing that stops a*

---

[39]I happened to be rewatching Woody Allen's 1997 film *Deconstructing Harry*, in which he descends into Hell in an elevator, and the floors (the underground levels) are announced. A voice says: "Eighth floor: fugitive war criminals, televangelists, and the NRA."

bad guy with a gun is a good guy with a gun." We all know how well that's worked out in the United States — no further comment.

Finally, Andreessen argues that, given the fierce competition with communist China, the only risky tactic on the part of the West would be to slow down AI research. I will again omit the detailed discussion, as my quotes from Andreessen's document are already too extensive.

In 2023, Andreessen apparently decided to share all of his views with the world. In October of this year, he published another programmatic document, not just a document, but a *Techno-Optimist Manifesto*. In this 5,200-word essay, he positioned himself as a proponent of "effective accelerationism," urging unfettered technological progress, which should be advanced without constraint. In the same year, Andreessen was interviewed by Lex Fridman, where he expressed the same views about AI that had already been presented in his "Why AI Will Save the World" blog post. He again said that close monitoring, regulation, and censorship of AI development would be counterproductive, even dangerous — "a slippery slope to Hell" (where, on the eighth floor, he would meet many of his own, if we were to believe Woody Allen — see the footnote).

In January 2025, Lex Fridman interviewed Marc Andreessen again on his podcast, and we could hear him say it again:

> I think the notion that AI will destroy humanity someday… is simply *not scientific*. It lacks a scientific foundation — it's a product of hysteria. We should *move forward* quickly, constrained only by the usual filters around *hateful content or protecting kids* — beyond that, *let it go full steam ahead.*

What can one say — this is the adviser who now has the 'ear' of the President of the United States. And what an ear… but that is another matter, which we will not go into here; it is evident enough as it is.

All of this is deeply discouraging: Any remaining hope for a cautious approach to AI development in the United States now seems to have vanished, at least for the next three years. And if there is no caution in the United States, there will be none anywhere. And yet, according to experts, these three years are the most decisive.

But on the other hand, was such hope ever real? After all, no one truly knows how to rein in an AI that has long since escaped every conceivable box and cage. We will return to this question in the Epilogue.

Geoffrey Hinton [*activity from 2023 to the present*]

To the reader of this book, Geoffrey Hinton is no stranger. We have spent many pages describing his foundational contributions to AI, and we understand why he is called the "Godfather of Deep Learning," and why he was awarded the 2024 Nobel Prize in Physics. Over a career spanning more than five decades, he has persistently championed artificial neural networks, often in the face of skepticism and neglect. In the 21st century, his stubborn faith in the power of neural-network-based AI has been rewarded a hundredfold. We have tried to describe in these pages how this was achieved and the brilliant, incredible results it has led to.

Born in 1947 in Wimbledon, London, Hinton is the great-great-grandson of the 19th-century logician George Boole, and the legacy of formal reasoning perhaps lingered in his intellectual DNA. After studying experimental psychology at Cambridge, he completed his Ph.D. in AI at the University of Edinburgh in 1978 His doctoral work focused on learning and mental representations — ideas that would evolve into his lifelong focus on distributed representations in neural networks.

Hinton worked at various institutions, including Carnegie Mellon University, the University of Toronto, and University College London. As we described in Chapter 4, in the 1980s, he co-invented the backpropagation algorithm, arguably the most influential learning algorithm in the history of neural networks. The 1986 *Nature* paper explaining this method was a turning point, but it would take another two decades before the field of deep learning caught up to the promise it embodied.

During the "AI winter," Hinton remained steadfast in his belief that deep (multilayered) neural networks could achieve remarkable results if given enough data and computing power. His persistence bore fruit in the 2000s and 2010s. In 2006, he and his students published a breakthrough paper on *Deep Belief Networks*, reviving interest in multi-layered neural architectures. Then, in 2012, his lab at the University of Toronto, led by students Alex Krizhevsky and Ilya Sutskever, stunned the machine vision community by winning the ImageNet competition with the deep convolutional neural network AlexNet, which ushered in the deep learning revolution.

In 2013, Hinton joined Google (initially on a part-time basis) to help bring neural networks to real-world applications. He later became a vice president and engineering fellow at *Google Brain* and was instrumental in projects involving speech recognition, image classification, and large-scale deep learning.

Throughout his distinguished career, Geoffrey Hinton has received numerous prestigious awards recognizing his foundational contributions to AI. In 2018, he was awarded the Turing Award — often called the "Nobel Prize of Computing" — alongside Yoshua Bengio and Yann LeCun for their work on deep learning. In 2024, he received the Nobel Prize in Physics, shared with John J. Hopfield, for their "foundational discoveries and inventions that enable machine learning with artificial neural networks," including the development of the Boltzmann machine and associative memory models.[40] Earlier honors include the IJCAI Award for Research Excellence (2005) and election as a Fellow of the Royal Society (1998).

In 2023, Hinton made headlines again — not for a new paper or model but for stepping down from Google and publicly voicing concerns about the existential risks posed by AGI. He expressed regret over parts of his life's work, warning that we may have created systems we no longer fully understand or control. This noble move reminds me of Russia's Andrey Sakharov, a brilliant physicist who was the leading force behind the Soviet hydrogen bomb but later retired from his classified government work and became the most vocal and brave opponent of the Soviet system.

Describing Hinton's views is a pleasure: He is clear and concise, and his wry British humor shines a warm light on the grimmest subjects. When he says something, we tend to believe him; if anyone knows what he is talking about when discussing AI, it is Geoffrey Hinton.

On May 1, 2023, an interview with Geoffrey Hinton appeared in *The New York Times*. It was entitled "The 'Godfather of AI' Leaves Google and Warns of Danger Ahead." I will take a few lines from that interview, which was conducted a few days after Hinton resigned from Google:

> A part of him, he said, now regrets his life's work. "I console myself with the normal excuse: If I hadn't done it, somebody else would have."

[40]This Nobel Prize surprised many physicists because Hopfield and Hinton did not contribute to physics per se; they only drew on certain physical ideas, such as Boltzmann statistics. If the Nobel Committee had a prize in computer science, Hinton and Hopfield would undoubtedly have received it. But in the absence of such an appropriate category, the prize committee chose to broaden its interpretation of what we consider physics.

We read in that interview about the evolution of Hinton's thinking:

> ...last year [*i.e., in 2022*], as Google and OpenAI built systems using much larger amounts of data, his view changed. He still believed the systems were inferior to the human brain in some ways, but he thought they were eclipsing human intelligence in others.
>
> Maybe what is going on in these systems," he said, "is actually a lot better than what is going on in the brain." As companies improve their AI systems, he believes, they become increasingly dangerous. "Look at how it was five years ago and how it is now," he said of AI technology. "Take the difference and propagate it forwards. That's scary."... Until last year, he said, Google acted as a "proper steward" for the technology, careful not to release something that might cause harm. But now that Microsoft has augmented its Bing search engine with a chatbot — challenging Google's core business — Google is racing to deploy the same kind of technology. The tech giants are locked in a competition that might be impossible to stop, Hinton said.

After briefly discussing the *immediate risks* of widespread false information on the internet and of upcoming job loss (of which Hinton said, "*It [AI] takes away the drudge work; it might take away more than that*"), the interview finally moved to discussing the appearance of ASI, which is the primary concern in Hinton's mind:

> ...he is worried that future versions of the technology pose a threat to humanity because they often learn unexpected behavior from the vast amounts of data they analyze. This becomes an issue, he said, as individuals and companies allow AI systems not only to generate their own computer code but actually to run that code on their own... The idea that this stuff could actually get smarter than people — a few people believed that...But most people thought it was way off. And I thought it was way off. I thought it was 30 to 50 years or even longer away. Obviously, I no longer think that."

On global regulation, Hinton is not optimistic, saying in that 2023 interview:

> But that [*global regulation*] may be impossible...Unlike with nuclear weapons, there is no way of knowing whether companies or countries are working on the technology in secret. The best hope is for the world's

leading scientists to collaborate on ways of controlling the technology. I don't think they should scale this up more until they have understood whether they can control it.

I watched numerous interviews with Geoffrey Hinton, including TV and podcast appearances, all taking place between 2023 and mid-2025. In some of them, he elaborated on the themes he had touched upon in his first 2023 interview with *The New York Times*. In others, he added new thoughts. One could also note a significant change in the level of urgency in Hinton's message. In May 2023, when speaking to *The New York Times,* he conveyed his concern in the softest possible manner. In June 2025, at the latest Hinton's appearance at the time of writing, his tone has changed. That discussion with Steven Bartlett was titled "Godfather of AI: 'I Tried to Warn Them, But We've Already Lost Control!" Here are a few quotes from that discussion:

Bartlett: Why did you believe that modeling it [*AI*] off the brain[41] was a more effective approach?

Hinton: It wasn't just me [who] believed it. Early on, von Neumann believed it, and Turing believed it, and if either of those had lived [*longer*], I think AI would have had a very different history; but they both died young.

Bartlett: In this season of your life, what mission are you on?

Hinton: My main mission now is to warn people how dangerous AI could be…I was quite slow to understand some of the risks…that other people recognized…20 years ago; I only recognized it [*the risk*] a few years ago…It's a very good question — how could you not have seen that? But remember — neural networks 20–30 years ago were very primitive in what they could do. They were nowhere near as good as humans…the idea that you have to now worry about it getting smarter than people — that seems silly then.

Bartlett: When did that change?

Hinton: It changed for the general population when ChatGPT came out. It changed for me when I realized that the kinds of digital intelligences

[41] Meaning, using artificial neural networks (ANNs) as opposed to symbolic AI.

> we're making have something that makes them far superior to the kind of biological intelligence we have. If I want to share information with you…your brain is trying to figure out "how can I [*the brain*] change the strength of connections between neurons so I might have put that word next"…so that's roughly what's going on in your brain…Nobody really knows for sure how the brain works, and nobody knows how it gets the information about whether you should increase the strength of a connection or decrease the strength of a connection…But what we do know now from AI is that if you could get information about whether to increase or decrease the connection strength…then we can learn incredible things, [*and*] that's what we're doing now with artificial neuronets.

Hinton expressed similar thoughts about the advantages of digital over organic intelligence in many of his interviews; let me summarize them here in my own words: When humans learn something, such as calculus, it takes years of hard work; we do it slowly and painstakingly — "one synaptic connection at a time," so to speak. And at the end, when you receive your Harvard diploma, only one person has acquired all that knowledge — you — assuming you have not forgotten half of it by the time of graduation. In contrast, digital intelligence can learn all of that nearly instantaneously by copying the weights all at once from another system, assuming such a knowledgeable other system exists. The digital memory will never fade, and AI can effortlessly multiply the number of knowledgeable entities, making any number — 1,000, a million, a billion — of copies of the system of weights that corresponds to understanding calculus and disseminating them widely through the internet. The advantage AI has over us in the learning routine is indeed tremendous. Later in the interview, Hinton elaborated on the subject of "why AI is superior to a human brain":

> …we're lucky if we're transferring 10 bits a second; these things are transferring trillions of bits a second. So they're billions of times better than us at sharing information, and that's because they're digital...We're analog, and you can't do that. Your brain's different from my brain, and if I could see the connection strengths between all your neurons, it wouldn't do me any good because my neurons work slightly differently and they're connected up slightly differently. So when you die, all your knowledge dies with you…Suppose you take these two digital

intelligences that are clones of each other, and you destroy the hardware they run on. As long as you've stored the connection strength somewhere, you can just build new hardware that executes the same instructions. So, it'll know how to use those connection strengths, and you've recreated that intelligence. So, they're immortal; we've actually solved the problem of immortality, but it's only for digital things.

Bartlett: ...what are the big concerns you have around the safety of AI?

Hinton: ...First of all, I want to make a distinction between two completely different kinds of risk. There are risks that come from people misusing AI, and that's most of the risks and all of the short-term risks. And then there are risks that come from AI getting super smart and deciding it doesn't need us. That is a real risk, and I talk mainly about that second risk...Lots of people say, "Is that a real risk?" and — yes, it is. Now we don't know how much of a risk it is; we've never been in that situation before; we've never had to deal with things smarter than us...so really we have no idea how to deal with it...and anybody who tells you they know just what's going to happen and how to deal with it — they're talking nonsense...My friend Yann LeCun says, "No, no, no — we built these things, we're always going to be in control; we'll build them to be obedient."...And other people like Yudkowsky say "No, no, no — these things are going to wipe us out for sure. If anybody builds it, it's going to wipe us all out."...I think both of those positions are extreme; it's very hard to estimate the probabilities in between... I often say, there is a 10% to 20% chance that they will wipe us out[42]... but that's just 'gut'.

Hinton then gave the reason he places that probability of extinction at 10–20% rather than at the *Laplacian 50%*[43]:

...we're still making them [*AIs*], and we're pretty ingenious, and the hope is that if enough smart people do enough research with enough resources, we'll figure out a way to build them so they'll never want to harm us...

---

[42] I hardly need to tell my readers that this is a very grim assessment.

[43] The *Laplacian rule* for unknown probability is to make it to be 0.5, or 50%.

Sometimes I think [*about*] the invention of the atomic bomb and how it compares [*to AI*]...a lot of people at that time thought our days are numbered...but we're still here...so the atomic bomb was really only good for one thing, [*that*] it was very obvious how it worked even if you hadn't had the pictures of Hiroshima and Nagasaki...

...we're not going to stop the development [*of AI*]...People say, "Well, why don't we just stop it now?" We're not going to stop it because it's too good for many things. Also we're not going to stop it because it's good for battle robots and none of the countries that sell weapons are going to want to stop it...the European regulations they have about AI...they're not designed to deal with most of the threats... [*they*] have a clause in them that none of these regulations apply to military uses of AI...Governments are willing to regulate companies and people, but they're not willing to regulate themselves...When OpenAI releases a new model or a new piece of software in America, they can't release it to Europe yet because of regulations...

At this point in history, when we're about to produce things more intelligent than ourselves, what we really need is a kind of World Government that [*is*] run by intelligent, thoughtful people, and that's not what we got...We've got capitalism, which is done very nicely by us to produce lots of goods and services...but these big companies, they're legally required to try and maximize profits, and that's not what you want from the people developing this stuff...

Between 2023 and 2024, they [*cyberattacks*] increased by about a factor of 12,200% and that's probably because these Large Language Models make it much easier to do *phishing attacks*[44]...and now with AI, they can clone my voice [*and*] my image....

Hinton then shared a story about how his name and image (I assume, AI-generated video) are being used to lure innocent people into buying cryptocurrency, and how all his attempts to shut that down on Meta so far have not been successful. And he added:

I have a smaller version of that... people now publish papers with me as one of the authors...in order [*to*] get lots of citations to themselves.

---

[44]*Phishing attacks* are fraudulent attempts to obtain sensitive information, such as passwords, credit card numbers, or login credentials, by impersonating a trustworthy entity.

> …maybe by 2030 they'll be creating new kinds of cyberattacks which no person ever thought of….

Then, Hinton and Bartlett went down the list of immediate AI risks, including biomedical risks, corrupt elections, and the influence of AI on social media. And what if some or all of these risks coincide?

> …you can get combinatorially many risks by combining these other risks… for example, you could get a super intelligent AI that decides to get rid of people, and the obvious way to do that is just to make one of these nasty viruses…[*or*] it could send a warning on the nuclear systems in America that there's a nuclear bomb coming from Russia or vice versa and one retaliates… there's no way we're going to prevent it getting rid of us if it wants to. We're not used to thinking about things smarter than us. *If you want to know what life's like when you're not the apex intelligence, ask a chicken.*
>
> We have one example of that [*the opposite situation*], which is mothers and babies. Evolution put a lot of work into that. Mothers are smarter than babies, but babies are in control, and they're in control because the mother just can't bear the sound of the baby crying…
>
> The analogy I often use is — forget about intelligence; think about physical strength. Suppose you have a nice little tiger cub; it's…really cute; it's very cuddly, very interesting to watch, except that you'd better be sure that when it grows up, it never wants to kill you.
>
> …You think it's possible to train a superintelligence? I don't think it's clear that we can…I think it might be hopeless, but I also think we might be able to…
>
> …we had no idea that this stuff was going to happen this fast. We thought we had plenty of time to worry about things like that…
>
> It's not like I knowingly did something thinking this might wipe us all out, but I'm going to do it anyway…I feel I have a duty now to talk about the risks…I think we need people to tell their governments to force the companies to use their resources to work on safety… and they're not doing much of that because you don't make profits that way.
>
> **Bartlett:**…are you at all hopeful that anything can be done to slow down the pace and acceleration of AI?

Hinton: Okay, there are two issues. One is: can you slow it down, and the other is can you make it so it will be safe in the end [*and*] it won't wipe us all out? I don't believe we're going to slow it down, and the reason...is because there's competition between countries and competition between companies within a country, and all of that is making it go faster and faster, and if the U.S. slowed it down, China wouldn't slow it down.

The threat of AI-induced joblessness

In the past, new technologies have come in, which didn't lead to joblessness; new jobs were created...But here I think...for mundane intellectual labor, AI is just going to replace everybody...this is a very different kind of technology if it can do all mundane human intellectual labor... you'd have to be very skilled to have a job that it couldn't just do...for many jobs that'll mean you need far fewer people.

...muscles have been replaced; now intelligence is being replaced; so what remains? Maybe, for a while, some kinds of creativity. But the whole idea of superintelligence is [*that*] nothing remains.

Difference between current AI and superintelligence

Bartlett: So, what's the difference between what we have now and superintelligence, because it seems to be really intelligent to me when I use ChatGPT or Gemini.

Hinton: AI is already better than us at a lot of things, in particular, areas like chess, for example...the same in Go. In terms of the amount of knowledge they have — something like GPT-4 knows thousands of times more than you do. There are a few areas in which your knowledge is better than theirs...superintelligence when it's better than us at all things. My guess is that between 10 and 20 years, we'll have superintelligence[45]...I haven't come to terms with what the development of superintelligence could do to my children's future. I'm okay, I'm 77; I'm going to be out of here soon, but for my children and my younger friends, my nephews and nieces and their children...it's very hard to come to terms with it emotionally.

---

[45] In that, Hinton's estimate is close to Kurzweil's, which was 2045 for the singularity.

How AI may widen the wealth inequality gap

> …in a society that shared out things fairly, if you get a big increase in productivity, everybody should be better off. But if you can replace lots of people with AIs, then the people who get replaced will be worse off, and the company that supplies the AIs will be much better off, and [*so will be*] the company that uses the AIs. So, it's going to increase the gap between the rich and the poor, and we know that if you look at that gap between the rich and the poor, that basically tells you how nice the society is. If you have a big gap, you get very nasty societies…The International Monetary Fund has expressed profound concerns that generative AI could cause massive labor disruptions and rising inequality and has called for policies that prevent this from happening…If AI can make everything much more efficient and get rid of people for most jobs, or have a person assisted by AI doing many people's work, it's not obvious what to do about it.

Later in this fairly long interview, Bartlett asked Hinton about human uniqueness and whether an AI can become conscious. Since I feel that such questions are not productive or even well defined, let me report on Hinton's reply very briefly: He said that we, humans, can have subjective experiences, and so can an AI (Hinton gave an example of that). He also said, "I don't think there's anything in principle that stops machines from being conscious." As for human uniqueness, Hinton noted that it had been debunked many times throughout history, and that this moment is no different.

At the end of the interview, Bartlett asked Hinton for his parting thoughts on AI and AI safety. Of course, Hinton could not have said anything he had not said before. Here is part of his answer:

> There's still a chance that we can figure out how to develop AI that won't want to take over from us, and because there's a chance, we should put enormous resources into trying to figure that out. Because if we don't, it's going to take over…I really don't know; I genuinely don't know…When I'm feeling slightly depressed, I think people are *toast*… While [*when*] I'm feeling cheerful, I think we'll figure out a way. (Geoffrey Hinton on a Steven Bartlett podcast, *The Diary of a CEO*, aired June 16, 2025)

Epilogue

An Evolutionary Perspective

My previous book (Gurvitch, 2023) traced the history of electronics — from the electric telegraph to the present day, when electronic technologies have become ubiquitous and indispensable. As we noted in the Prologue, electronics may be broadly defined as *electromagnetic technology devoted to the manipulation of information.*[1] This is a deliberately broad definition. It encompasses the multitude of devices invented to transmit, gather, measure, analyze, and process information — what we might call the *hardware* of the information age. These include the now-extinct telegraph and also the telephone, radio, television, computers, smartphones, oscilloscopes, radars, radio telescopes, and innumerable other devices. Each group contains subgroups, which are further subdivided into individual models, comprising a vast sea of electronic devices. Servers, data centers, and the global network of the internet are also part of this vast domain.

But this landscape also includes software — the programs and algorithms that power these devices. It encompasses everything: from Morse code and early programming languages to internet protocols and the learning algorithms of artificial neural networks described in this book. The

[1] In this definition, information is placed at the center. The fact that information processing is carried out mainly by means of electricity and optics (electromagnetism) is merely a statement of fact — an acknowledgment that this is the best foundation for information technology currently known to us.

entire emerging world of AI is likewise part of electronics in this broad sense.

Then there are *hybrid fields*, where electronics blends with other domains: driverless cars (electronics plus automotive engineering), robotics (electronics plus mechanics and optics), magnetic resonance imaging (electronics plus nuclear magnetic resonance), the global positioning system (electronics plus satellite networks), and positron emission tomography (electronics plus nuclear physics), to name a few.

In practice, we tend to focus on just a sliver of this vast terrain — consumer electronics, computing, radar, for example. In this book, we have focused on AI and its historical roots. But when we attempt to envision this entire domain as a unified field, it becomes overwhelming. The traditional name "electronics," now more than a century old, no longer seems adequate.

In earlier writings, I proposed a new term, *meta-electronics*, to describe this integrated, expansive domain where information and electromagnetism meet, mingle, and evolve. In the absence of a better name,[2] I will continue to use it here.

I have always been fascinated by the history of our highest technology — the repository of our best scientific ideas. But about a dozen years ago, after reading Richard Dawkins' *The Selfish Gene*, my fascination with electronics took an unexpected turn. I began to see it as something more: an embodiment of a new kind of evolution — one that parallels Darwinian evolution in the natural world.

To explore this idea properly, I wrote a rather long article (Gurvitch, 2021) and later an even longer book (Gurvitch, 2023). And now, I'm going to attempt the impossible: to condense it all into just four pages.

Drum roll, please! Here we go.

Following in the footsteps of biologists Hamilton and Wilson, a young British zoologist named Richard Dawkins wrote a book in the mid-1970s intended for a general readership. In this book, *The Selfish Gene*, Dawkins presents the *gene* as the linchpin of biological evolution. Gene, the essential biological *replicator*, is the ultimate beneficiary of natural selection. Living organisms are merely gene-carrying vehicles; they fulfill

---

[2] You might ask, why not *Information Technology*? After all, this name corresponds to the definition — *technology dealing with information*. However, unfortunately, the term information technology (IT) has already been used for a specific, small subset of our range, and it has stuck. There are university degrees given in IT; we cannot change that fact.

their roles and die, while genes live forever.[3] Darwinian natural selection, while acting on individual members of a species, actually chooses better genes.

Almost as an afterthought, at the end of his book, Dawkins proposed that we might find meaningful replicators in fields other than biology. He suggested that a new kind of gene-like replicator is at work in human culture, and he coined the term "meme" to describe this cultural replicator. Later, Dawkins' followers ascribed to memes considerably more significance, and the short-lived science of *memetics* was born. This application of genetic concepts to human culture met with severe opposition, partly due to cultural philosophers' visceral reactions to the invasion of biological categories into their sacred temple. However, some of that criticism was justified, based on the significant difference between precise genetic transmission in biology and demonstrably imprecise cultural transmission.

So, once I read about memes and learned about the criticism of memetics, I thought — okay, this is a nice idea, but it's being rejected because human cultural transmission is too amorphous. The ideas tend to morph in transmission; this is pretty far from the accurate replication we have in DNA and genes. But, I thought, I know of a field where replication is also happening but in a much better organized and controlled way — I thought of technology, specifically electronics or meta-electronics. The replication of technological ideas is more precise because electronics is an offshoot of exact sciences — it is predominantly an application of physics, chemistry, and materials science, with an admixture of neuroscience and psychology (though admittedly less exact), all underpinned by a good deal of mathematics. In that no-nonsense environment, mistakes are not tolerated. An imprecise transmission would be quickly extinguished because then the new products would not function properly.

What do we mean by *technical ideas*? That category should be narrowed down, lest we find ourselves in the same predicament as memetics. We do not want to talk about all technical ideas. No, sir. We will consider the replication of precise technical *designs essential for building new products*. And where are these designs recorded and expressed in understandable but concise language? In patents, of course, in patent *claims*. So, I thought, here are my new genes — *e-genes* — new technological replicators: essential, patentable technical designs that travel almost unchanged through product generations.

[3]There are genes that are known to be billions of years old.

How is the new product — let's say, a new model of a smartphone — constructed? Most of it is a new combination of old designs, the so-called *standard engineering*, which is analogous to the mixing of existing genes in biology, known as *recombination*.

But don't designs change? Yes, they do, occasionally. Significant new ideas and designs are as rare in technology as *beneficial mutations* are in biology. Thus, we will draw a parallel between newly appearing e-genes and mutations.

Armed with this new genetics, we can now attempt to develop a theory of evolution in the field of meta-electronics. To do this, we need to incorporate genetics into the Darwinian theory. What is the essence of the theory of evolution? It is its mechanism, its *algorithm*, which Darwin called *natural selection*. This algorithm operates in three stages: *variation, heredity*, and *selection*, the last one based on the principle that a contemporary of Darwin, Herbert Spencer, called "survival of the fittest."

I will spare the reader a detailed analysis of these three algorithmic steps. I will note, however, that in our case, the first step — variation — is not random, as in natural environments, but is somewhat directional, as in human-supervised *selection* or *domestication*. Engineers assume the role of selectionists: They design new products by recombining existing designs. The engineer acts like a farmer who wants to breed a new long-haired sheep or like a gardener who wants to produce a new frost-resistant variety of tomato.

To my satisfaction, I found that biologists, starting with Darwin himself, and philosophers of science, such as Daniel Dennett, have long recognized that directional variation (also known as *artificial selection*) is as legitimate as the random variation prevalent in the wild. However, that directional variation — we can call it *intelligent design* — produces much faster evolution. This was noted by Dawkins when he considered human culture, of which technology is a part.[4]

What remains is to provide examples of e-evolutionary processes and products that serve as good analogs of biological processes and products. It was done in Gurvitch (2021, 2023).

---

[4]The timescale of technology is compressed compared to the biological timescale by ten million times, 10^7. To grasp this difference, imagine that biological evolution took one year instead of 4 billion years. Then, the evolution of electronics happened in just three seconds.

Finally, anticipating your question: Why electronics or meta-electronics and not all of technology? Can't we observe evolutionary features in the manufacturing and distribution of cars, airplanes, shoes, watches, and other products? Why can't we view their development as a new form of non-biological evolution?

My answer: The evolutionary algorithm outside of biology works in many cases, but eventually, the development it supports hits the "ceiling" — *if* there is a ceiling. The development that has a *final goal* tends to slow down and saturate. Incremental improvements are still occurring: New cars and new shoes are still being manufactured, but these are hardly good examples of true evolution. Meta-electronics, on the other hand, is akin to biology in that it is decisively *non-teleological*. It has no final goal.

The ultimate goal of shoe manufacturing is to provide people with a means of walking that does not harm their feet. Car manufacturing: to provide a means of transportation on the surface of the Earth that is consistent with the existing infrastructure of roads and reasonable safety measures (such as speed limits). All other embellishments are secondary; they are designed primarily to outdo the competition. That is why shoe manufacturing has hardly changed at all in the last three or four hundred years, and car manufacturing has hardly changed in the last century and a half.

In meta-electronics — perhaps because it deals with information — there is no final goal. When people invented the telegraph, their goal was to provide instantaneous long-distance communication. But when astrophysicists constructed the first radio telescope, they had a different goal — to "listen" to the Universe through long-wavelength electromagnetic waves. There is no single goal that can be pointed to for the entire field of meta-electronics; only separate, local goals exist.

True non-biological evolutions, with all the "bells and whistles" of Darwinian evolution, are rare. As far as I know, they proceed in just two fields: *historical linguistics* and *meta-electronics*. If you doubt that conclusion, please read Gurvitch (2021, 2023), where this thesis was examined and proven with considerable rigor.

Evolutions, like elections, have consequences. One feature of a "true" evolution is its *autonomy*. Evolutions proceed without asking for permission; they do whatever they can do. This is the consequence of that stubborn Darwinian algorithm we discussed above. Once it starts cranking, it will continue, producing new generations of entities — biological species, electronic devices, or modified languages — adapted to their respective environments. Daniel Dennett called the Darwinian algorithm "universal acid" — it will eat through anything. In this sense, if we believe in the evolution of electronics, we must accept that stopping it isn't an option.

One of the signs that we are dealing with real evolution is the appearance of unwanted, unclaimed features that no one "ordered." In my previous book, I complained about certain aspects of today's consumer electronics:

> When we worry about AI, we envision one extreme version of 'getting out of hand,' but there are other modes of 'misbehaving.' There are viruses and other software sicknesses, which, like diseases, nobody ordered. There are also more subtle signs, such as the ever-increasing inconvenience of using consumer devices. This inconvenience becomes apparent when we compare modern versions with the older ones, for example, tablets with laptops. I miss that mouse; it was a good thing. I am uncomfortable making magician-like passes over the tablet's screen. Or take the ever-increasing difficulty of pushing the keys on tiny on-screen keyboards. Or the difficulty of setting an electronic clock in your car; how nice would it be to have a little knob that moves that dial; but no, it has to be done through a program so cumbersome that even the dealership mechanics do not quite know how to operate it.
>
> Why do we have to suffer? The answer we are getting (if someone is answering): you are suffering because evolution is driven by relentless miniaturization and, lately, unification. After Nigel, Stanley Tucci's character in *The Devil Wears Prada*, I am tempted to say, "And that's my problem because... Oh, wait. No, it's not my problem."
>
> Or consider the growing unruliness of the internet, which, on the one hand, brought us priceless Wikipedia, but on the other hand, is bringing misinformation, dangerous indoctrination, cyber-bullying, and terrorist recruitment.
>
> Are we still in control, or have we already lost it? Various phenomena observed at the interface between electronics and human society raise "red flags" almost daily; soon, it will look like a May Day parade in Moscow's Red Square in the Soviet Union. (Gurvitch, 2023, Vol. 2, 349–350)

You might ask: Why can't we stop a process we ourselves created? I believe the answer is complex. Part of it was discussed in Chapter 5 of this book, not by the author, but by a diverse group of thinkers and participants. One brief answer is that you cannot stop an evolution. Another brief answer is: If you were alone, perhaps you could stop it. But there is your neighbor across the street who might disagree with you. And together, you and your neighbor cannot stop it.

What good is a theory if it cannot predict anything? Can we make any predictions based on the evolutionary theory of meta-electronics? One trend that may be extrapolated into the future is the observed unification of diverse functions. This trend was not present at the start of electronics but has emerged over the last 20 or 30 years. The unification trend makes the entire process symmetrical: It all began with the *common ancestor* (the electric telegraph) and will end with a unified, all-encompassing device (or entity) in the distant future — something we might call a *common descendant*.

Of course, in biology, there was a common ancestor (called the *Last Universal Common Ancestor*, LUCA), but the common descendant is not compatible with the structure of the ever-branching evolutionary tree. This is not the case in electronics, where the speciation tree (which can be seen in Gurvitch, 2021, 2023) has many branches rejoining each other. The current example of a device that unifies previously separated functions is, of course, the smartphone.

A smartphone is the champion of unification. It has concentrated considerable power previously distributed among many devices, combining the functions of a camera, map, music player, alarm clock, computer, wallet, and, yes, a telephone — all in one pocket-sized device. That makes it indispensable. Many of us cannot live without it, especially young people. Try to deny a smartphone to a teenager, and you will feel this power applied to you — and then, God help you!

A similar trend exists in linguistics. Researchers predict that languages may eventually converge into a single global language (Pagel, 2011, 2012).

Let's take that trend seriously and envision its future. What is the logical outcome of technological unification? A single device that combines all functions: autonomous energy sources, advanced AI (possibly even artificial superintelligence, ASI), sensors, and emitters for every known energy form. The projection points to a godlike machine. If that idea is even roughly accurate, then peaceful coexistence with such a machine race might no longer be feasible.

Parting Thoughts

Is there anything hopeful to be said? I'd like to try. I can't end this book as a *eulogy for humanity*.

A few words, then, about the possible *limits of reason*. I suspect that beings or machines far more intelligent than our greatest minds might be impossible. Greater speed? Certainly — we've already seen that. But greater depth of thought? I'm not so sure. As I wrote in the Prologue, it's hard for me to imagine a mind surpassing that of Albert Einstein. Take his *principle of equivalence*, the insight from which the entire General Theory of Relativity developed. This insight came to Einstein one day in 1907, and it was so original and profound that later he called it "The happiest thought of my life." Will such moments one day seem like the musings of a child compared to the feats of superintelligence? Perhaps. But I wouldn't bet on it. Nor can I prove otherwise. Yet it seems to me that a supermind cannot be infinitely stronger than our best minds — that there is some natural limit here. Again, not in the speed of thought, but in its depth.

Another possible constraint on superintelligence emerged when we discussed Sam Harris's views. In that section, I noted that physics is the foundational science and that physics is *experimental*. You cannot derive the laws of nature through thought alone; you must pose questions to Nature and await her answers. Even the fastest mind must be patient with the laboratory. This truth, recognized since the time of Galileo and Newton, may remain true even in the age of machines.

If we combine these two principles — the limit to the depth of thought and the limit to the rate of acquiring new experimental knowledge — perhaps we can avoid being like ants or dogs in relation to the coming superintelligence. Maybe we will be to superintelligence what a typical scientist is to John von Neumann, and that won't be so scary.

So, what more can I say in closing? In this book we reviewed the history of computation from ancient times to the present, and we have seen that this long and laborious history in our times reached new heights which half a century ago were considered unattainable. The structure at the pinnacle of this development is strong AI. This structure resembles a gilded mousetrap, with the most delicious cheese in the world inside. And now, the whole bunch of us have already entered this mousetrap, racing toward the cheese.

To be honest, it's frightening — for our children, grandchildren, and the entire human culture we've spent 10,000 years building. Yet, perhaps the techno-optimists are correct. Maybe this all ends well. Possibly, we will merge with the machines and become *Homo Deus*, to use the term coined by Yuval Harari.[5] As Geoffrey Hinton said, "When I'm feeling cheerful, I think we'll figure out a way."

One thing is certain: we can't stop what's coming. In that, we're like Goethe's *Sorcerer's Apprentice*: "*Sir, I'm in desperate straits! The spirits I summoned — I can't get rid of them.*"

[5] Harari, while envisioning a merger of *Homo sapiens* with the thinking machines, isn't rejoicing with that prospect.

References

Aaronson, Scott (2024). A discussion with Brian Greene, "Straight Talk on Quantum Computing." *World Science Festival/YouTube*, Oct. 2024 [Online video].

Aaronson, Scott (2025). *Why Philosophers Should Care About Computational Complexity*, Ekkolapto Polymath Salon, *YouTube*, May 17, 2025 [Online video].

Allen, Woody (1966). Yes, but can the steam engine do this? *The New Yorker*, Oct. 8, 1966.

Anderson, James A., and Rosenfeld, Edward, Editors (1998). *Taking Notes. An Oral History of Neural Networks*. A Bradford Book, The MIT Press.

Aristotle. *Prior Analytics*, Book I (1984). In: *The Complete Works of Aristotle*, Vol. **1**, edited by Jonathan Barnes. Princeton: Princeton University Press.

Aschenbrenner, Leopold (2024). *Situation Awareness: The Dacade Ahead* https://situational-awareness.ai/

Ball, Philip (2018). *Beyond Weird*. The University of Chicago Press.

Barrat, James (2013). *Our Final Invention*. Thomas Dunne Books, St. Martin's Griffin, New York.

Bengio, Yoshua (2025, May 20). The catastrophic risks of AI — and a safer path. *TED Conferences* [Online video].

Bostrom, Nick (2014). *Superintelligence: Paths, Dangers, Strategies*. Oxford: Oxford University Press.

Butler, Samuel (1863). Darwin Among the Machines, *The Press*, 13 June 1863; reprinted in H. F. Jones, ed., in Vol. **1** of The Shrewsbury Edition of the Works of Samuel Butler. London: Jonathan Cape, 1923.

Butler, Samuel (1872). *Erewhon*. Edited by Peter Mudford. Penguin Classics, 1985.

Cai, Jin-Yi. (2023). Shor's Algorithm Does Not Factor Large Integers in the Presence of Noise. *Quantum Physics, arXiv: 2306.10072v1.*

Čapek, Karel. (2006). *R.U.R. (Rossum's Universal Robots).* Translated into English by David Wyllie, University of Adelaide (eBooks@Adelaide), 2006 (published online).

Conway, Flo and Jim Siegelman (2006). *Dark Hero of the Information Age — In Search of Norbert Wiener, The Father of Cybernetics*, Basic Books.

Davis, Edward A. (ed.) (1998). Nevill Mott: Reminiscences and Appreciations. London: Taylor & Francis.

Dennard, Robert H., Fritz H. Gaensslen, Hwa-Nien Yu, V. Leo Rideout, Ernest Bassous, and Andres R. LeBlanc (1974). Design of ion-implanted MOSFETs with very small physical dimensions. *IEEE Journal of Solid-State Circuits*, **9**(5), 256–268.

Dowling, Jonathan P. (2013). *Schrödinger's Killer App.* CRC Press, Boca Raton, London, New York.

Dyakonov, Michel I. (2024). Thirty Years of Quantum Computing. *Fluctuation and Noise Letters*, **23**(4), Article S2440065.

Dyson, George B. (1997). *Darwin Among the Machines: The Evolution of Global Intelligence*. Perseus Books.

Dyson, George B. (2012). *Turing's Cathedral: The Origins of the Digital Universe*. New York: Pantheon.

Geffer, Amanda (2015). The Man Who Tried to Redeem the World with Logic, *Nautilus web publication, Technology*, 29 January 2015.

Gertner, Jon (2012). *The Idea Factory: Bell Labs and the Great Age of American Innovation*. Penguin Books.

Gidney, Craig and Martin Ekera (2021). *How to factor 2048 bit RSA integers in 8 hours using 20 million noisy qubits*, ArXiv:1905.09749v3 [quant-ph], 13 April 2021.

Good, Irving J. (1965). Speculations Concerning the First Ultraintelligent Machine. In F. L. Alt & M. Rubinoff (Eds.), *Advances in Computers* (Vol. **6**, pp. 31–88), Academic Press.

Gurvitch, Michael (1983). Experimental evidence versus exchange theory of resistivity saturation, *Physical Review* **B 28**, 2, 544–549.

Gurvitch, Michael (2017). *The Trace That Launched a Thousand Chips*, in "Josephson Junctions: History, Devices, and Applications," Ed. by E. L. Wolf, G. B. Arnold, M. Gurvitch, and J. F. Zasadzinski, Pan Stanford Publishing.

Gurvitch, Michael (2021). The Darwin is in the Details, *Techne: Research in Philosophy and Technology*, **25**, 1, 26–71.

Gurvitch, Michael (2023). *Brave New e-World*, in two volumes, World Scientific Publishing.

Hao, Karen (2025). *Empire of AI: Dreams and Nightmares in Sam Altman's OpenAI.* Penguin Press, New York.

Harari, Yuval Noah (2015; 2017). *Homo Deus: A Brief History of Tomorrow*. Vintage.

Harari, Yuval Noah (2018). "Yuval Noah Harari on Big Data, Google, and the End of Free Will." *The Economist*, October 2018.

Harari, Yuval Noah (2019). *21 Lessons for the 21st Century*, Spiegel & Grau.

Harari, Yuval Noah (2020). Remarks at the World Economic Forum Annual Meeting, Davos, Switzerland, January 2020.

Harari, Yuval Noah (2021). Interview with Anderson Cooper, 60 Minutes, CBS News, May 16, 2021.

Harari, Yuval Noah (2024a). Interview on Amanpour & Company, PBS, September 16, 2024.

Harari, Yuval Noah (2024b). Interview on The Rich Roll Podcast, episode featuring Yuval Noah Harari, September 2024.

Harari, Yuval Noah (2025). Interview on The Diary of a CEO (hosted by Steven Bartlett), June 2025.

Harris, Sam (2016). *Can we Build AI without Losing Control Over it? TED*, September 29, 2016 [Online video].

Harris, Sam (2023). WARNING By Sam Harris: ChatGPT Could Be The Start Of The End. *The Diary of a CEO with Steven Bartlett*, 7 Aug 2023.

Hebb, Donald (1949). *The Organization of Behavior: A Neuropsychological Theory*, New York: Wiley and Sons.

Hinton, Geoffrey (2025). What is understanding? [Keynote lecture]. International Association for Safe & Ethical AI (IASEAI '25), Paris, France, February 2025.

Hopfield, J. J. (1983). Collective Properties of Neuronal Networks. *Presented at the Meeting of the Corporate Associates of the American Institute of Physics, Xerox Palo Alto Research Center*, CA. [Online video].

HP Wire report on the Willow Chip https://www.hpcwire.com/2024/12/09/google-debuts-new-quantum-chip-error-correction-breakthrough-and-roadmap-details/

Isaacson, Walter (2014). *The Innovators: How a Group of Inventors, Hackers, Geniuses, and Geeks Created the Digital Revolution* New York: Simon & Schuster.

Jannai, Daniel *et al.* (2023). *Human or Not? A Gamified Approach to the Turing Test*, arXiv:2305.20010v1 [cs.AI], 31 May 2023.

Jones, Cameron R. and Benjamin K. Bergen (2025). *Large Language Models Pass the Turing Test*, arXiv:2503.23674, submitted March 25, 2025.

Kahneman, Daniel (2011). *Thinking, Fast and Slow*. New York: Farrar, Straus and Giroux.

Kasparov, Garry, and Mig Greengard (2017). *Deep Thinking: Where Machine Intelligence Ends and Human Creativity Begins*. New York: PublicAffairs.

Kikuchi, Makoto (1983). *Japanese Electronics*, The Simul Press, Inc., Tokyo, Japan.

Kokotajlo, D., Alexander, S., Larsen, T., Lifland, E., & Dean, R. (2025, April 3). AI 2027: A scenario forecast for advanced artificial intelligence. *AI Futures Project*. https://ai-2027.com/

Kurzweil, Ray (1990). *The Age of Intelligent Machines*, Cambridge, Mass., MIT Press.

Kurzweil, Ray (1999). *The Age of Spiritual Machines: When Computers Exceed Human Intelligence*. New York: Viking.

Kurzweil, Ray (2005). *The Singularity Is Near: When Humans Transcend Biology*. New York: Viking.

Kurzweil, Ray (2024). *The Singularity Is Nearer: When We Merge with AI*. New York: Viking.

Leibniz, Gottfried Wilhelm (1714). *Letter to Nicolas Rémond* (July 1714). In *Letters to Nicolas Rémond*, edited by Leroy E. Loemker, Philosophical Papers and Letters, Vol. **2**, pages 654–660. Dordrecht: Springer.

Lem, Stanisław (First published in 1959 in Polish; in English 1974). The Investigation. Translated by Adele Milch. New York: The Seabury Press.

Lem, Stanisław (First published in 1964 in Polish; in English 2013), *Summa Technologiae*, Translated into English by Johanna Zylinska, University of Minnesota Press.

Lem, Stanisław (1999). *What an Incredible Time we are Living in*. Interview taken by V.I. Borisov, in Russian.

Lem, Stanisław (2022). *The Truth and Other Stories*. Translated by Antonia Lloyd-Jones. Cambridge, MA: The MIT Press.

Lighthill, Sir James (1973). *Artificial Intelligence: A General Survey*, commissioned by the Science Research Council, UK.

Manabrea, Luigi F. and Ada Lovelace (1843). *Sketch of the Analytical Engine Invented by Charles Babbage*, Bibliotheque Universelle de Geneve, October 1842, No. 82, with notes upon the Memoir by the Translator Ada Augusta, Countess of Lovelace.

McCulloch, Warren and Walter Pitts (1943). A Logical Calculus of the Ideas Immanent in Nervous Activity, *Bulletin of Mathematical Biophysics* **5**, 115–133.

Mei, Qiaozhu; Xie, Yutong; Yuan, Walter; and Jackson, Matthew O. (2024). A Turing test of whether AI chatbots are behaviorally similar to humans, *Proceedings of the National Academy of Sciences* (*PNAS*), 22 February 2024.

Mims, Christopher (2024). This AI Godfather Says AI Is Dumber Than a Pet Cat, *Wall Street Journal*, October 12, 2024.

Minsky, Marvin (1967). Computation: Finite and infinite machines. *Scientific American*, Vol. **217**, 6 (December 1967), 78–86.

Minsky, Marvin, and Seymour Papert (1969). *Perceptrons: An Introduction to Computational Geometry*. MIT Press, Cambridge, MA.

Moore, Gordon E. (1965). Cramming more components onto integrated circuits, *Electronics Magazine* **38**, 8 (April 19, 1965), 114–117.

Musk, Elon (2014). Remarks at the MIT Aeronautics and Astronautics Centennial Symposium, Massachusetts Institute of Technology, Cambridge, MA, October 24, 2014.

Musk, Elon (2017). Remarks at the National Governors Association Summer Meeting, Providence, Rhode Island, July 15, 2017.

Musk, Elon (2018a). Interview on The Joe Rogan Experience, Episode #1169, September 6, 2018.

Musk, Elon (2018b). Remarks at South by Southwest (SXSW), Austin, Texas, March 11, 2018.

Musk, Elon (2023). Interview on Twitter Spaces, 2023. (Exact date varies by session; commonly cited discussion on AI risk and development pace.)

Musk, Elon (2025a). Remarks at xAI live demonstration event unveiling Grok 4, July 9, 2025.

Musk, Elon (2025b). Post on X (formerly Twitter), July 22, 2025.

Pagel, Mark (2011). How language transformed humanity, Talk at *TED Global-2011*, Edinburgh, Scotland.

Pagel, Mark (2012). *Wired for Culture: Origins of the Human Social Mind*, Norton&Co.

Pitts, Walter (1942). Some observations on the simple neuron circuit. *Bulletin of Mathematical Biophysics* **4**, 121–129.

Pitts, Walter and Warren McCulloch (1947). How we know universals: the perception of auditory and visual forms. *Bulletin of Mathematical Biophysics*, **9**, 127–147.

Riordan, Michael and Lillian Hoddeson (1997). *Crystal Fire*, W. W. Norton & Company, New York. London.

Rosin, Robert (1976). In the *Encyclopedia of Computer Science* Edited by Anthony Ralston and Chester L. Meek. Petrocelli/Charter, New York.

Rumelhart, D. E., Hinton, G. E., and Williams, R. J. (1986). Learning representations by back-propagating errors. *Nature*, 323(6088), 533–536.

Rumelhart, D. E., McClelland, J. L., and the PDP Research Group (1986). *Parallel Distributed Processing: Explorations in the Microstructure of Cognition*, Vol. **1**: Foundations. MIT Press.

Russel, Stuart, and Peter Norvig (2010). *Artificial Intelligence: A Modern Approach*, Third Edition, Prentice Hall.

Samuel, Arthur (1959). Some Studies in Machine Learning Using the Game of Checkers," *IBM Journal of Research and Development* **3**, 210–229.

Shannon, Claude (1950). A Chess-Playing Machine, *Scientific American*, Vol. **182**, No. 2 (February 1950), pp. 48–51.

Smalheiser, N. R. (2000). *Walter Pitts. Perspectives in Biology and Medicine,* **43**(2), 217–226.

Suleyman, Mustafa, with Michael Bhaskar (2023). *The Coming Wave*, Crown.

Swedin, Eric, and David Ferro (2005). *Computers. The life story of a technology*, Greenwood Technographies, Westport, Conn., London.

Turing, Alan (1948). *Intelligent Machinery*, https://weightagnostic.github.io/papers/turing1948.pdf

Turing, Alan (1950). Computing Machinery and Intelligence, *Mind* Vol. **49**: 433–460.

Turing, Alan (1951). Can digital computers think? BBC Radio Lecture. Reprinted in B. Jack Copeland (Ed.), *The essential Turing: Seminal writings in computing, logic, philosophy, artificial intelligence, and artificial life* (pp. 482–486). Oxford University Press, 2004.

Ulam, Stanisław (1976). *Adventures of a Mathematician.* New York: Charles Scribner's Sons.

Vaswani, A., Shazeer, N., Parmar, N., Uszkoreit, J., Jones, L., Gomez, A. N., Kaiser, Ł., & Polosukhin, I. (2017). *Attention Is All You Need.* In *Advances in Neural Information Processing Systems*, **30**, 5998–6008.

Vinge, Vernor (1993). *The Coming Technological Singularity: How to Survive in the Post-Human Era.* Paper presented at the Vision-21 Symposium, NASA Lewis Research Center and the Ohio Aerospace Institute.

Von Neumann, John (1955). Can we survive technology? *Fortune*, Vol. **51**, no. 6, June 1955, pp. 106–108, 151, 154, 156.

Von Neumann (1949 lecture notes, published in 1966). *The Theory of Self-Reproducing Automata*, edited and completed by Arthur W. Burks, University of Illinois Press.

Weiss, Eric A. (1992). Arthur Lee Samuel (1901–90). *IEEE Annals of the History of Computing*, **14**(3), pp. 55–69.

Wiener, Norbert (1961; first edition in 1948). *Cybernetics: Or Control and Communication in the Animal and the Machine.* The M.I.T. Press, Cambridge, Massachusetts.

Wiener, Norbert (1960). Some moral and technical consequences of automation. *Science*, **131**(3410), 1355–1358, May 6, 1960.

Williams, Michael R. (1997). *A History of Computing Technology*, IEEE Computer Society Press, Los Alamitos, California.

Yampolskiy, Roman V. (2024). *AI: Unexplainable, Unpredictable, Uncontrollable*, CRC Press.

Yudkowsky, Eliezer (2001). *Creating Friendly AI 1.0: The Analysis and Design of Benevolent Goal Architectures*, MIRI publication.

Yampolskiy, Roman V. (2025), in Steven Bartlett podcast The Diary of a CEO, September 4, 2025.

Yudkowsky, E. (2022, June). *AGI Ruin: A List of Lethalities.* LessWrong. https://www.lesswrong.com/posts/uMQ3cqWDPHhjtiesc/agi-ruin-a-list-of-lethalities

Yudkowsky, Eliezer (2023). Shut It All Down, *TIME Magazine* (29 Mar 2023).

Index

Q

R

S

T

Made in the USA
Columbia, SC
18 September 2026

445e1ca8-d938-41b6-9636-77a34d1bd60eR01